SOUTH ATLANTIC PALEOCEANOGRAPHY

SOUTH ATLANTIC PALEOCEANOGRAPHY

Edited by K. J. HSÜ and H. J. WEISSERT

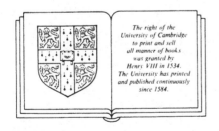

The right of the
University of Cambridge
to print and sell
all manner of books
was granted by
Henry VIII in 1534.
The University has printed
and published continuously
since 1584.

CAMBRIDGE UNIVERSITY PRESS

Cambridge

London New York New Rochelle

Melbourne Sydney

CAMBRIDGE UNIVERSITY PRESS
Cambridge, New York, Melbourne, Madrid, Cape Town, Singapore,
São Paulo, Delhi, Dubai, Tokyo

Cambridge University Press
The Edinburgh Building, Cambridge CB2 8RU, UK

Published in the United States of America by Cambridge University Press, New York

www.cambridge.org
Information on this title: www.cambridge.org/9780521129732

© Cambridge University Press 1985

First published 1985
This digitally printed version 2009

A catalogue record for this publication is available from the British Library

Library of Congress Catalogue Card Number: 85-412

ISBN 978-0-521-26609-3 Hardback
ISBN 978-0-521-12973-2 Paperback

Contents

Contributors

Benson, R.H., Smithsonian Institution, Washington, DC 20560, USA

Boersma, A., PO Box 404, Stony Point, New York, USA

Borella, P.E., Deep Sea Drilling Project, Scripps Institution of Oceanography, La Jolla, CA., USA

Chapman, R.E., Smithsonian Institution, Washington, DC 20560, USA

Dean, W., USGS, Denver, CO 80225, USA

Deck, L.T., Smithsonian Institution, Washington, DC 20560, USA

Diester-Haass, L., Fachrichtung Geographie, Univ. des Saarlands, Saarbrücken, BRD

Gardner, J., USGS, Menlo Park, CA 94025, USA

Gombos, A.M., Exxon Production Research Company, Houston, Texas 77001, USA

Hsü, K.J., Geological Institute, ETH-Z, CH-8092 Zürich, Switzerland

Hodell, D.A., Graduate School of Oceanography, Univ. Rhode Island, Narrangansett, RI 02882, USA

Jansen, J.H.F., Netherlands Inst. for Sea Research, NL-1790 Ab den Burg, Netherlands

Johnson, D.A., WHOI, Woods Hole, MA 02543, USA

Malmgren, B., Dept. of Paleontology, Univ. of Uppsala, Box 558 5-751 22 Uppsala, Sweden

McKenzie, J.A., Geological Institute, ETH-Z, CH-8092 Zürich, Switzerland

Moore, T.C., Exxon Production Research Company, Houston, Texas 77001, USA

Muza, J.P., Dept. of Geology, Florida State Univ., Tallahassee, FL 32302, USA

Oberhänsli, H., Geological Institute, ETH-Z, CH-8092 Zürich, Switzerland

Rabinowitz, P.D., Texas A & M University, College Station, Texas, USA

Shackleton, N.J., Godwin Laboratory, Univ. of Cambridge, Cambridge, UK

Thunell, R.C., Dept. of Geology, Univ. of South Carolina, Columbia, SC 29208, USA

Toumarkine, M., Geological Institute, ETH-Z, CH-8092 Zürich, Switzerland

Contributors

Vergnaud-Grazzini, C., Univ. Pierre et Marie Curie, Laboratoire de
   Geologie, 4, place jussieu, 75230 Paris, France
Weissert, H.J., Geological Institute, ETH-Z, CH-8092 Zürich, Switzerland
Williams, D.F., Dept. of Geology, Univ. of South Carolina, Columbia, SC
   29208, USA
Wise, S.W., Dept. of Geology, Florida State Univ., Tallahassee, FL 32302,
   USA
Wright, R., Exxon Production Research Company, Houston, Texas 77001,
   USA

Preface

Paleoceanography is a new science of the history of the oceans. This book describes what we have found out during the last decade about the past 100 million years of the history of the South Atlantic Ocean, thanks largely to drilling by *Glomar Challenger* during five expeditions in 1980. Paleotemperature studies have provided us with a history of climatic variations. Geochemistry of carbon isotopes had provided information on fertility of planktonic organisms and on intensity of oceanic over-turns. Correlation of sediment character to changes in oceanic chemistry and fertility has permitted interpretations of the variation of the level at which fossil skeletons became dissolved. All the authors are experts in their respective fields of specialization and almost all were participants in the 1980 expeditions to the South Atlantic. The book brings together the results of the major discoveries in one volume. This volume is a first modern regional synthesis of ocean history, and it serves as a methodological model for studies of the world's other oceans.

Acknowledgements

The editors are grateful to Ueli Briegel, who single-handedly organized the Paleoceanography Conference in Zürich. We wish to thank the authors of this volume for their contributions. We are indebted to W. H. Berger, M. B. Cita, L. Keigwin, K. Kelts, J. Kennett, K. Miller, L. Pratt, H. Thierstein, R. C. Thunell and T. Moore for reviewing the manuscripts. Finally, we would like to dedicate this volume to our colleague Hans Bolli, who contributed by establishing the biostratigraphic framework that made our paleoceanographical researches possible.

1

Introduction

K. J. HSÜ and H. J. WEISSERT

In the early days of the earth-science revolution, marine geologists were either geophysicists or micropaleontologists. They had all the fun, interpreting magnetic anomalies and dating the sediments above the ocean crust, and they were the heroes who upheld the seafloor spreading theory. Sedimentologists then had the thankless task of describing hundreds and hundreds of metres of pelagic oozes, which all looked alike – a routine that could be done by any high school graduate.

Micropaleontologists used to get together in international meetings called planktonic conferences, held every three years. The first was instrumental in straightening out the zonation of planktonic foraminifera, and the second the zonation of nannoplankton. At the third conference, in Kiel (1974), there was a shift in emphasis, and many micropaleontologists turned to isotope geochemistry. Analyses of the oxygen isotopes of foraminifera yielded paleotemperatures, though few seemed to understand what the carbon isotopes meant.

No more planktonic conferences have been held since then. Biostratigraphical zonation still had its problems, but the first flush of excitement was over. Paleoceanography was developing as a new earth-science discipline, though few thought of themselves as paleoceanographers.

When the idea of a meeting to compare notes and exchange ideas was first discussed, we thought of calling it the Fourth Planktonic Conference. However, on the advice of Bill Haq it became the First International Conference on Paleoceanography. This was held in 1983 under the sponsorship of the Geological Institute of the Swiss Federal Institute of Technology, the IUGS Commission on Marine Geology, and the Working Group 7 of the International Lithosphere Program.

The timing of the conference was influenced by the consideration that five deep-sea drilling cruises were sent to the South Atlantic in 1980. The preliminary results should have been completed after three years of post-cruise activities. The conference provided a forum for several key participants to report on their major findings. The exchange of ideas also contributed to a revision of their preliminary conclusions, previously published in the *Initial Reports of the Deep Sea Drilling Project*. A symposium on South Atlantic Paleoceanography was held during the conference, and the speakers were invited to contribute to this volume. We hope, therefore, that this present work achieves an up-to-date synthesis. All the significant results are now summarized in one volume, and we have been able to refer to discoveries made by preceding or successive cruises to broaden the basis of our interpretations. Like the analog volume *Indian Ocean Geology and Biostratigraphy*, which presented the results of six Indian Ocean cruises, Legs 22 to 27 (Heirtzler *et al.*, 1977), this volume gives the highlights of the five South Atlantic cruises, Legs 71 to 75.

The South Atlantic cruises were planned at the very start of the International Phase of Ocean Drilling (IPOD). The Planning Committee of the Joint Oceanographic Institutions for Deep Earth Sampling (JOIDES), reorganized the structure of advisory panels in 1975. Instead of geographical division (Pacific, Atlantic, and Indian Oceans), four thematic panels were constituted, namely the Ocean Crust, the Active Margin, the Passive Margin, and the Ocean Paleoenvironment Panels. Announcements were made to invite submissions of proposals to explore the deep-sea floor by drilling.

The idea to investigate South Atlantic paleoceanography first came up during the Leg 3 cruise of the Deep Sea Drilling Project. In 1968 the seafloor spreading theory of Vine and Matthews was still much debated. Maurice Ewing was skeptical. He had predicted a thickening of the sedimentary cover on the Mid-Atlantic Ridge in the direction away from the ridge-axis, because thicker sediments should be accumulated on older crust if the sedimentation rate has remained constant. The record of continuous seismic profiling showed, however, that it was not the case: there were no systematic changes in sediment thickness regionally, although the basinal sediments were always thicker than those on top or flanks of submarine elevations in each local area. Ewing cited his results to challenge the Vine and Matthews theory, and the Leg 3 drilling was to provide the first chance to test the revolutionary theory.

One of us (KJH) joined the Leg 3 cruise as a sedimentologist. The first objective at Site 14 of the South Atlantic was to determine the age of sediments above the crust of Anomaly 13; the theory had predicted a latest Eocene age of about 38 m.y. for the sediment at the base of a more than 100 m thick sequence. It was recalled that the first core was early Miocene, or some 20 m.y. in age, accumulated at a rate of less than half a meter per million years. At that rate, the age of the basal sediment should be Permian or Triassic. We began to regret that we had no conodont expert on board. To our surprise, the next series of cores, including some 100 m of nannofossil oozes, were all dated as Oligocene, and the last core yielded latest Eocene age as had been predicted. We were surprised by the non-uniform rate of sedimentation. The next drillsite was to ascertain the age of Anomaly 6, and the prediction was early Miocene (21 m.y.). After more than half of the 140 m thick sequence had been penetrated, the sediment encountered was still very late Miocene, i.e. 6 or 7 m.y. old. Just when we began to doubt the theory, the next core came up and contained a red clay of middle Miocene age. Soon the crust was reached and it was found to be early Miocene as predicted. So the story repeated itself at every drillsite of the cruise, and we could not predict the age of basal sediments at any site because of the unpredictable sedimentation rates. Ewing's puzzle was resolved: he was wrong in assuming a thicker sequence on older crusts, because the mid-Tertiary sediments have a negligible thickness in many parts of the Atlantic Ocean.

At the end of the Leg 3 cruise the pattern began to emerge that calcite dissolution was abnormally intense during the middle Miocene. Clinging to the Lyellian dogma and assuming constant calcite-compensation-depth (CCD) throughout geologic time, Hsü & Andrews (1970) postulated a varying paleodepth of the Mid-Atlantic Ridge to account for the variation of calcite-dissolution. Geophysical considerations proved this postulate untenable (Sclater & Francheteau, 1970). The scientific community soon came to a consensus that the CCD had varied, at times drastically, during the past; we had to adopt the 'Sclater curve' to back-track paleodepth and to reconstruct past CCD on the basis of such back-tracking information.

The presence of red clay in the middle Miocene was confirmed by several subsequent DSDP cruises to the Atlantic. We began to speak of a middle Miocene CCD-crisis, and the cause of rise in CCD during the middle Miocene became a theme of considerable controversy. A proposal was, therefore, sent to JOIDES by Hsü in 1975, to drill a transect of boreholes across the Mid-Atlantic Ridge along the 30°S parallel. This proposal was referred to the Ocean Paleoenvironment

Panel. Another idea that reached the OPP, submitted by W.W. Hay, was to study the Cretaceous anoxic events by drilling the Angola Basin.

A South Atlantic Working Group was constituted in 1975, and the membership included H. Bolli, J. Debyser, K. Hsü (Chairman), K. Perch-Nielsen, and W.B.F. Ryan. The Working Group met repeatedly, and the membership was gradually enlarged. Numerous new proposals were submitted but also centered around the two original themes. One particular suggestion was made by Ryan during the Edinburgh meeting of the Working Group in 1976; he presented a seismic profile across the Walvis Ridge and proposed to investigate the CCD-history and other paleoceanographical gradients there. Eventually the planning was finalized, a site survey was carried out, and the objectives were to be investigated during the Legs 73, 74, and 75, to the Mid-Atlantic Ridge, Walvis Ridge, and Angola Basin respectively.

The Rio Grande Rise was also first drilled during the Leg 3 cruise. The original aim was to obtain 100 m of Quaternary cores to study Pleistocene climatic changes, but the disturbed materials proved unsuitable for that purpose. We encountered difficulties in dynamic-positioning the drill vessel when drilling the hole on the flank of the Rise, sampled Plio-Quaternary foraminiferal sands, and found a large erosional surface spanning much of the mid-Tertiary, all testifying to strong bottom-current activities there. Subsequent deep-sea drilling cruises gathered unmistakable evidence for a mid-Tertiary epoch of current erosion on both margins of the Atlantic. The record spoke clearly for a past system of current circulation similar to the Antarctic Bottom Waters (AABW) of today, and the origin of the ancestral AABW must have been closely related to climatic changes of the Southern Oceans and to advance (northward) and retreat (southward) of the Antarctic Front.

Woody Wise submitted in 1976 a proposal to drill the Argentine Basin and the Falkland Plateau. Our Working Group then contacted Dave Johnson concerning an investigation of the Vema Channel, which connects the Argentine and the Brazil Basins. The possibility of scheduling cruises to the western South Atlantic did not open up, however, until the program of IPOD was expanded with a Phase-II extension. The South Atlantic Working Group was then split into two, with Hsü chairing the Southeastern and Kennett the Southwestern Atlantic Working Groups. Two more cruises were added to the South Atlantic Paleoceanography Project, scheduled to drill the Falkland Plateau in Leg 71 and the Vema Channel in Leg 72.

Originally scheduled for 1978, the drilling cruises were repeatedly delayed before they were realized in 1980. The delay turned out to be a lucky break, because the hydraulic-piston coring was developed in 1978 and first successfully tested at DSDP Site 480 on New Year's Day, 1979. The new development enabled us to use HPC during all South Atlantic Cruises. Leg 73, particularly, could be considered an HPC leg, when the rotary coring technique was used mainly to sample basement rocks. The recovery of undisturbed cores from continuous sequences permitted a correlation of Cenozoic magnetostratigraphy to seafloor anomalies on the one hand, and to biostratigraphy on the other. Prior to 1980, incontrovertible correlations were limited down to a level in the late Neogene (magnetostratigraphic Epoch 5 at about 5 m.y.), although various workers had attempted earlier Neogene and Paleogene corre-lations. The cores from Leg 73, we believe, yielded for the first time reliable correlations for the whole Cenozoic. The results are shown by Fig. 1. Numerical ages cited in the papers of this volume, especially in those authored by Leg 73 scientists, are based upon the correlations illustrated by Fig. 1.

This symposium volume includes 15 articles, all authored by speakers at the Symposium on South Atlantic Paleoceanography during the First International Conference on Paleoceanography. In the first paper (chapter 2) Malmgren studied *'Dissolution effects on size distribu-tion of recent planktonic foraminiferal species, South Atlantic Ocean'*. Malgrem documents that dissolution did not modify the mean test size of the various species analyzed. Jansen, in his study on *'Middle and Late Quaternary carbonate production and dissolution and paleoceanography of the eastern Angola Basin, South Atlantic Ocean'* (chapter 3), used the accumulation rates of $CaCO_3$ and the state of preservation of calcareous fossils as an indicator for paleoproductivity. He relates changing productivity to glacial–interglacial cycles. Diester-Haass studied vari-ations in radiolarian and diatom content in Quaternary sediments from the eastern Walvis Ridge which helped her to draw conclusions on *'Late Quaternary upwelling history off southwest Africa'* (chapter 4). In their abstract on the *'History of the Walvis Ridge: A précis of the results of DSDP Leg 74'* (chapter 5), Moore *et al.* summarize the information on the history of the Walvis Ridge gained from DSDP Leg 74. Dean and Gardner looked at *'Cyclic variations in calcium carbonate and organic carbon in Miocene to Holocene sediments, Walvis Ridge, South Atlantic Ocean'* (chapter 6). Climatically induced fluctuations of global sea-level had an impact on the deposition rate of terrigeneous detritic material in the area studied. In addition, changes in upwelling and productivity are reflected in the varying organic carbon contents of the sediments. The paleoceanography of the central South Atlantic is the theme of the next five papers. Weissert and Oberhänsli trace climate history back into the Pliocene. Their contribution (chapter 7) is entitled *'Pliocene oceanography and climate: An isotope record from the southwestern Angola Basin'*. The late Miocene is the subject of the paper by McKenzie and Oberhänsli (chapter 8). They studied *'Paleoceanographic expressions of the Messinian salinity crisis'* in sediments cored during DSDP Leg 73.

Fig. 1. A Cenozoic time scale of magnetostratigraphy, based on Leg 73 results (in three parts).

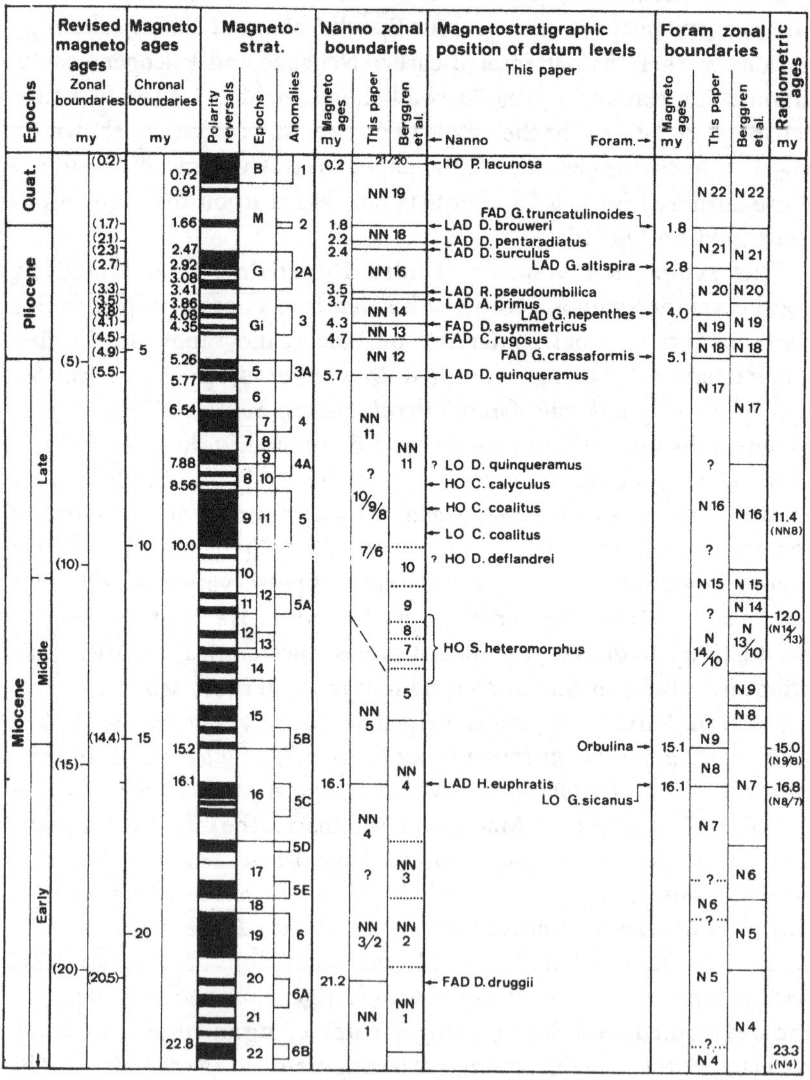

Epochs	Revised magneto ages — Zonal boundaries (my)	Magneto ages — Chronal boundaries (my)	Magneto-strat. (Polarity reversals / Epochs / Anomalies)	Nanno zonal boundaries (Magneto ages my / This paper / Berggren et al.)	Magnetostratigraphic position of datum levels — This paper (← Nanno / Foram →)	Foram zonal boundaries (Magneto ages my / Th.s paper / Berggren et al.)	Radiometric ages (my)
Quat.	(0.2)	0.72 / 0.91	B — 1	0.2 / 21/20	← HO P. lacunosa		
				NN 19			
Pliocene	(1.7)	1.66	M — 2	/ NN 18	FAD G. truncatulinoides ← LAD D. brouweri	N 22 / N 22 / 1.8	
	(2.1)(2.3)	2.47		1.8 2.2 2.4	← LAD D. pentaradiatus ← LAD D. surculus	N 21 / N 21 / 2.8	
	(2.7)	2.92 / 3.08	G — 2A	/ NN 16	LAD G. altispira →		
	(3.3)(3.5)(3.8)(3.8)(4.1)	3.41 / 3.86 / 4.08 / 4.35	Gi — 3	3.5 3.7 / NN 14 / NN 13	← LAD R. pseudoumbilica ← LAD A. primus / LAD G. nepenthes → ← FAD D. asymmetricus	N 20 / N 20 / 4.0 N 19 / N 19	
	(4.5)(4.9)(5)	5.26		4.3 4.7 / NN 12	← FAD C. rugosus / FAD G. crassaformis →	N 18 / N 18 / 5.1	
	(5.5) — 5	5.77	5 — 3A	5.7	← LAD D. quinqueramus	N 17 / N 17	
Miocene — Late		6.54	6 / 7 — 4 / 8 9 / 8 10 — 4A	NN 11 / NN 11	? LO D. quinqueramus ← HO C. calyculus	?	
		7.88 / 8.56	9 11 — 5	? 10/9/8	← HO C. coalitus ← LO C. coalitus	N 16 / N 16 / 11.4 (NN8)	
	(10) — 10	10.0	10 / 11 12 — 5A	7/6 / 10	? HO D. deflandrei	? N 15 / N 15	
Miocene — Middle			12 13 / 14	9 / 8 / 7	HO S. heteromorphus	? N 14 N 14/10 / 13/10 / 12.0 (N14/13) N 9	
	(14.4) — 15	15.2	15 — 5B	NN 5 / 5		N 8 ? N 9 / 15.1	
	(15)	16.1	16 — 5C	16.1 / NN 4	Orbulina → ← LAD H. euphratis LO G. sicanus →	N 8 / N 7 / 16.1 / 15.0 (N9/8) / 16.8 (N8/7)	
Miocene — Early			5D	NN 4		N 7	
			17 — 5E	? / NN 3		... ? ... / N 6	
			18			N 6 ... ? ...	
	— 20		19 — 6	NN 3/2 / NN 2		N 5	
	(20)(20.5)		20 — 6A	21.2	← FAD D. druggii	N 5	
		22.8	21 / 22 — 6B	NN 1 / NN 1		N 4 ... ? ... / N 4 / 23.3 (N4)	

They conclude that the late Miocene glaciation in the Southern Hemisphere preceded the desiccation event in the Mediterranean. Oberhänsli and Toumarkine present information on 'The Paleogene oxygen and carbon isotope history of Sites 522, 523, and 524 from the central South Atlantic' (chapter 9). They analyzed samples from Late Paleocene–Early Eocene and from Middle Eocene deposits. They further discuss paleoceanographic changes at the Eocene/Oligocene boundary. Hsü and Wright traced the 'History of calcite dissolution of the South Atlantic Ocean' (chapter 10). In their calcite-dissolution

Epochs	Revised magneto ages — Zonal boundaries my	Magneto ages — Chronal boundaries my	Magneto-strat. (Polarity reversals / Epochs / Chrones or Anomalies)	Nanno zonal boundaries — Magneto ages my	Nanno — This paper	Nanno — Berggren et al.	Magnetostratigraphic position of datum levels (This paper) — ←Nanno / Foram→	Foram zonal boundaries — Magneto ages my	Foram — This paper	Foram — Berggren et al.	Radiometric ages my
Oligocene — Late	(22.7) (22.9)	23.6 23.9 24.3	23 | 6C	23.8	NP 25		←LAD D. bisectus / LAD D. scrippsae / FAD Globigerinoides spp→	24.0			23.0 (Mi/O2)
	(25)	25	24						P 22	P 22	
		25.9	C-7		NP 25						
		26.8	C-7A			NP 24			P 21		
	(27.6)	27.3	C-8								
		28.7	C-9	28.9	NP 24		←LAD S. distentus / LAD G. opima→	28.9			
	(29.5) (30)	30 30.3							P 21		
				30.9	NP 23		←FAD S. ciperoensis		G. amp.		
	(31.2)	31.9	C-10								
		33.3 33.7	C-11			NP 23	FAD G. opima→	32.7	P. mic.	G. amp.	
Oligocene — Early	(33.7) (33.8) (34.6) (35)	35	C-12	35.4	22	NP 22	LAD Pseudohastigerina / ←LAD R. umbilica	35.3	P. mic. / C. chi.	C. chi.	
		36.2		36.0	NP 21		←LAD C. formosus				35.5 (NP21)
	(35.6) (36.0)	36.8	C-13	37.2 37.6	20 / NP 19	NP 21 / NP 21 / NP 20	←LAD D. saipanensis / ←FAD S. pseudoradians / LAD G. cocoaensis→		T. cer.	T. cer.	36.5 / 35–37 (O1/5Q) / 36.7
Eocene — Late	(37.2) (37.5)	38.2 39.1	C-15	38.9	NP 18	NP 19	←FAD I. recurvus	39.0	G. sem.	G. sem.	38.5 (Jacks)
	(39.0)	40 40.6	C-16	40.9	NP 17	18	FAD C. omaruensis / LAD C. grandis		T.		37–41 (NP18)
	(40)	42.5	C-17		NP 17	NP 17			P 14		38.5–42.7 (P14/13) / 38–43.5 (NP17/16)
Eocene — Middle	(41.4)		C-18	43.4			←LAD C. solitus / LAD T. frontosa→	43.4	T. rohri / P 13		
		44.8	C-19		NP 16	NP 16			T. / P 12 / pos.		
		45 46.0									

record they could recognize long-term variations of the order of 10^7 years and second-order fluctuations with a periodicity of 10^4–10^5 years, which they could relate to climatic change. Hsü, McKenzie, and Weissert summarize information on the *'Cenozoic carbon-isotope record in South Atlantic sediments'* (chapter 11). In a last contribution on the eastern South Atlantic, Jansen deals with *'Hiatuses in Mesozoic and Cenozoic sediments of the Zaire–Congo continental shelf, slope, and deep-sea fan'* (chapter 12). Jansen correlates hiatuses in shelf and slope sediments of the African margin with denudation surfaces on the

Epochs	Revised magneto ages Zonal boundaries (my)	Magneto ages Chronal boundaries (my)	Polarity reversals	Chrones	Nanno zonal boundaries Magneto ages (my)	This paper	Berggren et al.	Magnetostratigraphic position of datum levels This paper (← Nanno Foram →)	Foram zonal boundaries Magneto ages (my)	This paper	Berggren et al.	Radiometric ages (my)
Eocene — Middle	(45)	47.6		C-20		NP 15	NP 15				P 11		47.2 (P 12)
	(47.1)	50				NP 14		FAD T. possagnoensis →	49.4	P 10 (T. frontosa)		
				C-21		NP 13				P 9		
Eocene — Early	(50)					NP 12				? P 8		
						11				P 7		48.5 (P 6)
		55		C-22		NP 10						
				C-23		13 NP 9		} LAD T. orthostylus		? P 6		
						12 ?				P 5		49.5B (NP 12–11)
	(55)	[55.6]		C-24	58.1	11 NP 8				? P 4		53–55 (Eo.–Pal.)
						10 NP 7	NP 9	← LAD F. tympaniformis				
		60				NP 9 NP 6					P 4	
	(57.6) (57.8)	60.2		C-25	60.6	NP 5 NP 4		FAD P. pseudomenardii / ← FAD D. multiradiatus	60.4		P 3	58 (P 3)
Paleocene — Late	(59.0)			C-26	61.8	NP 8/7/6 NP 3		← FAD A. kleinpelli		P 3		
	(60)					NP 5		FAD G. angulata / ← FAD E. tympaniformis / ← FAD E. macellus		P 2		
	[60.6] [60.9]	63.9		C-27	63.5 63.8	NP 3		FAD G. daubjergensis →	63.7			
Paleocene — Early	(61.8)	65.0		C-28		NP 2			64.7	P 2 P 1		
	(62.5)	66.0		C-29	65.5 66.3 66.5	NP 2	NP 1	← FAD C. danicus / FAD G. eugubina / ← FAD C. tenuis	66.5	P 1		
	[63.3] [63.5]	67.0		C-30								
Cretaceous	(65)	68.5 69.3 70		C-31								

African continent. Williams, Thunell, Hodell, and Vergnaud-Grazzini present a *'Synthesis of Late Cretaceous, Tertiary, and Quaternary stable isotope records of the South Atlantic based on Leg 72 DSDP core material'* (chapter 13). They evaluate how the southwestern Atlantic Ocean responded to climatic changes at the Cretaceous–Tertiary boundary, at the Eocene–Oligocene boundary, and in Late Neogene times. Johnson, in his paper on *'Abyssal teleconnections II. Initiation of Antarctic Bottom Water flow in the southwestern Atlantic'* (chapter 14), discusses the evolution of AABW. He explains why the initiation of AABW flow sometime during the Oligocene was governed by the thermal isolation of Antarctica, by injection of high-salinity water, and by tectonic factors controlling the circulation pattern. Wise, Gombos, and Muza reconstructed the *'Cenozoic evolution of polar water masses, southwest Atlantic Ocean'* (chapter 15). Based on faunistic data such as changes in radiolarian fauna, and on oxygen and carbon isotope date they can draw a picture of the Cenozoic high latitude South Atlantic. Finally, Benson, Chapman, and Deck give *'Evidence from the Ostracoda of major events in the South Atlantic and world-wide over the past 80 million years'*. With their data they can show the conversion of a latitudinal, salinity driven, oceanic system to a longitudinal, thermally driven, strongly two-layered ocean system.

REFERENCES

Berggren, W.A. & Van Couvering, J.A. 1974. The late Neogene. *Palaeogeogr. Palaeoclimatol. Palaeoecol.*, 16, 1–216.

Heirtzler, J.R., Bolli, H.M., Davis, T.A., Saunders, J.B. & Sclater, J.G., 1977. *Indian Ocean Geology and Biostratigraphy*, Washington, DC, American Geophysical Union, 616 pp.

Hsü, K.J. & Andrews, J.E. 1970. Lithology. In: Maxwell, A.E. *et al.*, *Initial Reports Deep Sea Drilling Project*, 3, Washington, DC, US Government Printing Office, pp. 445–453.

Sclater, J.G. & Francheteau, J. 1970. The implication of terrestrial heat flow observations on current tectonic and geochemical models of the crust and the upper mantle of the Earth. *Royal Astron. Soc. Geophys. J.*, 20, 509–542.

2

Dissolution effects on size distribution of recent planktonic foraminiferal species, South Atlantic Ocean

B. MALMGREN

Abstract

The effects of dissolution on mean test size of various planktonic foraminiferal species were studied. Seven species or varieties of modern planktonic foraminifera were analysed from surface sediments from the Maurice Ewing Bank. These species included *Neogloboquadrina pachyderma* sinistral and dextral forms, *Globorotalia inflata, Globigerina bulloides, G. quinqueloba, Globigerinita glutinata,* and *G. uvula.* A dissolution index was determined quantitatively from relative abundances of fragments of planktonic foraminifera, benthonic foraminifera, radiolarians, mineral grains, the $CaCO_3$ content, and the ratio of the dissolution-resistant *G. inflata* to the more susceptible *G. bulloides.* The index was used to divide the samples into two groups, less dissolved and more dissolved, respectively. There is no difference in mean size between the two sets indicating that dissolution does not modify the mean test size of any of these species. Examination of the shapes of the size distribution curves also does not point to any preferential removal of either large or small specimens within each species.

Introduction

The shells of planktonic foraminifera are made of calcite, which is subject to dissolution in many regions of the ocean. The susceptibility to dissolution varies among species; shallow-living forms like *Globigerinoides* are the most susceptible, whereas deeper-dwelling forms, for example, *Globorotalia*, are more tolerant. The order of susceptibility has been discussed in several articles (Berger, 1968, 1970; Parker & Berger, 1971; Adelseck, 1978; Thunell & Honjo, 1981; and Malmgren, 1983).

Few studies have been carried out to determine the effects of dissolution on size distribution in planktonic foraminiferal species, although size variation in planktonic foraminifera may be useful in paleoceanographic reconstructions and evolutionary studies. For example, Stone (1956), Kennett (1968), Bé et al. (1973), Hecht (1974, 1976), and Malmgren & Kennett (1976, 1977) studied size changes in some modern planktonic foraminiferal species in relation to the physical properties of the sea water. In general, mean sizes are largest in areas where species are best adapted; warm-water species are largest in warm waters and cool-water species are largest in cool waters (Hecht, 1975; Malmgren & Kennett, 1976). Such relationships between environment and mean size in planktonic foraminifera have been employed by Malmgren & Kennett (1978) and Bé (1976) to analyse Quaternary paleoclimatic history. If size is to be used as a paleoclimatic index, we must know how dissolution influences it. Dissolution is known to preferentially remove small species, and thus to concentrate large ones (Parker & Berger, 1971; Berger et al., 1982). However, little is known about the *intraspecific* changes in size due to dissolution. Does dissolution mainly affect small specimens within a species, causing an increase in mean size, or are large specimens the ones that are dissolved, causing a decrease in mean size? Is there a similar response in all species?

In order to answer these questions, we analysed seven cool-water species of planktonic foraminifera from surface-sediment samples from the Maurice Ewing Bank (eastern Falkland Plateau), South Atlantic Ocean. These species were ranked by Malmgren (1983) in the following order from the most resistant to the most susceptible: *Neogloboquadrina pachyderma* sinistral form, *Globorotalia inflata*, *Globigerina bulloides* and *Globigerinita glutinata* (the same susceptibility), *N. pachyderma* dextral form, *Globigerinita uvula*, and *Globigerina quinqueloba*. In addition, *Globorotalia truncatulinoides*, which is not included in the size analyses, was ranked between *G. inflata* and *G. bulloides*.

The Maurice Ewing Bank is ideal for dissolution studies of planktonic foraminifera, since carbonate material is available from a wide range of water depths. Also, there are no major planktonic foraminiferal biogeographic boundaries across this narrow latitudinal interval. Statistical tests of relative abundances of the species in five samples from each of the areas to the south and to the north of the subantarctic boundary of the Polar Front (Fig. 1) that exhibited the least degree of dissolution on the basis of the dissolution index formulated here, showed that no differences exist (Malmgren, 1983).

Material and methods

Thirty-two core-top samples (topmost 1 to 2 cm), collected during *Eltanin* cruises 9 and 22, *Islas Orcadas* cruises 7 and 11, *Vema* cruises 22 and 31, and *R. D. Conrad* cruise 16 to the Maurice Ewing Bank were analyzed (Fig. 1; Table 1). Most samples were from north of the Polar Front. The samples ranged in water depth between 1200 and 3500 m. The present calcium carbonate compensation depth (CCD) in this area is at about 3500 m (Berger & Winterer, 1974). Biostratigraphic analysis indicates that all samples are of Holocene age (Malmgren, 1983).

Between 10 and 132 specimens per species were picked randomly from the > 90 μm fraction of each sample (Table 2). Sufficient sample sizes (more than 10 specimens) could not always be obtained for dissolution-susceptible species (*G. glutinata, N. pachyderma* dextral form, *G. uvula,* and *G. quinqueloba*) in highly dissolved samples (Table 2). Size was measured, using a light microscope, as the maximum width of the test (in micrometres). The width of the test was used as a measure of size instead of the length of the test, because length is much affected by size variation in the highly variable final chamber. Only undamaged specimens were included in the analysis. Specimens that showed evidence of having lost the final chamber were not included. Careful inspections of the area around the aperture revealed whether an additional chamber had been present.

The following sediment particles (> 90 μm fraction) were also counted in aliquots of the samples:

  (1) whole tests of planktonic foraminifera;

  (2) fragments of planktonic foraminifera (unidentifiable remains of planktonic foraminifera);

Fig. 1. Locations of core-top samples from the Maurice Ewing Bank used in this study. Samples are listed in Table 1.

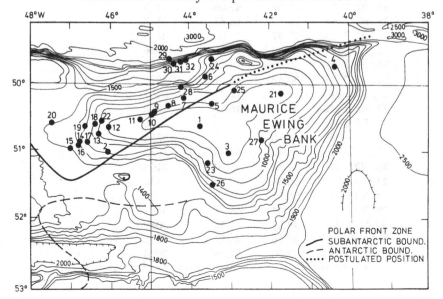

(3) benthonic foraminifera;
(4) radiolarians;
(5) mineral grains (mostly quartz).

Each particle found in the aliquots, consisting of between 450 and 1500 grains per sample, was referred to one of these categories, and the relative abundance of each category was determined.

In addition, the CaCO$_3$ content was analysed from the bulk sediment (data are from Cronblad & Malmgren, 1981). The ratio of the relative

Table 1. *Locations of sites from the Maurice Ewing Bank used in this study.*

Core	Lat. (°S)	Long. (°W)	Water depth (m)
1. E 9–2	50.62	43.60	1272
2. E22–10	50.98	46.10	2110
3. E22–11	51.00	43.05	1340
4. IO 7–1	49.69	40.40	2090
5. IO 7–40	50.31	43.42	1605
6. IO 7–42	49.87	43.64	2621
7. IO 7–43	50.22	44.16	1713
8. IO 7–44	50.31	44.54	1651
9. IO 7–45	50.42	44.88	1621
10. IO 7–46	50.47	44.96	1599
11. IO 7–47	50.56	45.31	1517
12. IO 7–48	50.65	46.09	1493
13. IO 7–49	50.74	46.34	1784
14. IO 7–50	50.86	46.77	2344
15. IO 7–51	50.96	47.04	2547
16. IO 7–52	50.92	46.83	2558
17. IO 7–53	50.87	46.62	2229
18. IO 7–54	50.60	46.39	1856
19. IO 7–55	50.63	46.65	2255
20. IO 7–57	50.59	47.52	2525
21. IO11–9	50.16	42.29	1441
22. RC16–108	50.59	46.26	1650
23. RC16–111	51.23	43.44	1716
24. V22–90	49.60	42.75	2288
25. V22–92	50.10	42.93	1686
26. V22–94	51.52	43.50	2164
27. V31–73	50.88	42.25	1732
28. V31–75	50.17	44.28	1772
29. V31–79	49.64	44.51	3449
30. V31–80	49.67	44.43	3089
31. V31–81	49.66	44.12	2952
32. V31–83	49.60	44.05	3078

abundances of *G. inflata* (dissolution resistant) and *G. bulloides* (relatively more susceptible) was also estimated from census data based on more than 300 specimens of planktonic foraminifera per sample.

In order to establish a dissolution ranking index, the relative abundances of fragments of planktonic foraminifera, benthonic foraminifera, radiolarians, and mineral grains, $CaCO_3$ content, and the *G. inflata/G. bulloides* ratio, were entered as variables into a principal components analysis (Malmgren, 1983). The rationale of formulating a single dissolution index from these variables is that the amount of their variation due to dissolution may be isolated quantitatively. Each parameter, taken separately, is a potential dissolution index, but their variation may partially be due to factors other than dissolution. The $CaCO_3$ content may also be affected by productivity of calcareous plankton in the water column and dilution by noncarbonate material. The degree of test fragmentation has been related directly to dissolution (Berger, 1970), but may also to some degree reflect mechanical breakage during bioturbation (Thunell, 1976).

Benthonic foraminifera are generally more dissolution resistant than planktonic forms (Berger, 1973), and an increase in benthonics may indicate enhanced dissolution. However, variation in benthonic abundance could also reflect changing bottom-water conditions (Thunell, 1976) or variation in productivity. Similarly, radiolarians generally should be more abundant in strongly dissolved samples, but their abundance may increase in the sediments by increased productivity at the Polar Front.

The *G. inflata/G. bulloides* ratio should not be much affected by possible differential response of these species to environmental change

Table 2. *Number of samples in which mean size was determined for different species, and means and ranges of sample sizes.*

	Number of samples	Sample size (number of specimens)	
		Mean	Range
N. pachyderma (sinistral form)	32	82	51–132
G. inflata	32	38	10–62
G. bulloides	32	46	31–73
G. glutinata	22	19	10–62
N. pachyderma (dextral form)	22	21	12–36
G. uvula	29	38	11–64
G. quinqueloba	26	39	10–75

across this narrow latitudinal belt. *G. inflata* is known to be more resistant to dissolution than *G. bulloides* (Berger, 1970), and this ratio should, therefore, represent a sensitive dissolution index in this area.

The first principal component displays a negative correlation between CaCO$_3$ content on the one hand and relative abundances of fragments of planktonic foraminifera, radiolarians, and mineral grains, and the *G. inflata/G. bulloides* ratio on the other (Table 3), and is interpreted as a dissolution index. Benthonic foraminiferal frequencies are not dissolution-dependent in this area.

The scores of individual samples on this dissolution index (coordinates on first principal component axis) indicate their degree of dissolution. A low score marks moderate dissolution and a high score marks stronger dissolution. As expected, scores are generally higher in samples from greater water depths, but they are also pronounced in some samples from shallower water (Fig. 2).

Results

In order to determine dissolution effects on the test size distribution of the various species, mean sizes were plotted against the dissolution index (Fig. 3). If mean sizes were modified by dissolution, there should be distinct differences between samples exhibiting moderate dissolution (low scores) and samples exhibiting stronger dissolution (higher scores). The plot shows that mean sizes vary greatly

Table 3. *Derivation of the dissolution ranking index from a principal components analysis of six potential dissolution-related variables. The first principal component, accounting for 45% of the variability in these variables, is interpreted as isolating the portions of their variation that are due to dissolution.*

Variable	First principal component loading
% CaCO$_3$	− 0.92[a]
% fragments of planktonic foraminifera	+ 0.54[b]
% benthonic foraminifera	− 0.21
% radiolarians	+ 0.39[c]
% mineral grains	+ 0.94[a]
G. inflata/G. bulloides ratio	+ 0.69[a]

[a] $p \leqslant 0.001$
[b] $p \leqslant 0.01$
[c] $p \leqslant 0.05$

among the samples in most species, but that there is no obvious systematic change relating to dissolution in any of them.

The dissolution effects were tested statistically by dividing the samples into two groups, the one group composed of samples that are moderately affected by dissolution (negative scores) and the other of more strongly dissolved samples (positive scores), and computing grand mean sizes for each species in the respective group. The first group consists of 17 samples and the other of 15 samples. If dissolution had had an effect, the grand means should be different between the dissolution regimes. Student's *t*-tests indicate that the differences in grand means are far from reaching statistical significance (Table 4). Consequently, there is no evidence that mean sizes are affected by dissolution.

Fig. 2. Relationship between dissolution indices and water depths. Dissolution is stronger for higher scores.

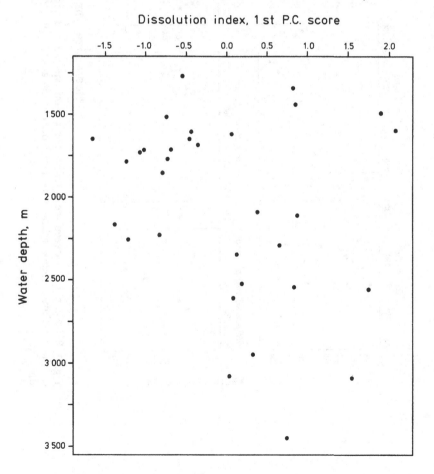

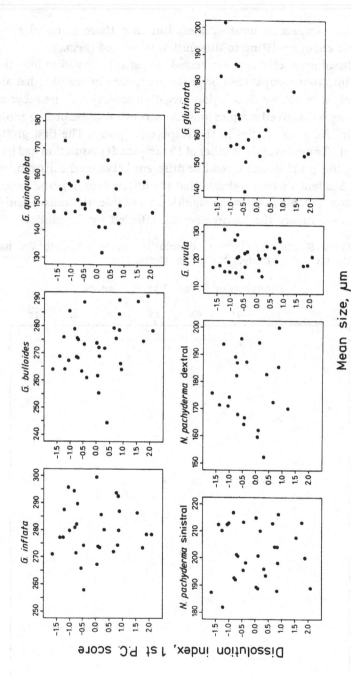

Mean size, μm

Fig. 3. Relationships between mean size of the various species and degree of dissolution. A low score on the dissolution index indicates moderate dissolution and a high score indicates stronger dissolution.

However, the fact that mean sizes are not different does not necessarily exclude the possibility that the *shapes* of the size distribution curves may differ between dissolution regimes. Thus there may be a bias toward either small or large specimens in the size distributions in strongly dissolved samples despite similar mean sizes. Fig. 4 shows histograms of the size distributions in each of the three samples that show the minimum and maximum dissolution effects (the three lowest and highest scores, respectively). In a few cases, the most strongly dissolved samples were replaced by other samples ranking below these on the dissolution index, because sufficient samples were not obtained (Fig. 4).

The histograms do not show any tendency for more pronounced skewness with enhanced dissolution that could be accounted for by selective removal of either small or large specimens. Normality tests of the distributions confirm that most of them, as is usually the case (Malmgren, 1979), conform to the log-normal distribution with regard to skewness. Only two samples, the sinistral variety of *N. pachyderma* in V22–94 and *G. inflata* in I07–52 deviated significantly from log-normality. Both these showed negative skewness (see Fig. 4).

In *G. glutinata*, there may appear to be smaller amounts of larger specimens (> 220 μm) in strongly dissolved samples, but this is probably due to a shift of mean sizes toward smaller values in the samples plotted here. The fact that the *grand means* do not differ

Table 4. *Mean sizes ($\bar{x}$; in micrometres) of various species in two dissolution regimes (samples showing moderate and stronger signs of dissolution, respectively) and t-tests of differences in means. None of the t-values are statistically significant, indicating that there are no differences in mean size. N is the number of samples and d.f. is the degrees of freedom.*

	Moderate dissolution		Stronger dissolution		t	d.f.
	N	$\bar{x}$	N	$\bar{x}$		
N. pachyderma (sinistral form)	15	203	17	202	0.27	30
G. inflata	15	279	17	281	0.41	30
G. bulloides	15	272	17	274	0.48	30
G. glutinata	15	167	7	161	0.87	20
N. pachyderma (dextral form)	14	178	8	176	0.36	20
G. uvula	15	120	14	121	0.51	27
G. quinqueloba	15	153	11	149	1.29	24

between dissolution regimes and also the log-normality of the distributions point to a definite lack of dissolution influence on the size of this species.

In conclusion, the test size of these cool-water species does not seem to be changed in any nonrandom way by dissolution on the sea floor. This is in contradiction to some previous studies of the relationship between dissolution and size in warm-water planktonic foraminiferal species, but the results of these studies are not conclusive. Berger

Fig. 4. Histograms showing size distributions of the various species in each of three samples exhibiting moderate and stronger dissolution effects (the three smallest and largest scores, respectively, on the dissolution index). Samples displaying moderate dissolution are plotted in the uppermost three rows; those displaying stronger dissolution are plotted in the lowermost three rows (D.I. shows the score on the dissolution index). The strongest dissolved samples were replaced in a few cases by samples with less dissolution, because sample sizes of susceptible species were not sufficient. Arrows mark the mean sizes. Histograms were constructed to contain an average of four specimens per interval.

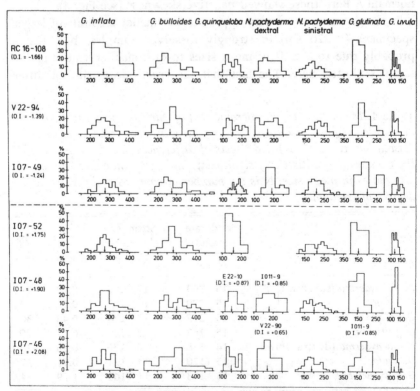

(1970) observed in his buoy experiments that small specimens of *Globigerinoides ruber* were more resistant to dissolution than larger specimens, and the same conclusion was reached by Adelseck (1978) in his laboratory experiments on *G. ruber*. Hecht *et al.* (1975) carried out experimental dissolution studies on various warm-water species (among them *G. ruber*). In their first experiment, they compared dissolution effects on size in the fractions greater and less than 250 μm, and found that the small-size fraction tended to dissolve considerably more quickly than the large-size fraction, which is in contrast to the studies by Berger (1970) and Adelseck (1978). Adelseck, however, pointed out the possibility that larger specimens of *G. ruber* may seem to be less resistant to dissolution only because they possess a fragile and easily dissolvable final chamber, which is not fully developed in ontogenetically younger specimens.

The second experiment of Hecht *et al.* (1975) was made on three species (*G. ruber, G. trilobus = G. sacculifer*, and *Orbulina universa*), in order to evaluate changes within the larger-size fraction (> 250 μm). They noted decreases in mean size with enhanced dissolution, but the validity of this observation is not clear, since no statistical tests were employed. Standard deviations were not given, but assuming a co-efficient of variation (V) of 20%, it appears that the only statistically significant change is in *G. ruber* (unbroken tests). This again supports the contention that specimens in the larger size ranges are removed selectively by dissolution in this species.

It follows that more studies are needed to fully understand how dissolution affects the test size distribution in planktonic foraminifera. High-latitude forms may differ from low-latitude forms in their response to dissolution. It may be that high-latitude forms are already preadopted with respect to size, because of higher dissolution in the water column as living forms. In any event, the lack of dissolution effects on size in cool-water species suggests that mean size in those species that closely follow modern oceanographic gradients are potentially useful Quaternary paleoenvironmental indices in sediments from cool waters (cf. Malmgren & Kennett, 1978).

Acknowledgements
I thank W. H. Berger, Scripps Institution of Oceanography, J. P. Kennett, University of Rhode Island, and J. Backman, University of Stockholm, for comments on the manuscript; D. Cassidy, Antarctic Research Facility, Florida State University, for providing the *Eltanin* and *Islas Orcadas* material; F. McCoy, Lamont-Doherty Geological

Observatory of Columbia University, for providing the *Vema* and *Conrad* material; B. Säwensten, University of Stockholm, for secretarial assistance; and S. Jevall and I. Arnström, University of Stockholm, for drafting assistance. This research was supported by a grant from the Swedish Natural Science Research Council (G 4076–101).

REFERENCES

Adelseck, C.G. Jr 1978. Dissolution of deep-sea carbonate: preliminary calibration of preservational and morphologic aspects. *Deep-Sea Res.*, **25**, 1167–1185.

Bé, A.W.H. 1976. Subtropical Convergence fluctuations and Quaternary climates in the middle latitudes of the Indian Ocean. *Science*, **194**, 419–422.

Bé, A.W.H., Harrison, S.M. & Lott, L. 1973. *Orbulina universa* d'Orbigny in the Indian Ocean. *Micropaleontol.*, **19**, 150–192.

Berger, W.H. 1968. Planktonic foraminifera: selective solution and paleoclimatic interpretation. *Deep-Sea Res.*, **15**, 31–43.

Berger, W.H. 1970. Planktonic foraminifera: selective solution and the lysocline. *Mar. Geol.*, **8**, 111–138.

Berger, W.H. 1973. Deep-sea carbonates: Pleistocene dissolution cycles. *J. Foram. Res.*, **3**, 187–195.

Berger, W.H. & Winterer, E.L. 1974. Plate stratigraphy and the fluctuating carbonate line. In: Pelagic Sediments: On Land and under the Sea, ed. K.J. Hsü & H.C. Jenkyns, *Int. Ass. Sedimentol., Spec. Publ.*, **1**, 11–48.

Berger W.H., Bonneau, M.-C. & Parker, F.L. 1982. Foraminifera on the deep-sea floor: lysocline and dissolution rate. *Oceanol. Acta*, **5**, 249–258.

Cronblad, H.G. & Malmgren, B.A. 1981. Climatically controlled variation of Sr and Mg in Quaternary planktonic foraminifera. *Nature*, **291** (5810), 61–64.

Hecht, A.D. 1974. Intraspecific variation in Recent populations of *Globigerinoides ruber* and *Globigerinoides trilobus* and their applications to paleoenvironmental analysis. *J. Paleontol.*, **48**, 1217–1234.

Hecht, A.D. 1975. An ecologic model of planktonic foraminifera. *Geol. Soc. Am. Proc.*, **7**, 1107–1108.

Hecht, A.D. 1976. An ecologic model for test size variation in Recent planktonic foraminifera: applications to the fossil record. *J. Foram. Res.*, **6**, 295–311.

Hecht, A.D., Eslinger, E.V. & Garmon, L.B. 1975. Experimental studies on the dissolution of planktonic foraminifera. *Cushman Found. Foram. Res., Spec. Publ.*, **13**, 56–69.

Kennett, J.P. 1968. Latitudinal variation in *Globigerina pachyderma* (Ehrenberg) in surface sediment samples of the southwest Pacific Ocean. *Micropaleontol.*, **14**, 305–318.

Malmgren, B.A. 1979. Multivariate normality tests of planktonic foraminiferal data. *J. Int. Assoc. Math. Geol.*, **11**, 285–297.

Malmgren, B.A. 1983. Ranking of dissolution susceptibility of planktonic foraminifera at high latitudes of the South Atlantic Ocean. *Mar. Micropaleontol.*, **8**, 183–191.

Malmgren, B.A. & Kennett, J.P. 1976. Biometric analysis of phenotypic variation in Recent *Globigerina bulloides* d'Orbigny in the southern Indian Ocean. *Mar. Micropaleontol.*, **1**, 3–25.

Malmgren, B.A. & Kennett, J.P. 1977. Biometric differentiation between Recent *Globigerina bulloides* and *Globigerina falconensis* in the southern Indian Ocean. *J. Foram. Res.*, **7**, 130–148.

Malmgren, B.A. & Kennett, J.P. 1978. Late Quaternary paleoclimatic applications of mean size variations in *Globigerina bulloides* d'Orbigny in the southern Indian Ocean. *J. Paleontol.*, **52**, 1195–1207.

Parker, F.L. & Berger, W.H. 1971. Faunal and solution patterns of planktonic
  foraminifera in surface sediments of the South Pacific. *Deep-Sea Res.*, **18**, 73–107.
Stone, S.W. 1956. Some ecologic data relating to pelagic foraminifera.
  Micropaleontol., **2**, 361–370.
Thunell, R.C. 1976. Calcium carbonate dissolution history in Late Quaternary deep-
  sea sediments, western Gulf of Mexico. *Quatern. Res.*, **6**, 281–297.
Thunell, R.C. & Honjo, S. 1981. Calcite dissolution and the modification of
  planktonic foraminiferal assemblages. *Mar. Micropaleontol.*, **6**, 169–182.

3

Middle and Late Quaternary carbonate production and dissolution, and paleoceanography of the eastern Angola Basin, South Atlantic Ocean

J. H. F. JANSEN

Abstract

A detailed study of the contents and accumulation rates of calcium carbonate in relation to the state of preservation of the calcareous skeletons in deep-sea sediments from the Zaire (Congo) fan indicates that, during the Middle and Late Quaternary, carbonate accumulation was lower during glacials than interglacials in the equatorial Atlantic. The interglacial maxima show a higher accumulation due to increased production and not to better preservation. Non-carbonate sedimentation rates are higher during glacial than interglacial periods and emphasize the downcore fluctuations of the carbonate content caused by the variations of the carbonate accumulation.

In the sediments below the present, and probably also the interglacial, river plume, strong carbonate production is attributed to enlarged zooplankton activity in an area of normal oceanic phytoplankton productivity. This special Zaire effect may have existed during glacials as well, although then phytoplankton production increased because of the occupation from the south by waters of greater oceanic productivity, due to a slight shift of the main branch of an intensified Benguela Current. This also initiated coastal upwelling as is indicated by extra carbonate production in the near-shelf zone during glacials. Here the combination of sedimentation and carbonate accumulation rates caused a carbonate, and probably also a preservation peak at 14 000 yBP which is only a local coincidence without any paleoceanographical meaning.

The glacial CCD and lysocline were about 1000 m higher than the present levels at 5600 and 4800 m. The interglacial levels were nearly as high as the glacial, caused by postdepositional carbonate dissolution below the glacial lysocline at c. 4000 m. The data from the Angola Basin sediments provide no evidence that glacial carbonate dissolution was controlled by an inflow of Antarctic Bottom Water.

Introduction

The Quaternary sediments of the Zaire (Congo) deep-sea fan in the eastern Angola Basin contain many continuous hemipelagic sections. Their relatively rapid sedimentation provides a high resolution of the various regional and ocean-wide marine and terrestrial events and offers the opportunity to study in detail these paleoceanographic and climatic events. Because of its large size the Zaire fan, extending to more than 1000 km from the African coast, covers different marine environments. This study is an attempt to characterize the various oceanic processes in the eastern Angola Basin and to distinguish local from regional and ocean-wide phenomena on the basis of their sedimentary record. Special attention is paid to the production and dissolution of carbonate organisms in relation to paleoceanography, using detailed calculations of sedimentation and carbonate accumulation rates.

The material studied consists of 33 piston cores of 4–17 m length,

Table 1. *Locations, corrected water depths, and length of five pistoncores from the eastern Angola Basin.*

Core number	Core length	Latitude	Longitude	Corrected water depth
T78–33	1012 cm	5°11.0′S	7°58.0′E	4120 m
T78–45	1081 cm	7°47.8′S	10°07.0′E	4070 m
T78–46	1102 cm	6°50.1′S	10°45.3′E	2100 m
T78–49	1028 cm	3°57.8′S	8°03.7′E	4340 m
T80–7	1716 cm	7°00.0′S	9°01.9′E	3946 m

Fig. 1. Map of the study area with core locations. The numerals indicate the cores T78–33, –45, –46, –49, and T80–7.

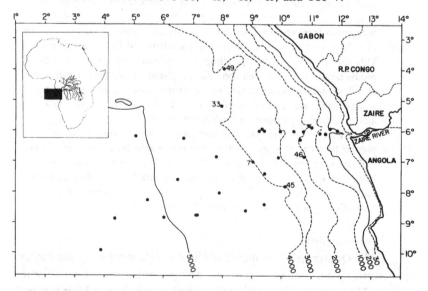

taken in the area of the Zaire deep-sea fan during two expeditions by the Netherlands Institute for Sea Research with the MV *Tyro* in 1978 and 1980 (Fig. 1 and Table 1). The cores are described in Jansen *et al.* (1984b).

Sedimentation and carbonate accumulation

Stratigraphy

The Quaternary hemipelagic sediments of the Zaire fan show fluctuations in calcium carbonate content that are very typical of the entire Atlantic Ocean: high carbonate during warm periods and low carbonate during cold periods (Arrhenius, 1952). The fluctuations are generally attributed to variations in postdepositional dissolution, but there is a debate over whether they are due to increased or decreased production of cold and erosive bottom waters during glacial stages. For a recent discussion see Volat *et al.* (1980) and Olausson (1981). The

Table 2. *Conventional radiocarbon ages for several intervals in five pistoncores from the eastern Angola Basin.*

| Core no. | Depth in core (cm) | Carbonate | | Organic carbon | |
		Date no.	Age (years)	Date no	Age (years)
T78–33	2–22	GrN–11417	3310 ± 70	–	–
	48–75	GrN–11416	9360 ± 120	GrN–11425	13890 ± 170
	262–282	–	–	GrN–11784	26700 ± 750
T78–45	6–18	GrN–11421	4060 ± 100	GrN–11427	5040 ± 130
	35–55	–	–	GrN–11426	11600 ± 150
	212–223	–	–	GrN–11773	26200 ± 650
	290–300	–	–	GrN–11772	30400 ± 1100
T78–46	7–17	GrN–10763	$94.3 \pm 1.8\%$	GrN–10775	1920 ± 70
	102–130	GrN–10939	5930 ± 160	GrN–10935	6890 ± 70
	269–282	GrN–10764	13840 ± 110	GrN–10776	13160 ± 130
	312–332	GrN–10938	15430 ± 270	GrN–10934	15170 ± 70
	464–486	GrN–10937	18550 ± 300	GrN–10933	19990 ± 110
	710–722	–	–	GrN–10930	25370 ± 350
	807–829	GrN–10765	$27480 ^{+2100}_{-1700}$	GrN–10777	$27970 ^{+2100}_{-1700}$
	962–970	–	–	GrN–11774	$35400 ^{+2700}_{-2000}$
T78–49	19–31	GrN–11423	4460 ± 100	–	–
	61–70	GrN–11422	9000 ± 130	–	–
	250–262	–	–	GrN–11776	26900 ± 650
	358–381	–	–	GrN–11775	30350 ± 650
T80–7	18–32	GrN–11415	5980 ± 80	GrN–11424	6960 ± 150
	63–82	–	–	GrN–11778	13520 ± 200

fluctuations are used to define a carbonate stratigraphy with stages named in Arabic numerals conforming to the oxygen isotope stages by Emiliani (1955) and Shackleton & Opdyke (1973) (Jansen et al., 1984b). Even numerals are given to cold periods and odd numerals to warm periods. Its time scale is valid for the last 40 000 years as is demonstrated by a series of radiocarbon datings (Figs. 2 and 8; Table 2; Jansen et al., 1984b). This means that the scale can be applied to the older core intervals, the oldest core (T78–38 at 5490m water depth) reaching back to about half a million years ago. The stratigraphy has been tested and confirmed by investigations of microfossils (Bjørklund & Jansen, 1984; de Ruiter et al., 1984; Mikkelsen, 1984; Zachariasse et al., 1984) ^{230}Th-excess dating (Jansen et al., 1984b) and stable isotopes (Olausson, 1984). Fig. 2 shows the stratigraphy for the representative hemipelagic core T80-7. The ages older than 40 000 years are mainly adopted from Shackleton & Opdyke (1976) and Berggren et al. (1980). A detailed account of the core stratigraphy is given in Jansen et al. (1984b).

Fig. 2. Carbonate fluctuations, carbonate stages, and ages in the Zaire deep-sea fan as represented by core T80–7. Ages are given in thousands of years. Two radiocarbon dates (years) are given in roman numerals (carbonate date) or italics (organic carbon date).

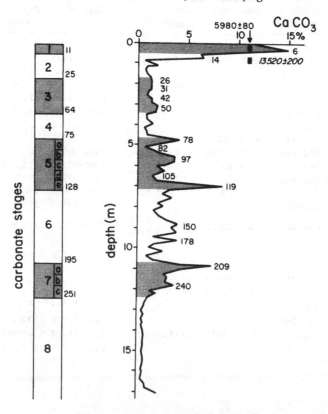

Downcore fluctuations

Calcium carbonate content of sediment is always the result of the combined influence of biogenic carbonate production, carbonate dissolution in the sea water and on the sea floor, and dilution by non-carbonate material. The detailed stratigraphy of our cores allows the determination of non-carbonate sedimentation rates and carbonate accumulation rates (Jansen *et al.*, 1984b). This is done with the aim of distinguishing the effect of dilution from those of production and dissolution. Gardner (1975), who related the length of glacial and interglacial sections to their carbonate percentages, calculated that the downcore variations in carbonate are controlled by all three factors.

We have calculated the carbonate accumulation rates for the well documented hemipelagic cores in terms of weight quantities. This is done using the water and carbonate contents which have been measured downcore at distances of 20 cm or less. Detailed information could be obtained from five cores. It appears that the highest peaks of the interglacial carbonate stages 1, 5e, and 7a all fall in a period of maximum accumulation (Fig. 3). Other, minor peaks are sometimes also related to high accumulation, but usually the data are not accurate enough for this observation. In the non-carbonate sedimentation

Fig. 3. Cumulative carbonate accumulation of four representative cores T78–33, T78–45, T78–49, and T80–7. The slopes of the curves represent the accumulation rates. Carbonate stages are indicated.

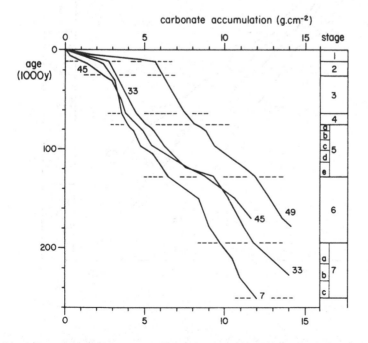

pattern the high rates generally correspond to the carbonate minima (stages 2, 4, 5b, and 6), while most low rates coincide with carbonate peaks (Fig. 4). It follows that the downcore carbonate pattern is attributed to variations in carbonate production and dissolution, which together controlled the accumulation. The opposite pattern of terrigenous non-carbonate sedimentation emphasizes this effect.

The shallower cores (<3800 m water depth) have a deviating carbonate pattern (Fig. 8, core T78–46). They demonstrate a carbonate peak at c. 14000 yBP, while the Holocene peak at 6000 yBP is suppressed. The carbonate accumulation curve of T78–46 shows that this peak falls in a period of strong accumulation from 27000 to c. 12500 yBP, reaching a maximum between 18600 and 15400 yBP (Fig. 5). The non-carbonate sedimentation rate, however, decreased around 14500 yBP following the general postglacial pattern of the area. This caused a relative enrichment with carbonate at 14500 yBP that ended at 12500 yBP by decreasing accumulation, and produced a carbonate peak at

Fig. 4. Cumulative non-carbonate sedimentation of four representative cores T78–33, T78–45, T78–49, and T80–7. The slopes of the curves represent the sedimentation rates. Carbonate stages are indicated.

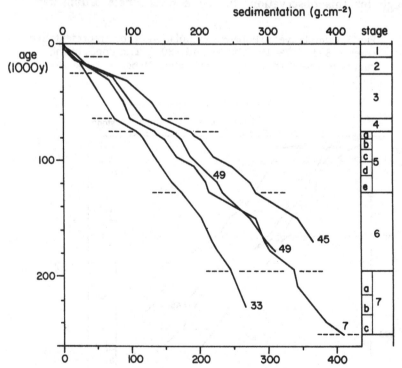

14 000 yBP which is only the result of the coincidence of two independent events.

Regional distribution

For the sedimentation and carbonate accumulation rates of the best documented cores we have found that rather constant proportions occur between the rates of characteristic periods like interglacial and glacial stages, interglacial maxima, and glacial minima of the last 200 000 years. The proportions for the interglacial maximum, average, and glacial minimum rates for carbonate are 1.6:1.0:0.6, and for non-carbonate 0.6:1.0:1.5 (Jansen *et al.*, 1984b). It follows that the maps of the averaged rates (Figs. 6 and 7) can be used for the different time intervals by applying these proportions.

The non-carbonate sedimentation map demonstrates two general trends. The first trend of isolines follows the isobaths and includes a decrease from the continent towards deeper waters. It suggests that in the area where it is most dominant, the continental shelf is an important sediment source. A second trend shows an extensive lobe of high sedimentation rates stretching from the canyon oceanwards. In the

Fig. 5. Cumulative carbonate accumulation and non-carbonate sedimentation of core T78–46. The slopes of the curves represent the accumulation and sedimentation rates. Carbonate stages are indicated.

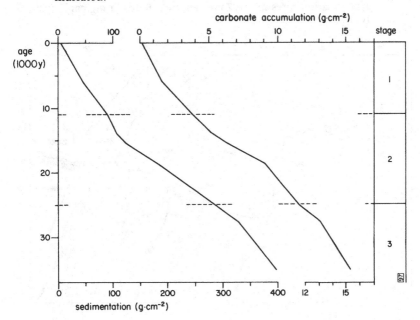

Fig. 6. Averaged Middle and Late Quaternary non-carbonate sedimentation rates in the Zaire fan area. Rates in g.cm^{-2} per 1000y.

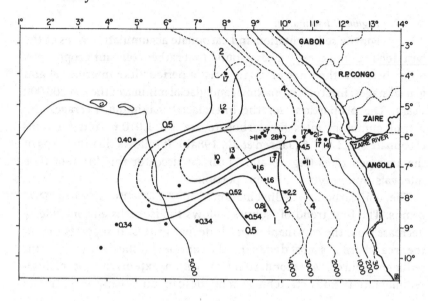

Fig. 7. Averaged Middle and Late Quaternary carbonate accumulation rates in the Zaire fan area. Rates in mg.cm^{-2} per 1000y.

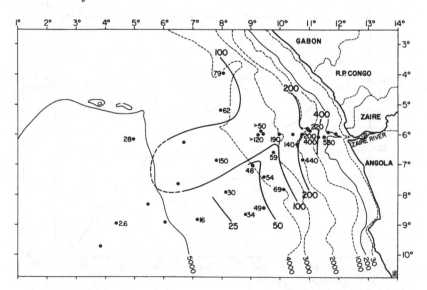

central fan area it coincides with an area of Late Quaternary turbidites (Van Weering & Jansen, unpublished results). We conclude that the lobe is caused by sedimentation processes of the Zaire deep-sea fan.

The lines of equal carbonate accumulation rates also follow the isobaths, due to the well known effect of stronger dissolution with increased waterdepth (Berger, 1973). The relatively high rates in the central area point to the presence of a lobe of high carbonate accumulation as well, with a distribution pattern comparable to that of the non-carbonate sedimentation. Absolute numbers of foraminifera and coccoliths per gram sediment are low in the lobe (Zachariasse *et al.*, 1984) but this is easily explained by the large differences in non-carbonate sedimentation inside and outside the lobe.

Carbonate production and dissolution

Downcore fluctuations

If the carbonate accumulation in a sediment is known, the state of preservation of the carbonate skeletons may give an insight into the attribution to it of carbonate production and dissolution. The calcareous microfossils from several cores have been investigated by Zachariasse *et al.* (1984) and it appears that the carbonate peaks, which generally are also carbonate accumulation peaks, are associated with preservation peaks in the coccolith flora (Fig. 8). This indicates that dissolution plays an important role in the carbonate accumulation processes in the eastern equatorial Atlantic Ocean. On the other hand, the prominent carbonate peaks of the stages 1, 5e, and 7a are distinguished from other peaks by increased carbonate accumulation, but not by their coccolith dissolution idexes. Since this indicates that the differences in dissolution between the peaks are very limited, the prominent peaks represent periods of increased production of carbonate by coccolithophores and foraminifera.

A special case is formed by the cores from the shallower area above 3800 m water depth, as represented by core T78–46, where a period of strong carbonate accumulation occurred from 27 000 to c. 12 500 yBP (Fig. 5). Since the preservation of the carbonate skeletons in the interval 27 000–15 400 yBP is not better but even worse than in the intervals above and below (Fig. 8), it also represents a period of increased carbonate production lasting till 15 400 yBP or somewhat later. Its influence, however, was limited to a small zone off the African coast for about 150 nautical miles wide. A coccolith preservation peak synchronizes with the carbonate peak at 14 000 yBP in core T78–46. The decrease of the carbonate accumulation at c. 12 500 yBP, apart from other possibilities, may be connected to a decrease of the carbonate production around that time. If this is true it would have lowered the amount of carbonate that was disposable to dissolution by carbon dioxide and would have caused a preservation low. Together with the general postglacial improvement of the preservation this may have resulted in a local preservation peak at 14 000 yBP, which then is also a coincidence, like the carbonate peak.

Fig. 8. Carbonate contents, coccolith dissolution indexes (from Zachariasse *et al.*, 1984) and carbonate stages of four cores from the Zaire deep-sea fan. Radiocarbon datings of carbonate are given in roman numerals and of organic carbon in italics.

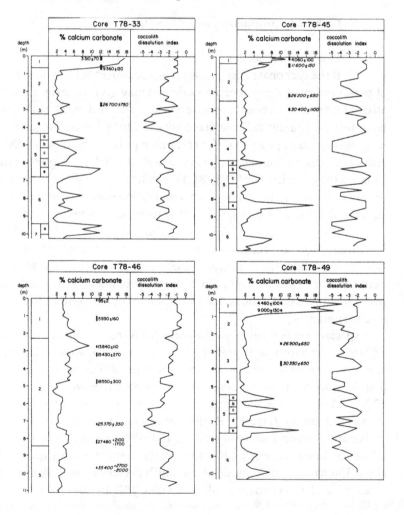

Regional distribution

The preservation pattern of the Holocene core tops shows that the carbonate dissolution is more severe in the lobe of strong carbonate accumulation than around it (Zachariasse *et al.*, 1984). Following the same reasoning as above, increased carbonate production must also have caused the high accumulation rates here.

Paleoceanography

We deduce above the following major events which closely mirror oceanographic processes in the eastern equatorial Atlantic Ocean.

(1) During the last *c.* 200 000 years a lobe of high carbonate accumulation was present in the central area of the Zaire fan, which for the Holocene is due to strong carbonate production.

(2) The interglacial maxima (carbonate stages 1, 5e, 7a) are also characterized by increased carbonate production.

(3) In a zone *c.* 150 nautical miles wide along the African coast, strong carbonate production occurred from 27 000 to *c.* 15 000 yBP with a possible maximum between 18 600 and 15 400 yBP.

Hydrography and recent and interglacial sediments

A major feature in the Angola Basin is a plume of low salinity Zaire water extending westward till *c.* 800 km from the river mouth (Fig. 9A). Below the plume and around it a confrontation of three different water masses, the northward-moving Benguela Current, deflected westward at 20°S by the eastward Equatorial Under Current and South Equatorial Counter Current, causes a very complicated pattern of fronts, gyres, and domes. This brings nutrient-rich shallow subsurface water into the photic zone especially south of the Zaire fan, resulting in an increase in primary production (van Bennekom & Berger, 1984).

The Zaire River is distinguished from other large rivers by the presence of a deep canyon-head in its river mouth, giving ocean water easy access to the estuary which is very narrow (4 to 5 km wide). This forces a rapid outflow of river water in a thin, sharply bounded turbid surface layer, 10 m thick (Eisma & van Bennekom, 1978) and causes a displacement of the phytoplankton production to a narrow zone outside the estuary in waters with salinities of about 30‰ (van Bennekom *et al.*, 1978; Cadée, 1978, 1984). This phytoplankton forms the source for a large plume with high values of dissolved organic carbon (DOC) which covers roughly the lobe of high carbonate accumulation. As one of the possible mechanisms for the conversion of phytoplankton into dis-

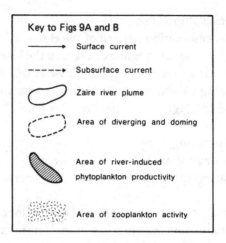

Key to Figs 9A and B

→   Surface current

--→   Subsurface current

⬭   Zaire river plume

Area of diverging and doming

Area of river-induced
phytoplankton productivity

Area of zooplankton activity

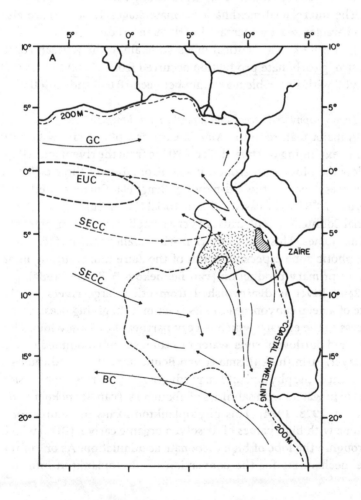

solved organic carbon Cadée mentions excretion by zooplankton. Thus the DOC zone probably marks an area with an active zooplankton population but without high phytoplankton productivity (Cadée, 1984).

The carbonate in the sediments consists mainly of coccoliths and to a lesser extent foraminifera shells. Foraminifera are animals while the majority of coccolithophores are grazed by zooplankton and reach the deep-sea floor mainly via rapidly sinking faecal pellets (Honjo, 1976). This means that a carbonate rain towards the ocean floor is a sign of

Fig. 9. Generalized oceanic circulation in the eastern Angola Basin.
A. Present circulation, mainly after van Bennekom & Berger (1984).
B. Glacial circulation.
BC: Benguela Current, GC: Guinea Current, SECC: South Equatorial Counter Current, EUC: Equatorial Under Current.

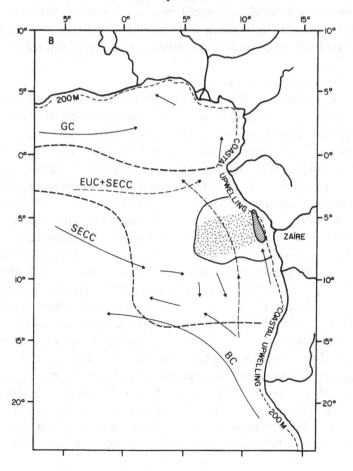

zooplankton consumption rather than phytoplankton production. Usually both features are coincidentally high, but the Zaire plume with its large DOC contents forms an exception to this rule. Here the source of food of the zooplankton is probably phytoplankton produced in the narrow zone off the river mouth. The ocean in the plume area is deeper than 2000 m, so that it can be expected that the zooplankton utilize all available phytoplankton and only little organic matter escapes from the pelagic system to be deposited on the ocean floor. This is in accordance with the measurements made by Menzel & Ryther (1961) who calculated that in the pelagic system of the Sargasso Sea the respiration of zooplankton equals the primary productivity. We infer from this that the Holocene lobe of high carbonate production in the Zaire fan is caused by increased zooplankton production which leads to the sinking of carbonate skeletons. Like the Holocene, the two interglacial maxima documented in our cores are distinguished by high carbonate production. Since it is generally accepted that their climates also much resembled the present climate, we assume that the Holocene oceanographic situation is representative of both interglacial maxima.

Glacial sediments and paleoceanography

The position of the accumulation lobe remained the same during interglacial–glacial oscillations in the oceanic regime, although it is not known whether it was also a lobe of glacial carbonate production. On the other hand, there are no reasons to believe that the major controlling conditions of the characteristic Zaire effect, the dimensions of the estuary and the easy access of ocean water, were very different from present-day conditions. The deep and narrow valley on the Zaire mouth continues across the shelf and is still a narrow canyon at the shelf break at 120 m waterdepth (Eisma & van Bennekom, 1978; Jansen et al., 1984a). A glacial sea level fall of c. 110 m (Giresse et al., 1981) would have caused an equally narrow and even deeper estuary near the shelf edge, with an easier access of ocean water. It is therefore probable that the effect existed during glacials as well. In contrast to the interglacials there is diatom and radiolarian assembly evidence which indicates that nutrient-rich conditions existed during glacials (Bjørklund & Jansen, 1984; Mikkelsen, 1984) pointing to a larger phytoplankton production. This suggests that the fan area became influenced by the front and gyre system which at present is situated south of it. This is in accordance with reconstructions from foraminifera and radiolarian associations in the Guinea Basin and Angola Basin by Gardner & Hays (1976) and Morley & Hays (1979) who concluded that the Benguela Current intensified during the glacial maximum of 18 000 yBP. The stronger Benguela Current may have pushed northwards the front and gyre system of the interacting water masses (Fig. 9B).

The coastal zone of glacial carbonate production indicates that coastal upwelling accompanied the intensified Benguela Current, analogous to the present coastal upwelling of cold Benguela Current water further south off the Namib desert as described by Schuette & Schrader (1981). This interpretation, already proposed by van Zinderen-Bakker (1967), is corroborated by the signals of glacial aridity in deep-sea sediments (Jansen *et al.*, 1984b) at the shelf (Caratini & Giresse, 1979; Giresse & LeRibault, 1981; Giresse *et al.*, 1981) and the African continent (see Jansen *et al.*, 1984b). The fall at 15 400 yBP in carbonate production recorded in core T78–46 coincides with a drastic change from a subantarctic planktonic foraminiferal association indicative of eutrophic waters to a tropical–subtropical assemblage (Zachariasse *et al.* 1984). Apparently, the influence of the glacial Benguela Current decreased at the latitude of the Zaire mouth between 15 400 and 14 500 yBP. A short but intense return of subantarctic species occurred at approximately 11 400 yBP, and has been correlated by Olausson (1984) and Zachariasse *et al.* (1984) to a cold event dated in the North Atlantic Ocean at *c.* 10 500 yBP. It represents a short excursion to the north of the Benguela Current coastal upwelling system.

The switch toward tropical and subtropical microfossils around 15 000 yBP is almost simultaneous with the start of the last pluvial with accompanying high lake levels in tropical Africa around 13 000 yBP (Street & Grove, 1976) and is in agreement with the scarce information from the Zaire river area about the transition to moister conditions (de Ploey, 1965; Livingstone, 1975; Nicholson & Flohn, 1980). Late Quaternary lake levels of Lake Tanganyika, observed by Livingstone (1975), show a postglacial rise that was interrupted by a transitory fall between 11 000 and 10 000 yBP. This indicates an arid episode which is probably connected to the short oceanic shift at 11 400 yBP mirrored in core T78–46.

The Benguela Current

Transfer function analyses of radiolarians from surface sediments in the South Atlantic by Morley & Hays (1979) identified a gyre margin assemblage which is associated with the waters of the Benguela Current. They reconstructed a paleotemperature pattern of the last glacial maximum at 18 000 yBP with a tongue of cold Benguela Current water in the Guinea Basin at 0° latitude. Comparable work on foraminifera indicates the same tongue (Gardner & Hays, 1976). The radiolarian data, however, allow us to draw a similar Holocene tongue as well, with the contribution of this gyre margin assemblage differing only in being smaller than at 18 000 years ago. It is situated at the same place and has the same northern boundary with the assemblage attributed to the waters of the warm Guinea Current. However, in the present-day water mass above this assumed Holocene tongue, Benguela Current water is not recognized by its physical and chemical properties (van Bennekom & Berger, 1984). Moreover, if a much stronger Benguela Current had intruded into the Guinea Basin, why then would it not have driven back or even displaced the Guinea Current. We suggest, therefore, that not only is temperature a controlling factor of the gyre margin assemblage, but that the state of fertility is also a highly important property. If this is true, the contribution of this assemblage to the radiolarians of the sediments in the Guinea Basin is probably due to increased primary production caused by oceanic divergence and doming, similar to the present situation in the area south of the Zaire fan (Fig. 9A). The glacial intensification of the Benguela Current may have enhanced the diverging forces in the borderland of the two currents, resulting in an increase in fertility and the contribution of the gyre margin radiolarian assemblage associated with it. The westward deflection of the main branch moved probably only a few degrees to the north which is in accordance with the assumed displacement of the southern area of the oceanic doming towards the Zaire fan region (Fig. 9B).

(Paleo) carbonate compensation depth and lysocline

The carbonate compensation depth (CCD) and lysocline can be determined in the classical way with a plot of the calcium carbonate content of surface sediments versus water depth, giving the usual very scattered diagram known from many publications (see, for example, the plot of the carbonate contents in the samples of this study in Jansen et al. (1984b). A clearer picture can be obtained by utilizing carbonate accumulation rates instead of percentages and by using only carefully selected (hemi) pelagic cores with no signs of reworked material. Since the dilution effect is eliminated, the accumulation rates show less variation and allow us to draw conclusions from a limited amount of data. Fig. 10 shows the results for the cores from the northeast Angola Basin which are described in Jansen et al. (1984b). A present CCD of 5600 m is in accordance with data from the same latitude in the Angola Basin by Thunell (1982), who placed the CCD below 5400 m. A plot by

Ellis & Moore (1973) of percentages from the entire Angola Basin indicating a CCD of 5800–6000 m is also a good agreement with our value which is not a maximum for the basin. Fig. 10 indicates a lysocline between 4500 and 4900 m, that fits well with the data of Berger (1968) giving a depth for the Zaire fan area of 4700–4800 m. The plot by Thunell (1982) demonstrated a sharp increase in the rate of dissolution at *c.* 4800 m and data from the whole Angola Basin collected by Ellis & Moore (1973) and Biscaye *et al.* (1976) indicated the lysocline at 4800–5000 m. Accordingly, the present carbonate lysocline is placed at 4800 m.

The positions of the CCD and lysocline are also easily identified for the different climatic periods. The climate optimums of stages 5e and 7a suggest a CCD of 4500–4700 m and a lysocline at *c.* 4000 m, while the glacial maxima of stages 2 and 4 and the beginning and end of stage 6 have had a CCD of 4300–4500 m and a lysocline at 3800–4000 m. This is a

Fig. 10. Relationship between carbonate accumulation rates and water depth for the Holocene (black dots), the interglacial maximum stages 5e and 7a (open circles), and the glacial minimum stages 2, 4, and 6 (triangles) on the northeastern Angola Basin.

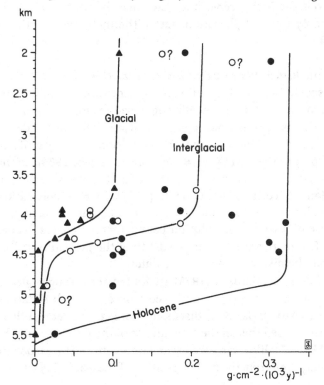

perplexing result since the levels of the CCD and lysocline during the climatic optima approximate more closely those of the glacials rather than the recent levels. On a worldwide basis the oceanic as well as the continental climatic regimes of these interglacial periods very much resemble the present situation instead of the glacials. The high interglacial levels must be the result of postsedimentary carbonate dissolution. During the subsequent glacial periods in the regions deeper than 4000 m, the position of the interglacial sediments was only 25 cm or less below the sea floor and could have become exposed to glacial hydrographic conditions due to benthic mixing. This applies particu-larly to the depths below the level where carbonate dissolution starts, in other words the depths below the glacial lysocline. As a result the apparent interglacial CCD and lysocline tend to approximate the depths of the subsequent glacial episode: during the glacial climatic minima the CCD rose 900–1100 m and the lysocline c. 800 m in the eastern Angola Basin.

This rise casts doubt on the calculations of interglacial CCD and lysocline values for the Atlantic Ocean, and on glacial values for the Indian and Pacific Oceans with their opposite carbonate signals, not only throughout the Pleistocene, but also since the Middle Miocene when presumably the first glaciation started (Kennett, 1982; Keller & Barron, 1983).

Antarctic Bottom Water and carbonate dissolution

As discussed above, dissolution is a major factor controlling the accumulation of carbonate in hemipelagic sediments. Within the Angola Basin there is no water mass boundary around 4800 m, the depth of the present lysocline. The oceanographic conditions are very uniform in the deeper part of the basin (van Bennekom & Berger, 1984), so the lysocline must be related to a transition level of carbonate saturation to undersaturation. According to Thunell (1982) such a correlation cannot be found, but his saturation values do not exclude a transition at 4800 m because of a lack of dissolution parameters for waters deeper than 4400 m. Therefore we attribute the depth of the present lysocline to the regional chemistry of the water column.

Bornhold (1973) and Gardner (1975) attributed the stronger glacial dissolution to an increase in the inflow of corrosive, CO_2-rich Antarctic Bottom Water (AABW). Gardner, discussing cores from among others the Guinea Basin, based this on the coincidence of the lysocline and the top of the AABW. AABW was thus assumed to control the carbonate dissolution today, and from the higher position of the glacial lysocline it was deduced that the top of the AABW had risen in the Guinea Basin during glacials. Bornhold studied cores from the Angola Basin. Van Bennekom & Berger (1984), however, demonstrated that AABW is only recognizable in a thin layer near the sills of the Romanche Trench in the north and Walvis Ridge in the south. It quickly loses its identity in the Angola Basin by vertical mixing with North Atlantic Deep Water. This eliminates the AABW as a cause of actual carbonate dissolution in the Angola Basin. Consequently the data from the Angola Basin provided no information on glacial dissolution by AABW.

Conclusions

The Quaternary carbonate fluctuations, characteristic of the Atlantic Ocean, are due to variations in carbonate accumulation. Their pattern is emphasized by the opposite pattern of non-carbonate sedimentation causing extra dilution with terrigenous matter during cold periods.

The carbonate accumulation rates are largely controlled by dissolution processes. The relationships with the state of preservation of the calcareous microfossils indicate extra carbonate production during the Holocene and interglacial maxima of stages 1, 5e, and 7a.

For the Holocene, and presumably also for the interglacials, this production is attributed to increased activity of zooplankton in the Zaire river plume, which is not accompanied by high phytoplankton productivity in the same area. This is related to hydrographic features connected to the deep incision of the Zaire canyon into the river mouth.

The high zooplankton activity probably existed during glacials as well. Moreover, the plume area became occupied from the south by a complex water mass of three interacting ocean currents in which nutrient-rich subsurface water is brought into the photic zone by diverging and doming, causing increased phytoplankton production.

In the shallower coastal zone high carbonate accumulation occurred during glacial stage 2. Based on the state of preservation we infer that it is the result of increased carbonate production. In combination with the non-carbonate sedimentation pattern it created peaks in the carbonate content and preservation, only as a coincidence of two independent processes. These peaks, both at 14 000 yBP, do not reflect any

44 J. H. F. Jansen

oceanographic event. The production, between 27 000 and c. 15 000 yBP, originates from glacial coastal upwelling in relation to an intensified Benguela Current, which also had a postglacial excursion at c. 11 400 yBP.

During the last glacial maximum the Benguela Current shifted only a few degrees to the north. The presence of gyre margin radiolarian assemblages in the Guinea Basin and the northern Angola Basin (Morley & Hays, 1979) reflects its nutrient content rather than its temperature.

The present CCD is 5600 m and the lysocline lies at 4800 m at the latitude of the Zaire fan, while the glacial values are 4300–4500 m and 3800–4000 m, respectively. The levels of the interglacial maxima, however, approximate the glacial levels. This is caused by postsedimentary carbonate dissolution at depths below 4000 m, and casts doubt on all paleo-CCD and lysocline values after the first glaciation of the Middle Miocene.

In the Angola Basin carbonate dissolution is controlled by the regional chemistry of the deeper waters. No evidence is available from the Angola Basin for a glacial inflow of Antarctic Bottom Water.

Acknowledgements

Thanks are due to Bert Aggenbach for the drawings and to Lidy Everhardus for typing the manuscript. I am indebted to W. G. Mook, Isotope Physics Laboratory, University of Groningen for the ¹⁴C datings. This work was financially supported by the Netherlands Council of Oceanic Research, who chartered the MV *Tyro* for the expeditions.

REFERENCES

Arrhenius, G. 1952. Sediment cores from the east Pacific. *Rep. Swed. Deep Sea Exped. 1947–1948*, 5, 1–228.

Bé, A. H. W., Damuth, J. E., Lott, L. & Free, R. 1976. Late Quaternary climatic record in western equatorial Atlantic sediment. *Geol. Soc. Am. Mem.*, 145, 165–200.

Bennekom, A. J. van & Berger, G. W. 1984. Hydrography and silica budget of the Angola Basin. *Neth. J. Sea Res.*, 17 (2–4), 149–200.

Bennekom, A. J. van, Berger, G. W., Helder, W. & de Vries, R. T. P. 1978. Nutrient distribution in the Zaire estuary and river plume. *Neth. J. Sea Res.*, 12 (3/4), 296–323.

Berger, W. H. 1968. Planktonic Foraminifera: selective solution and paleoclimatic interpretation. *Deep-Sea Res.*, 15 (1), 31–43.

Berger, W. H. 1973. Deep-sea carbonates: Pleistocene dissolution cycles. *J. Foram. Res.*, 3 (4), 187–195.

Berggren, W.A., Burckle, L.H., Cita, M.B., Cooke, H.B.S., Funnell, B.M., Gartner, S., Hays, J.D., Kennett, J.P., Opdyke, N.D., Pastouret, L., Shackleton, N.J. & Takayanagi, Y. 1980. Towards a Quaternary time scale. *Quat. Res.*, 13 (3), 277–302.

Biscaye, P.E., Kolla, V. & Turekian, K.K. 1976. Distribution of calcium carbonate in surface sediments of the Atlantic Ocean. *J. Geophys. Res.*, 81 (5), 2595–2603.

Bjørklund, K.R. & Jansen, J.H.F. 1984. Radiolaria distribution in Middle and late Quaternary sediments and palaeoceanography in the eastern Angola Basin. *Neth. J. Sea Res.*, 17 (2–4), 299–312.

Bornhold, B.D. 1973. Late Quaternary sedimentation in the eastern Angola Basin. *Techn. Rep. Woods Hole WHOI*, 73 (8), 1–213.

Cadée, G.C. 1978. Primary production and chlorophyll in the Zaire river, estuary and plume. *Neth. J. Sea Res.*, 12 (3/4), 368–381.

Cadée, G.C. 1984. Particulate and dissolved organic carbon and chlorophyll a in the Zaire river, estuary and plume. *Neth. J. Sea Res.*, 17 (2–4), 426–440.

Caratini, I.C & Giresse, P. 1979. Palynological contribution to the study of continental and marine environments in the Congo at the end of the Quaternary. *C.R. Hébd. Scéances Acad. Sci.*, Sér. D, 288 (4), 379–382.

Eisma, D. & van Bennekom, A.J. 1978. The Zaire river and estuary and the Zaire outflow in the Atlantic Ocean. *Neth. J. Sea Res.*, 12 (3/4), 255–272.

Ellis, D.B. & Moore, T.C. Jr 1973. Calcium carbonate, opal, and quartz in Holocene pelagic sediments and the calcite compensation level in the South Atlantic Ocean. *J. Mar. Res.*, 31 (3), 210–227.

Emiliani, C. 1955. Pleistocene temperatures. *J. Geol.*, 63 (6), 538–578.

Emiliani, C. 1966. Paleotemperature analysis of Caribbean cores P6304–8 and P6304–9 and a generalized temperature curve for the past 425,000 years. *J. Geol.*, 74 (2), 109–124.

Gardner, J.V. 1975. Late Pleistocene carbonate dissolution cycles in the eastern equatorial Atlantic. In: W.V. Sliter, A.H.W. Bé & W.H. Berger (eds.), Dissolution of deep-sea carbonates. *Cushman Found. Foram. Res. Spec. Publ.*, 13, 129–141.

Gardner, J.V. & Hays, J.D. 1976. Responses of sea-surface temperature and circulation to global climatic change during the past 200,000 years in the eastern equatorial Atlantic Ocean. *Geol. Soc. Am. Mem.*, 145, 221–246.

Giresse, P. & LeRibault, L. 1981. Contribution de l'étude exoscopique des quartz à la reconstitution paléogéographique des derniers épisodes du Quaternaire littoral du Congo. *Quat. Res.*, 15 (1), 86–100.

Giresse, P., Jansen, J.H.F., Kouyoumontzakis, G. & Moguedet, G. 1981. Les fonds du plateau continental congolais et le delta sous-marin du fleuve Congo. *Trav. Doc. Office de la Recherche Scientifique et Technique Outre-Mer, Paris*, 138, 13–45.

Honjo, S. 1976. Coccoliths: production, transportation and sedimentation. *Mar. Micropaleontol.*, 1 (1), 65–79.

Jansen, J.H.F., Giresse, P. & Moguedet, G. 1984a. Structural and sedimentary geology of the Congo and southern Gabon continental shelf, a seismic and acoustic reflection survey. *Neth. J. Sea Res.*, 17 (2–4), 364–384.

Jansen, J.H.F., van Weering, T.C.E., Gieles, R. & van Iperen, J. 1984b. Middle and Late Quaternary oceanography and climatology of the Zaire–Congo fan and the adjacent eastern Angola Basin. *Neth. J. Sea Res.*, 17 (2–4), 201–249.

Keller, G. & Barron, J.A. 1983. Paleoceanographic implications of Miocene deep-sea hiatuses. *Geol. Soc. Am. Bull.*, 94 (5), 590–613.

Kennett, J.P. 1982. *Marine Geology.* Prentice Hall Inc., Englewood Cliffs, 813 pp.

Ledbetter, M.T. 1979. Fluctuations of Antarctic Bottom Water velocity in the Vema Channel during the last 160,000 years. *Mar. Geol.*, 33 (1/2), 71–89.

Livingstone, D.A. 1975. Late Quaternary climatic change in Africa. In: R.F. Johnston, P.W. Frank & C.D. Michener (eds.), *Annual Review of Ecology and Systematics*, vol. 6. Annual Rev. Inc., Palo Alto Calif., pp.249–280.

Menzel, D.W. & Ryther, J.H. 1961. Zooplankton in the Sargasso Sea off Bermuda and its relation to organic production. *J. du Conseil*, 26, 250–258.

Mikkelsen, N. 1984. Diatoms in the Zaire deep-sea fan and Pleistocene palaeoclimatic trends in the Angola Basin and west equatorial Africa. *Neth. J. Sea Res.*, **17** (2–4), 280–292.

Morley, J.J. & Hays, J.D. 1979. Comparison of glacial and interglacial oceanographic conditions in the South Atlantic from variations in calcium carbonate and radiolarian distributions. *Quat. Res.*, **12** (3), 396–408.

Nicholson, S.E. & Flohn, H. 1980. African environmental and climatic changes and the general atmospheric circulation in late Pleistocene and Holocene. *Climatic Change*, **2** (4), 313–348.

Olausson, E. 1981. Dissolution and carbonate fluctuations in Pleistocene deep-sea cores: a reply. *Mar. Geol.*, **41** (3/4), 339–342.

Olausson, E. 1984. Oxygen and carbon isotope analyses of a Late Quaternary core in the Zaire (Congo) fan. *Neth. J. Sea Res.*, **17** (2–4), 276–279.

Ploey, J. de, 1965. Position géomorphologique, génèse et chronologie de certains dépôts superficiels au Congo occidental. *Quaternaria*, **9**, 131–154.

Prell, W.L. & Hays, J.D. 1976. Late Pleistocene faunal and temperature patterns of the Colombia Basin, Caribbean Sea. *Geol. Soc. Am. Mem.*, **145**, 201–220.

Ruiter, R.S.C. de, Jansen, J.H.F. & van Weering, T.C.E. 1983. Middle and Late Quaternary silicoflagellates from the Angola Basin. *Neth. J. Sea Res.*, **17** (2–4), 293–298.

Schuette, G. & Schrader, H. 1981. Diatoms in surface sediments: a reflection of coastal upwelling. In: F.A. Richards (ed.), *Coastal Upwelling*. Am. Geophys. Union, Washington DC, pp. 372–380.

Shackleton, N.J. & Opdyke, N.D. 1973. Oxygen isotope and paleomagnetic stratigraphy of equatorial Pacific core V28–238: oxygen isotope temperatures and ice volumes on a 10^5 year and 10^6 year scale. *Quat. Res.*, **3** (1), 39–55.

Shackleton, N.J. & Opdyke, N.D. 1976. Oxygen-isotope and paleomagnetic stratigraphy of Pacific core V28–39 late Pliocene to latest Pleistocene. *Geol. Soc. Am. Mem.*, **145**, 449–464.

Street, F.A. & Grove, A.T. 1976. Environmental and climatic implications of late Quaternary lake-level fluctuations in Africa. *Nature*, **261** (5559), 385–390.

Thunell, R.C. 1982. Carbonate dissolution and abyssal hydrography in the Atlantic Ocean. *Mar. Geol.*, **47** (3/4), 165–180.

Volat, J.-L., Pastouret, L. & Vergnaud-Grazzini, C. 1980 Dissolution and carbonate fluctuations in Pleistocene deep-sea cores: a review. *Mar. Geol.*, **34** (1/2), 1–28.

Zachariasse, W.J., Schmidt, R.R. & van Leeuwen, R.J.W. 1984. Distribution of Foraminifera and calcareous nannoplankton in Quaternary sediments of the eastern Angola Basin in response to climatic and oceanic fluctuations. *Neth. J. Sea Res.*, **17** (2–4), 250–275.

Zinderen-Bakker, E.M. van, 1967. Upper Pleistocene and Holocene stratigraphy and ecology on the basis of vegetation changes in sub-Saharan Africa. In: W.W. Bishop & J.D. Clark (eds.), *Background to Evolution in Africa*. Univ. Chicago Press, Chicago, pp. 125–146.

4

Late Quaternary upwelling history off southwest Africa (DSDP Leg 75, HPC 532)

L. DIESTER-HAASS

Abstract

A course fraction analysis was performed on the uppermost four cores of hydraulic piston core 532. Variations in opal (radiolarian and diatom) content are interpreted in terms of varying upwelling influence. During interglacial periods there was a higher opal content than during the glacial periods. The reason has been found in the Benguela Current, which, during interglacials turns westward in the area of 20°S, carrying eddies of upwelled water to the west, whereas in glacial periods the Benguela Current flowed parallel to the coast until about 17°S, without leaving an upwelling signal in Site 532 in 20°S.

Introduction

During DSDP Leg 75 a hydraulic piston core (HPC) was taken on the eastern Walvis Ridge off south-west Africa, at 1341 m water depth, about 180 km offshore (10°31.13E, 19°44.61S). Abundant siliceous biogenous particles, besides calcareous shells, occurred at this site and the question arose as to the influence of coastal upwelling, although this site is outside the area of intense coastal upwelling. Today, upwelling is concentrated on the inner shelf and over the shelf edge (Summerhayes, 1983).

Are these siliceous remains, mainly radiolaria and diatoms, due to upwelling influence? Which mechanism is responsible for indicators of upwelling influence on the upper continental slope in the Walvis Ridge Terrace area? What about temporal variations in upwelling influence in Site 532?

Materials and methods

The questions above were studied by means of a coarse fraction analysis carried out on 20 cm-spaced samples from cores 1–4 of HPC 532.

The samples were prepared for the coarse fraction analysis (> 40 μm) according to the procedure described in Diester-Haass (1975). Within the sand fraction, five subfractions were obtained (40–63, 63–125, 125–250, 250–500 and 500–1000 μm) and in each fraction about 800 grains counted, if present. Various biogenous, terrigenous and autigenous components were distinguished (see Diester-Haass, 1985).

The stratigraphic control was performed by means of an oxygen isotope curve for cores 1–3, which was established by Nick Shackleton in Cambridge.

4   Fig. 1. Results of HPC 532. Left-hand side column: Core, section, and depth in metres below surface. Small horizontal lines indicate position of investigated samples. CC = core catcher.
   A. Oxygen isotope curve and oxygen isotope stages (established by N. Shackleton, Cambridge).
   B. Composition of the sand fraction: radiolaria, benthos, planktonic foraminifera and others.
   C. Percentage of radiolaria and diatoms in the 40–63 μm fraction.
   D. Percentage of terrigenous matter in the 40–63 μm and > 63μm fractions.

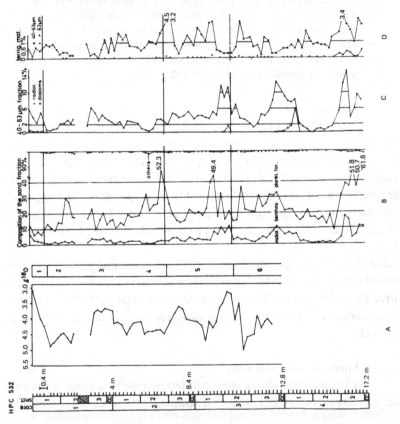

Results

The data presented are from the uppermost four cores, that is oxygen isotope stages 1–6, and perhaps stage 7 in the lowest part (Fig. 1A).

The amount of sand fraction ($> 63 \mu m$) decreases regularly from 35% in the Holocene to $< 5\%$ in stage 6. The 40–63 μm fraction forms a constant amount of 2–6%

E. Fragmentation of planktonic foraminifera, calculated as fragmented tests/(fragmented tests + whole tests) × 100.

F. Benthos/plankton foraminiferal ratio, calculated as benthonic foraminifera (benthonic foraminifera + planktonic foraminifera) × 100.

G. Radiolarian/planktonic foraminiferal ratio, calculated as radiolaria/(radiolaria + planktonic foraminifera) × 100.

H. Radiolarian/benthonic foraminiferal ratio, calculated as (radiolaria/(radiolaria + benthonic foraminifera)) × 100.

I. Percentage of CaCO$_3$ in the sediments.

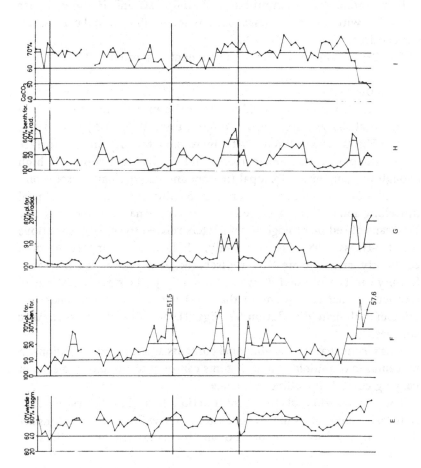

The composition of the sand fraction (Fig. 1B) shows widely varying amounts of benthos and planktonic foraminifera which form the most important part of the sand fraction. Radiolaria form up to 10% of the fraction.

Within the 40–63 µm fraction, radiolaria form up to 14% and sometimes diatoms occur (Fig. 1C). Terrigenous matter occurs in very small amounts in the >40 µm fractions (Fig. 1D), its maxima being related to maxima in benthos (cf. Figs. 1B and D).

The ratio of fragmented versus whole tests of planktonic foraminifera – as a measure of calcium carbonate dissolution (Thunell, 1976) – shows strong variations (Fig. 1E) from <40% up to 70% fragments of the sum whole tests plus fragments. Maxima in fragmentation are related to maxima in benthos (Fig. 1B) and to maxima in benthos/plankton foraminiferal ratios (Fig. 1F).

Both the radiolarian/planktonic foraminiferal ratio and the radiolarian/benthos foraminiferal ratio (Figs. 1G and H show similar variations with maxima in warm periods and minima in cool periods (cf. oxygen isotope curve, Fig. 1A).

The CaCO₃ content of the total sediment is plotted in Fig. 1I.

Discussion

The opal content of sediments can be an indicator of fertility in continental margin areas (Goll & Bjorklund, 1974; Maynard, 1976; Diester-Haass, 1977, 1983). With increasing fertility in the surface water, diatom production increases and finally opal supply is high enough to compensate for opal (diatom and radiolarian) dissolution.

Until now, a distinction between fertility produced by coastal upwelling and fertility produced by river supply has only been possible by means of sedimentological parameters related to climatic variations (Diester-Haass, 1983). In our study, all factors are in favour of a constantly arid climate. River discharge as a reason for increased fertility can be excluded (Diester-Haass, 1985). The small variations in terrigenous matter amount in the >40 µm fractions are related to calcium carbonate dissolution (cf. Figs. 1D and E) and not to climatic changes.

This means that the varying opal contents as shown by variations in percentages of radiolaria and diatoms can be interpreted by means of varying coastal upwelling influence.

The composition of the sand fraction (Fig. 1B) shows varying radiolarian amounts, but it is not clear as to how far these variations are due to varying dilution by benthos and by planktonic foraminifera. So

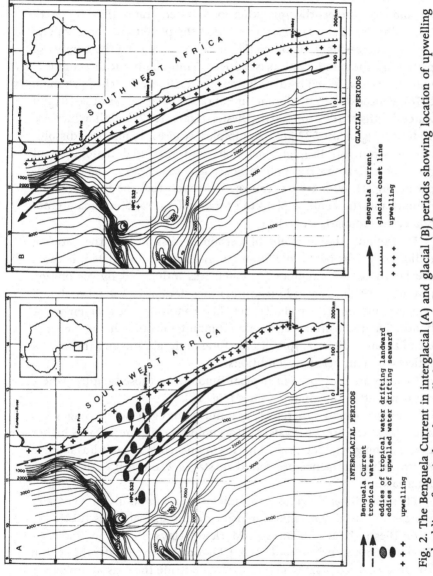

Fig. 2. The Benguela Current in interglacial (A) and glacial (B) periods showing location of upwelling and eddies of tropical and upwelled water (based on results of Hart & Currie, 1960 and van Zinderen-Bakker, 1976).

the ratio of radiolaria versus planktonic foraminifera has been calculated: it shows higher radiolarian proportions in warm intervals than in cold ones. But planktonic foraminifera are subject to dissolution, as one can see from the fragmentation curve of planktonic foraminifera (Fig. 1E) and from the benthos/plankton foraminiferal ratios (Fig. 1F). With increasing calcium carbonate dissolution the planktonic foraminiferal tests are increasingly dissolved (Berger et al., 1982) and the number of planktonic foraminiferal tests is reduced. Thus benthonic foraminifera, which are less affected by dissolution, are relatively enriched (Thunell, 1976). Benthos/plankton foraminiferal ratios have values of 5–10% in water depths of 1000 to 1500 m (Diester-Haass, 1982) and they increase up to > 50% in Site 532 in those samples, where foraminiferal dissolution is strongest. It might be, that with increased dissolution of planktonic foraminifera, radiolaria are also relatively enriched. Comparing Fig. 1E with 1G one can see, however, that the fragmentation minima and the benthos/plankton foraminiferal ratio minima are not related to radiolarian maxima.

To exclude the problem of carbonate dissolution, the ratio of radiolaria versus benthonic foraminifera has been calculated, because the benthonic foraminifera are much less influenced by dissolution (Thunell, 1976). The radiolarian/benthos foraminiferal ratio should remain constant under similar fertility conditions in a given water depth. The radiolarian/benthonic foraminiferal ratio shows the same trend as the radiolarian/planktonic foraminiferal ratio and thus really reflects varying fertility conditions.

Diatoms are only present in small amounts in those samples, where radiolaria attain maximum values (Fig. 1C).

One can conclude, therefore, that during oxygen isotope stages 1, 3 and 5 (i.e. the warm intervals) opal accumulation in Site 532 was higher than during glacial intervals. Minima in opal content and thus minima in upwelling influence are found in stages 2 and 4 (i.e. the coldest intervals).

These findings seem to contradict the assumptions of generally increased coastal upwelling intensity during glacial times (CLIMAP, 1976; Diester-Haass, 1978, 1983; Morley & Hays, 1979; van Zinderen-Bakker, 1980; Sarnthein et al., 1982). But one has to consider the special conditions of Site 532, which is situated outside the coastal upwelling area, 180 km offshore. The site is situated in an area where the Benguela Current turns westward from its SE–NW direction (Hart & Currie, 1960; van Zinderen-Bakker, 1976; Martin, 1981) (Fig. 2A). From the north, tropical water masses shift southward until about 20°S. In this area of 20°S, eddies of tropical water drifting towards the coast and eddies of upwelled water drifting from the near-coastal area of upwelling seaward are generated (Hart & Currie, 1960; Hagen *et al.*, 1981). The seaward-drifting eddies of upwelled water contain biogenous opal skeletons (Hart & Currie, 1960, Goll & Bjørklund, 1974; Schuette & Schrader, 1981) and thus are responsible for the relatively high opal content in sediments from interglacial periods in Site 532.

During glacial periods, however, the Benguela Current did not turn westward in the area of Site 532, but it flowed northward until about 17°S parallel to the coast (van Zinderen-Bakker, 1976) (Fig. 2B). Coastal upwelling most probably continued to occur on the glacial inner shelf, but there is no reflection of this upwelling in the glacial sediments of Site 532, because no westward-turning current carried the signal to the Walvis Ridge Terrace.

These results do not allow any conclusion as to the intensity of the near-coastal upwelling phenomena during glacial periods.

First results on the Early Quaternary and Pliocene sediments of HPC 532 reveal the same phenomenon of increased interglacial upwelling influence related to good carbonate preservation, and low opal content in colder periods related to strong carbonate dissolution. The only difference is that the opal amount increased considerably in the Early Quaternary warm intervals compared to the Late Quaternary ones (Diester-Haass *et al.*, 1985).

Conclusions

Results of a coarse fraction analysis performed on the uppermost four cores of HPC 532 revealed:

(1) Strongly varying amounts of benthos, planktonic foraminifera and radiolaria.

(2) Higher radiolarian and diatom contents in interglacial than in glacial periods.

(3) A varying opal content which is not a result of changes in dissolution and thus dilution by calcium carbonate. $CaCO_3$ dissolution maxima, as revealed by high fragmentation of planktonic foraminifera and high benthos/plankton foraminiferal ratios are not related to opal maxima.

The high interglacial (oxgyen isotope stages 1, 3, and 5) opal content is explained by the influence of the Benguela Current, which turns westward in the area of Site 532 in interglacial periods. It transports eddies of upwelled water from the upwelling centre on the shelf to the Walvis Ridge Terrace.

The low glacial (oxygen isotope stages 2, 4, and 6) opal content is explained by the Benguela Current, which flowed parallel to the coast until about 17°S in glacial periods, without turning westward in the area of the Walvis Ridge.

Acknowledgements

The samples were provided by the US National Science Foundation, Grant OCE 80–22 109 and OCE 78–25 448. I thank N. Shackleton of Cambridge University for having done the oxygen isotope study. Thanks are due to J. Gardner, P. Meyers and P. Rothe for helpful discussions. I gratefully acknowledge the laboratory work done by M. Killinger and E. Hartmann in Mannheim and I thank P. Rothe for providing working facilities. The financial support of the Deutsche Forschungsgemeinschaft is gratefully acknowledged.

REFERENCES

Berger, W.H., Bonneau, M.-C. & Parker, F.L. 1982. Foraminifera on the deep-sea floor, lysocline and dissolution rate. *Oceanol. Acta*, 5 (2), 249–258.

CLIMAP Project members, 1976. The surface of the ice-age earth. *Science*, 191, 1131–1144.

Diester-Haass, L. 1975. Sedimentation and climate in the Late Quaternary between Senegal and the Cape Verde Islands. *'Meteor' Forsch. Ergebn.*, C, 20, 1–32.

Diester-Haass, L. 1977. Radiolarian/planktonic foraminiferal ratios in a coastal upwelling region. *J. Foram. Res.*, 7 (1), 26–33.

Diester-Haass, L. 1978. Sediments as indicators of upwelling. In: *Upwelling Ecosystems*, ed. R. Boje & M. Tomczak, Springer, Berlin, Heidelberg, pp. 261–281.

Diester-Haass, L. 1982. Indicators of water depth in bottom sediments of the continental margin off West Africa. *Mar. Geol.*, 49, 311–326.

Diester-Haass, L. 1983. Differentiation of high oceanic fertility in marine sediments caused by coastal upwelling and/or river discharge off Northwest Africa during the Late Quaternary. In: *Coastal Upwelling. Its Sediment Record*, ed. J. Thiede & E. Suess, NATO Conf. Ser., IV, 10 b, Plenum Press, New York, London, pp. 399–419.

Diester-Haass, L. 1985. Late Quaternary sedimentation on the Eastern Walvis Ridge, SE Atlantic (HPC 532, IPOD Leg 75, and neighboured piston cores). *Mar. Geol.*, (in press).

Diester-Haass, L., Meyers, P. and Rothe, P. 1985. Origin of light-dark sediment cycles in HPC 532B (IPOD LEG 75) on the Walvis Ridge Terrace (SW Africa). *Marine Geology*, in press.

Goll, R.M. & Bjørklund, K.R. 1974. Radiolaria in surface sediments of the South Atlantic. *Micropal.*, 20 (1), 38–75.

Hagen, E., Schemainda, R., Michelsen, N. Postel, L. Schulz, S. & Below, M. 1981. Zur küstensenkrechten Struktur des Kaltwasserauftriebs vor der Küste Namibias. *Geodät., geophys. Veröff.*, series IV, part 36, 99 pp.

Hart, T.J. & Currie, R.T. 1960. The Benguela Current. *Discovery Reports*, 31, 123–298.

Martin, R.A. 1981. Benthonic Foraminifera from the Orange–Lüderitz shelf, Southern African Continental Margin. *Bull. Mar. Geoscience Unit*, No. 11, Dept. of Geol., Univ. Cape Town, 82 pp.

Maynard, N. 1976. Relationship between diatoms in surface sediments of the Atlantic Ocean and the biological and physical oceanography of overlying waters. *Paleobiology*, **2**, 99–121.

Morley, J.J. & Hays, J.D. 1979. Comparison of glacial and interglacial oceanographic conditions in the South Atlantic from variations in calcium carbonate and radiolarian distributions. *Quat. Res.*, **12**, 396–408.

Sarnthein, M., Thiede, J., Pflaumann, U., Erlenkeuser, H., Fütterer, D., Koopmann, B., Lange, H. & Seibold, E. 1982. Atmospheric and oceanic circulation patterns off NW Africa during the past 25 million years. In: *Geology of the Northwest African Continental Margin*, ed. U. von Rad, K. Hinz, M. Sarnthein & E. Seibold, Springer, pp. 545–604.

Schuette, G. & Schrader, H.-J. 1981. Diatom taphocoenoses in the coastal upwelling area off South West Africa. *Mar. Micropal.*, **6**, 131–155.

Summerhayes, C.P. 1983. Sedimentation of organic matter in upwelling regimes. In: *Coastal Upwelling. Its Sediment Record*, ed. J. Thiede & E. Suess, NATO Conf. Ser., IV, 10 b, Plenum Press, New York, London, pp. 29–72.

Thunell, R.C. 1976. Calcium carbonate dissolution history in Late Quaternary deep-sea sediments, Western Gulf of Mexico. *Quat. Res.*, **6**, 281–297.

van Zinderen-Bakker, E.M. 1976. The evolution of Late Quaternary paleoclimates of Southern Africa. *Paleoecology of Africa*, vol. IX, Balkema, Cape Town, pp. 160–202.

van Zinderen-Bakker, E.M. 1980. Comparison of Late Quaternary climatic evolutions in the Sahara and the Namib–Kalahari region. *Paleoecology of Africa*, vol. XII, Balkema, Rotterdam, pp. 381–394.

5

History of the Walvis Ridge. A précis of the results of DSDP Leg 74

T. C. MOORE, Jr, P. D. RABINOWITZ, P. E. BORELLA,
N. J. SHACKLETON and A. BOERSMA

Five sites were drilled along a transect of the Walvis Ridge. The basement rocks range in age from 69 to 71 Ma; the deeper sites are slightly younger, in agreement with the seafloor-spreading magnetic lineations. Geophysical and petrological evidence indicates that the Walvis Ridge was formed at a mid-ocean ridge at anomalously shallow elevations. The basement complex associated with the relatively smooth acoustic basement in the area consists of pillowed basalt and massive flows alternating with nannofossil chalk and limestone that contain a significant volcanogenic component. Basalts are quartz tholeiites at the ridge crest and olivine tholeiites downslope. The sediment sections are dominated by carbonate oozes and chalks; volcanogenic material, probably derived from sources on the Walvis Ridge, is common in the lower parts of the sediment columns (up through lower Paleocene sediments).

Paleodepth estimates based on the benthic fauna are consistent with a normal crustal-cooling rate of subsidence of the Walvis Ridge. The shallowest site in the transect sank below sea level in the late Paleocene, and benthic fauna indicate a rapid lowering of the sea level in the mid-Oligocene.

Average accumulation rates during the Cenozoic indicate three peaks in the rate of supply of carbonate to the seafloor: early Pliocene, late middle Miocene, and late Paleocene to early Eocene. Carbonate accumulation rates for the rest of the Cenozoic averaged 1 g/cm² per 1000 y. These peaks in accumulation are found at all sites (Fig. 1 and 2) at depths ranging from 1000 m to about 4000 m; thus, they are thought to result from increased productivity, not from changes in dissolution or winnowing.

Dissolution had a marked effect on sediment accumulation in the deeper sites, particularly during the mid-early Miocene, late Oligocene, and middle to late Eocene. Changes in accumulation rates with depth (Fig. 2) indicate that the upper part of the water column had a greater degree of undersaturation with respect to carbonate during times of high productivity. Even when the carbonate compensation depth (CCD) was below 4400 m, a significant amount of carbonate was dissolved at the shallower sites. This is evidenced by the higher accumulation rates of

Fig. 1. Map (inset) and transect locations of sites of DSDP Leg 74

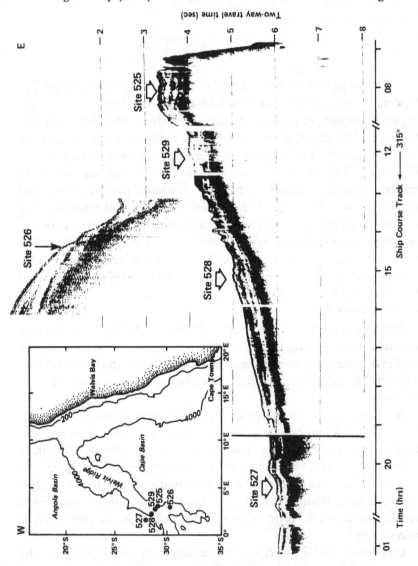

coarse-grained carbonate at the shallowest site (526, 1054 m) compared with the adjacent deeper site (525, 2467 m).

From late Paleocene to late Oligocene times there is evidence for a mid-depth zone (2000–2500 m) of poor preservation. During the mid to late Eocene, a shallow CCD virtually merged with this mid-depth zone of

Fig. 2. A carbonate accumulation rate (mg/cm² per 1000y) of the ≤63 (calcareous nannofossil) size fraction versus age. Data were plotted at 1 Ma increments for each site at the paleodepth of the site. A simple thermal cooling curve was used to estimate paleodepth. The 0, 50, 500, 1000, and 2000 mg/cm² per 1000y contours are shown. Regions of no fine-grained carbonate accumulation are shown in dark shading; regions of low carbonate accumulation (≤50 mg/cm² per 1000y) are shown in light shading.
B. Carbonate accumulation (mg/cm² per 1000y) of the >63 μm (foraminifera) size fraction. Plotted as in Fig. 2A. Contours are 0, 50, 100, 200, 400, 600, and 800 mg/cm² per 1000y. Regions with no coarse-grained carbonate accumulation are shown in dark shading. Accumulation rate data are from Shackleton et al. (1984). The sharp maximum in accumulation at the Oligocene–Miocene boundary is largely a result of an error in the time scale as applied to the calcareous nannofossil stratigraphy.

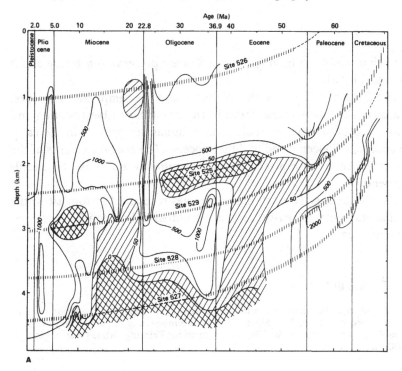

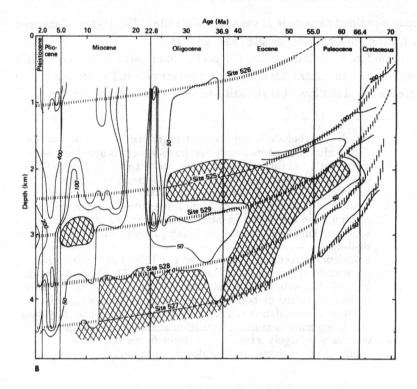

corrosion, resulting in very poor carbonate preservation below about 2000 m (Fig. 2).

The flora and fauna of the Walvis Ridge are temperate in nature. Warmer water faunas are found in the uppermost Maestrichtian and lower Eocene sediments; cooler water faunas are present in the early Paleocene, Oligocene, and middle Miocene. The boreal elements of the lower Pliocene are replaced by more temperate forms in the middle Pliocene.

The Cretaceous/Tertiary boundary was recovered in four sites; the sediments contain well-preserved nannofossils but poorly preserved foraminifera.

REFERENCES

Shackleton, N. J. and members of the shipboard party. 1984. Accumulation rates in Leg 74 sediments. In: T. C. Moore Jr, P. D. Rabinowitz, *et al.*, *Initial Reports of the Deep Sea Drilling Project*, **74**, US Government Printing Office (Washington, DC), pp. 621–44.

6

Cyclic variations in calcium carbonate and organic carbon in Miocene to Holocene sediments, Walvis Ridge, South Atlantic Ocean

W. DEAN and J. GARDNER

Abstract

The entire upper Miocene to Holocene sedimentary sequence recovered in a hydraulic piston core at DSDP Site 532 on Walvis Ridge shows distinct cycles in amount of $CaCO_3$ that correlate with dark and light cycles of sediment color. The average periodicities of the carbonate cycles for the Quaternary, upper Pliocene, and lower Pliocene are about 35, 46, and 28 ky, respectively, with an overall average of about 36 ky for the last 5 my. Most minima in carbonate abundance correspond to dark parts of color cycles, and most maxima correspond to the lightest parts of color cycles. The darker parts of color cycles usually contain higher concentrations of organic carbon, but organic carbon does not follow the color cycles in detail. Organic-carbon cycles were analyzed only for the last 2.5 my, and for this interval they have an average periodicity of about 34 ky.

The carbonate and color cycles persist through more than 5 my during which major changes in relative proportions of siliceous–biogenic, calcareous–biogenic, and terrigenous–clastic components occurred in response to climate change and to the waxing and waning of the Benguela Current upwelling system off southwest Africa. The cyclic nature of these sediments probably is the result of dilution by terrigenous clastic material and not dissolution of carbonate. We believe that the forcing mechanism that produced the cyclicity was external to the area of Site 532. Because of the similarity among the periodicities of the Walvis Ridge cycles and those of carbonate cycles in the northeastern Atlantic, Caribbean, and eastern equatorial Pacific, these cycles probably are responses to global events. We conclude that fluctuations in global sea level with an average period of about 36 ky during the last 5 my caused variations in influx of terrigenous clastic material from the African continental margin.

Introduction

Walvis Ridge is a structural spur that projects southwestward from the continental margin of southwest Africa off Namibia, and is beneath the cold, nutrient-rich, Benguela Current upwelling system (Fig. 1). Site 532 of the Deep Sea Drilling Project (DSDP) is located on the eastern part of Walvis Ridge at a water depth of 1331 m near DSDP Site 362, which was drilled with rotary methods. Hydraulic piston coring at Site 532 recovered a nearly continuous, undisturbed 291-m sediment section of late Miocene to Holocene age. The section was divided into three lithostratigraphic sub-units (units 1a–c in Fig. 2) on the basis of relative proportions of siliceous and calcareous microfossils (mostly diatoms and nannofossils) and nonbiogenic material (mostly clay) (Hay, Sibuet *et al.*, 1982). Shipboard descriptions of the section at Site 532 noted cycles of dark and light sediment (Fig. 3) that contain abrupt changes in abundances of foraminifers and diatoms (Hay, Sibuet *et al.*, 1984), but these changes in microfossil abundance are not systematic within the color cycles. We thought initially that the changes in color and abundances of siliceous and calcareous microfossils might reflect changes in productivity associated with the intensity of the Benguela Current upwelling system. We therefore chose this site to construct detailed stratigraphies of carbonate and organic carbon to see if there are cyclic variations that can be correlated with variations in other sediment parameters.

Fig. 1. Map of the continental margin off southwest Africa showing locations of DSDP Sites 362 and 532.

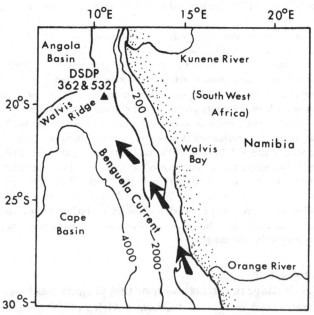

Description of the cycles

Color cycles

The most noticeable characteristic of the sediment at Site 532 is the cyclic dark and light variations in sediment color (Fig. 3). The contacts between light and dark interbeds usually are gradational over about 10 to 20 cm. We determined the periodicities of the color cycles by counting the number of cycles between time datums established by nannofossil zones (Gardner *et al.*, 1984). The color cycles have periodicities of 55, 58, and 30 ky for the Quaternary, upper Pliocene, and lower Pliocene, respectively.

The lighter parts of most cycles contain more $CaCO_3$ (Fig. 4), and the darker parts contain more clay. The darker parts of the cycles usually contain more organic carbon although there are commonly high concentrations of organic carbon that do not appear to correspond to color changes (Fig. 4). The average concentrations of $CaCO_3$ and organic carbon in 85 light-colored beds are 51.5% and 2.07%, respectively (Meyers *et al.*, 1984). The average concentrations of $CaCO_3$ and organic carbon in 65 dark-colored beds are 31.8% and 3.71%, respectively. Smear-slide estimates of biogenic components show that variations in abundances of siliceous and calcareous microfossils do not appear to be systematically related to color variations (Gardner *et al.*, 1984).

Carbonate and organic-carbon cycles

The entire section in Hole 532 was sampled at 20-cm intervals for analyses of $CaCO_3$. This sampling resulted in 1057 samples with an average sampling interval of about 5000 y/sample. All samples were analyzed for $CaCO_3$ using the gasometric technique of Hülsemann (1966). One hundred and fifty shipboard analyses of $CaCO_3$ by the carbonate bomb method (Meyers *et al.*, 1984) from Holes 532 and 532B were merged with the larger gasometric data set. These 150 shipboard samples were collected from 85 light-colored and 65 dark-colored beds and are plotted separately by bed color in Fig. 4. Four hundred and twenty of the 20-cm carbonate samples between 0 and 109 m also were analyzed for organic carbon by the LECO carbon analyzer method. The only analyses of organic carbon that we have for the interval between 109 and 250 m are from Meyers *et al.* (1984) (Fig. 4). The raw data for $CaCO_3$ and organic carbon are available in Gardner *et al.* (1984).

The results of analyses of $CaCO_3$ and organic carbon, plotted in Figs. 2 and 4, show that the average concentration of $CaCO_3$ tends to decrease from about 60% at the bottom of the section to about 25% at 55 m, and

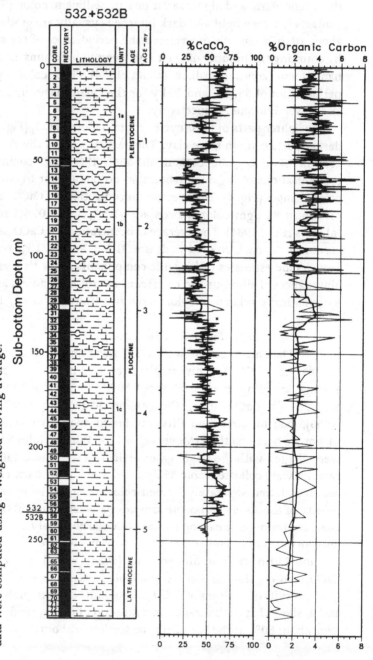

Fig. 2. Lithology, age, % CaCO₃, % organic carbon, % total nonbiogenic material, % total calcareous biogenic components, % total siliceous biogenic components, % diatoms, and % biogenic silica in sediments recovered at DSDP Site 532. Percentages of biogenic and nonbiogenic components are from smear-slide estimates (Hay, Sibuet *et al.*, 1984). Percent biogenic silica is from Dean & Parduhn (1984). Percentages of CaCO₃ and organic carbon are from Gardner *et al.* (1984). Smoothed curves for % CaCO₃, % organic carbon, and the smear-slide data were computed using a weighted moving average.

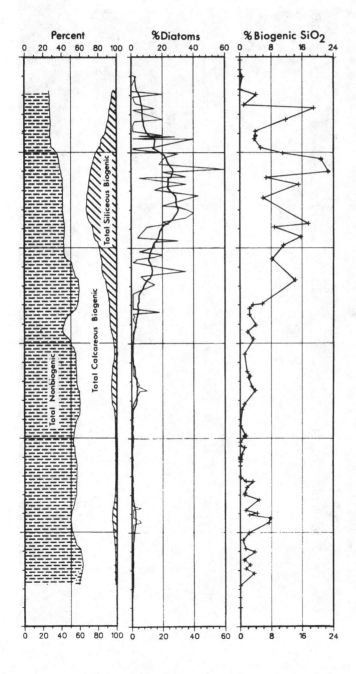

Fig. 3. Photograph of Cores 9 and 10, Hole 532, showing cyclic dark and light variations in sediment color.

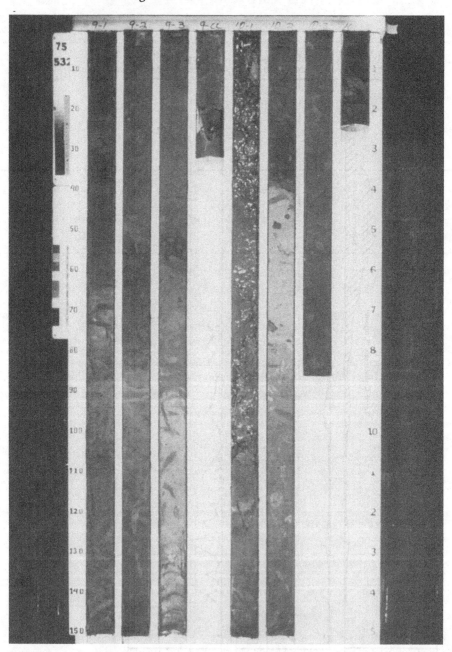

Fig. 4. Plots of percent CaCO₃ and organic carbon (C-org) in samples from 85 light-colored beds and 65 dark-colored beds from Site 532. Data are from Meyers *et al.* (1984).

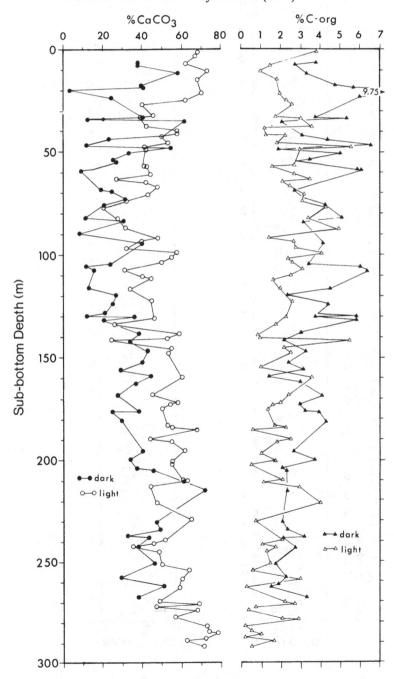

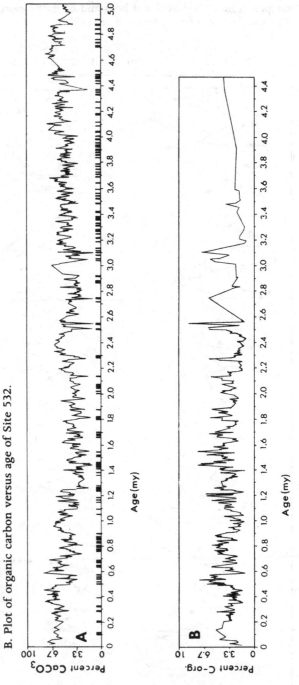

Fig. 5. A. Plot of CaCO$_3$ versus age for Site 532. Black bars at the bottom of the plot correspond to locations of dark-colored beds.
B. Plot of organic carbon versus age of Site 532.

then increase to almost 70% at the top of the section. Stratigraphic variations in organic-carbon concentration are roughly opposite to those of $CaCO_3$; organic carbon increases from about 2% at the bottom of the section to a maximum of almost 4% at about 55 m. The most striking features of the detailed profiles (Fig. 2), however, are the high-frequency fluctuations in amounts of both $CaCO_3$ and organic carbon.

We define a $CaCO_3$ cycle as a section of the detailed $CaCO_3$ profile (Fig. 2) betweeen successive $CaCO_3$ minima. We define an organic-carbon cycle as the section between successive maxima on the organic-carbon profile from 0 to 109 m (0 to 2.4 my). The color cycles were defined by shipboard visual core descriptions.

Successive age datums based on nannofossil zonation (Gardner *et al.*, 1984) were used to calculate and average sedimentation rate between datums. Each sample was then assigned an age by linear extrapolation. The resulting plots of percent $CaCO_3$ and organic carbon versus age are shown in Figs. 5A and B. Any age scale for determining sedimentary cycles is only as good as the method of dating, which in our case was nannofossil zonation (Hay, Sibuet *et al.*, 1984). All nannofossil zones are present, but there could still be minor hiatuses that are not recognized. Therefore, the sedimentation rates used to extrapolate between age datums should be considered minimum rates. Also, we have assumed a constant sedimentation rate between nannofossil age datums, which is unlikely; the sedimentation rate undoubtedly varied, particularly between deposition of light and dark beds. By taking an average sedimentation rate between age datums, we probably averaged 'stair-steps' in the true sedimentation-rate curve. The age scale probably is the largest source of error in all of our calculations of periodicities.

We subdivided the age axes of these plots into 0.5-my intervals and counted the number of cycles in each 0.5-my segment. We calculated the average percentages of $CaCO_3$, organic carbon, and biogenic silica for each 0.5-my segment (Table 1), and the periodicities of $CaCO_3$, organic carbon, and dark beds (Table 2).

The periodicities of the $CaCO_3$ cycles for each 0.5-my interval (Table 2) range from 26 ky to 49 ky, and average 36 ky. Most of the $CaCO_3$ minima correspond to the dark parts of color cycles (Fig. 5A), and most of the maxima correspond to the lightest parts of color cycles. The periodicities of the organic-carbon cycles for each 0.5-my interval (Table 2) range from 28 ky to about 36 ky, with an average of about 34 ky, for the last 2.5 my. Although most dark beds contain high concentrations of organic carbon of marine origin (Meyers *et al.*, 1984), the relationship between organic-carbon concentration and dark and

Table 1. *Basic statistics for concentrations of CaCO₃, organic carbon, and biogenic SiO₂ for 0.5-my intervals between 0 and 5.0 my in Hole 532. Data for CaCO₃ and organic carbon are from Gardner et al. (1984). Date for biogenic SiO₂ are from Dean & Parduhn (1984).*

	CaCO₃			Organic carbon			Biogenic SiO₂	
0.5-my interval	Number of analyses	Mean (%)	Standard deviation (%)	Number of analyses	Mean (%)	Standard deviation (%)	Number of analyses	Mean (%)
0.0–0.5	85	62.4	7.1	81	3.05	0.84	3	0.18
0.5–1.0	113	50.0	11.2	108	3.72	1.37	6	6.62
1.0–1.5	110	38.9	11.9	110	3.56	1.43	9	12.8
1.5–2.0	117	35.0	9.0	84	3.73	1.18	3	9.17
2.0–2.5	111	44.0	10.9	78	2.82	1.30	4	13.3
2.5–3.0	89	36.4	10.6	27	3.54	1.86	2	11.2
3.0–3.5	117	44.6	11.2	29	3.04	1.75	7	3.24
3.5–4.0	160	48.2	7.4	20	2.52	1.00	6	2.43
4.0–4.5	140	54.3	7.4	20	2.52	1.00	6	2.43
4.5–5.0	130	58.0	8.9	10	2.19	0.93	14	3.15

Table 2. *Periodicities of CaCO₃, color, and organic-carbon cycles for each 0.5-my interval between 0 and 5.0 my in Hole 532.*

	CaCO₃ cycles			Color cycles			Organic-C cycles		
Time interval (my)	Number of cycles	Period (ky/ cycle)	Mean	Number of cycles	Period (ky/ cycle)	Mean	Number of cycles	Period (ky/ cycle)	Mean
0–0.5	14.5	34.5		6	83.3		18	27.8	
0.5–1.0	14	35.7	35.1	11	45.4	55.1	18	27.8	31.2
1.0–1.5	16	31.7		12	41.7		15	33.3	
1.5–2.0	13	38.5		10	50.0		14	35.7	
2.0–2.5	10	48.8		8	62.5		11.5	43.5	
2.5–3.0	10.5	40.5[a]	45.6	5	85[a]	58.4	–	–	43.5
3.0–3.5	10.5	47.6		18	27.8		–	–	
3.5–4.0	19	26.3		18	27.8		–	–	
4.0–4.5	16	25.6[b]	28.4	14	29[b]	30.1	–	–	
4.5–5.0	9	33.3[c]		9	33[c]		–	–	
Grand means			36.3			48.6			33.6

[a]Based on number of cycles within 0.425 my; rest of section missing due to incomplete recovery.
[b]Based on number of cycles within 0.41 my; rest of section missing due to incomplete recovery.
[c]Based on number of cycles within 0.3 my; rest of section missing due to incomplete recovery.
Dashes (–) indicate insufficient data to determine periodicity.

light beds is not so clear as for CaCO$_3$; many of the organic-carbon maxima do not appear to correspond to a dark bed (Fig. 5B). As a result, the organic-carbon cycles have a higher frequency (shorter period) than those of the carbonate and color cycles (Table 2).

The CaCO$_3$ and organic-carbon profiles (Fig. 2) also show lower frequency cycles with an average thickness of about 6 m, which corresponds to an average periodicity of about 130 ky. These longer cycles are best seen in the smoothed curves for parts of the profiles between 40 and 80 m (Figs. 2 and 6).

Origin of the cycles

Cyclic variations in relative abundances of carbonate and noncarbonate components may be due to carbonate dissolution, changes in productivity of calcareous organisms, or dilution of carbonate by noncarbonate material (e.g. Gardner, 1975; Dean *et al.*, 1977, 1981). We concluded from shipboard investigations that carbonate dissolution probably was not an important cause of the cycles at Site 532 because the site is only 1331 m deep and the calcareous microfossils are moderately to well preserved with no apparent signs of dissolution (Hay, Sibuet *et al.*, 1984). In order to further test for carbonate dissolution, samples were collected from within a well-developed cycle (Core 10, Sections 2 and 3; Fig. 3) to determine if calcareous nannofossils showed any evidence of dissolution, either by difference in degree of preservation or by differences in relative abundance of coccoliths and the more dissolution-resistant discoasters. Both of these indicators have been used to identify Tertiary carbonate dissolution cycles in the eastern equatorial Atlantic (Dean *et al.*, 1981). The nannofossils from the cycle at Site 532 did not show any indications of differential dissolution (H. Stradner, personal communication, 1982) which we take as further evidence that carbonate dissolution was not a cause for the cycles.

Changes in periodicities of carbonate, organic carbon, and color with time are plotted in Fig. 7. Periodicities of carbonate cycles from DSDP Sites 502 and 503 in the western Caribbean and eastern equatorial Pacific, respectively (Gardner, 1982), also are plotted in Fig. 7. The carbonate cycles at all three sites have periods that are mostly between 30 and 50 ky with overall averages of about 40 ky. Notice that the changes in periodicity with time also are similar at all three sites, with maximum periodicities during the late Pliocene (Fig. 7). The ranges of periodicities of the carbonate cycles at all three sites also are similar to the range of periodicities of carbonate cycles of Oligocene and early Miocene age on the Sierra Leone Rise in the eastern equatorial Atlantic

Fig. 6. Smoothed curves of % CaCO₃ and % organic carbon from Fig. 2.

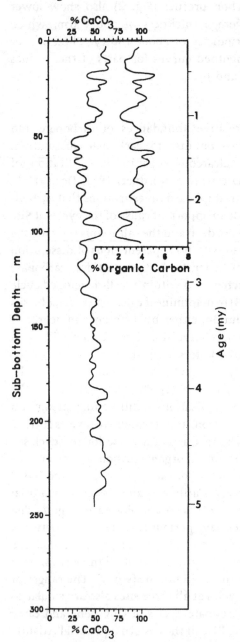

(Dean *et al.*, 1981). If changes in productivity of calcareous organisms were a primary cause of the carbonate cycles at Site 532, we would expect these changes to be more in response to local conditions of circulation along the African continental margin rather than a response to global changes. The similarities of magnitude and changes with time of periodicities of carbonate cycles from widely separated areas (Fig. 7), however, suggest that the forcing mechanism for the carbonate cycles is global and not local in extent, and that the cycles probably are not due to variations in productivity of calcareous organisms.

If variations in carbonate dissolution or productivity are not the main cause of the cycles, this leaves dilution by noncarbonate material as the most likely cause. Intuitively we can see dilution of $CaCO_3$ by fine-grained terrigenous siliciclastic material as a very likely control of the carbonate cycles at Site 532 because the area is near the continental margin of Africa, a continent with a history of abundant but variable

Fig. 7. Plots of periodicities of carbonate cycles, averaged at 0.5-my intervals, from Site 532, Site 502 in the western Caribbean, and Site 503 in the eastern equatorial Pacific. Periodicities of color and organic-carbon cycles from Site 532 also are plotted.

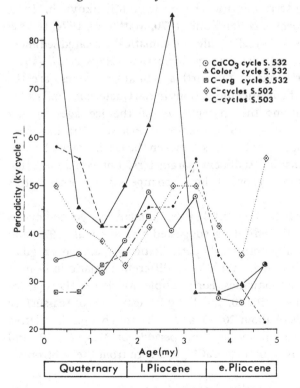

supply of terrigenous material to the Atlantic Ocean. Dilution has been demonstrated as a mechanism for reducing the carbonate content of sediments off northwest Africa (e.g. Gardner, 1975; Diester-Haass, 1976; Chamley et al., 1977). Semi-quantitative estimates of clay abundance from smear-slides show that clay content of dark interbeds usually is about 25% compared with about 10% in the lightest parts of light, high-$CaCO_3$ interbeds (Hay, Sibuet et al., 1984). This difference represents a 150% increase in clay content between lighter and darker interbeds, which suggests that there was at least some dilution by terrigenous clastics.

Assuming that the carbonate cycles at Site 532 and elsewhere were caused by global events, as the periodicities suggest, and assuming that variation in supply of terrigenous clastics was the main cause of the cycles, than we need a cyclic driving mechanism that could pulse terrigenous clastics with the appropriate periodicity to produce the cycles. The only mechanism that we can think of that could do this is climatic change. Climatic changes related to the earth's orbital cycles of precession, obliquity, and eccentricity (with periods of about 23, 41, and 100 ky, respectively) have been suggested as the main driving forces behind long-term climatic change (e.g. Milankovitch, 1930; Anderson, 1964; Broecker & Van Donk, 1970; Matthews, 1974; Hays et al., 1976). Matthews (1974) called inferred climatically related sedimentary cycles the 'tuning forks' of geologic time. Hays et al. (1976) concluded that variations in the earth's orbital parameters are the fundamental cause of climatic fluctuations in the Quaternary and called these orbital variations the 'pacemakers of the ice ages'. Other investigators (e.g. Emiliani, 1978; Kominz & Pisias, 1979), however, agreed that there appears to be a link between the earth's orbital cycles and climatically sensitive sediment parameters, but concluded that much of the sedimentary record of climate change is due to other factors that are more random.

The relationships between orbital cycles, climate change, and cycles of sedimentary climatic indicators, suggested by Hays et al. (1976) and numerous other investigators, imply periodicities that are more regular than are actually observed. In addition, different climatic indicators may have different periodicities. For example, Morley & Hays (1978) found periods of about 150, 30, and 19 ky for estimates of sea-surface temperature, periods of about 20, 15, and 12 ky for changes in relative abundance of radiolarian species, and periods of 37.5, 23, 19, and 14.5 ky for variations in percent $CaCO_3$ in a core from the subtropical Atlantic Ocean.

In spite of all of the variability described above, and in spite of all of the potential sources of error in determining the periodicities of sedimentary variables that may be related to climate, carbonate cycles with periods between 20 and 100 ky are characteristic of most Pleistocene deep-sea carbonate sediments. Climate-related sedimentary cycles with periodicities that range from 20 to 100 ky are not unique to the Quaternary, but extend well back into the Tertiary (e.g. Dean *et al.*, 1977, 1981; Clifton, 1980), Mesozoic (e.g. Fischer, 1964; Arthur & Fischer, 1977; Dean *et al.*, 1977; Arthur, 1979; McCave, 1979; Arthur *et al.*, 1984) and even the Paleozoic (e.g. Anderson, 1982; Schwartzacher & Fischer, 1982).

Assuming that there is a connection between variations in the earth's orbital parameters and climate, what are the connections between climate and sediment composition? One of the best-established links between climate and a sediment compositional variable is that between global ice volume, the oxygen isotopic composition of the ocean, and the oxygen isotopic composition of both planktonic and benthic foraminifers. Studies of cyclic fluctuations in the oxygen isotopic composition of foraminifers from Quaternary deep-sea sediments have related these cycles to variations in global ice volume and hence to changes in global sea level. (Shackleton, 1967). These studies have been extended into the Pliocene (Prell, 1982), and arguments have been presented that oxygen-isotope cycles throughout the Tertiary are related to fluctuations in global ice volumes (Matthews & Poore, 1980). These cycles in oxygen-isotopic composition have strong periodicities of 20 to 100 ky.

Conclusions

On the basis of the preceding discussion, we interpret the carbonate and color cycles at Site 532 to be the result mainly of fluctuations of terrigenous detrital clastic material. The pulsing of terrigenous clastics was caused by changes in eustatic sea level, which in turn, were caused by changes in global ice volume. During eustatic low-stands, terrigenous sediment was fed directly to the area of the shelf break, and sediment that initially was stored on submerged shelves during eustatic high-stands was shed and deposited offshore, resulting in an increase in the flux of terrigenous material to dilute the 'pelagic rain' of the calcareous and siliceous biogenic debris.

In general, there is an inverse relationship between $CaCO_3$ and organic carbon, both on the scale of individual dark–light cycles, and on the scale of the entire sequence (Figs. 2–6). Most dark beds correspond to an organic-carbon maximum, but there are many organic-carbon maxima that do not correspond to dark beds. This suggests that other factors may have affected the organic-carbon cycles. Variations in amount of organic carbon that accumulated at Site 532 most likely were caused by cyclic fluctuations in the intensity of upwelling and resulting productivity along the continental margin of southwest Africa. If our interpretation about the relationships among global ice volume, sea level, and influx of terrigenous clastics are correct, then periods of greater clastic influx (lower $CaCO_3$, greater ice volume, lower sea level) would also tend to be periods of more vigorous circulation, upwelling, and productivity. This would explain the usual coincidence of high concentrations of marine organic carbon in dark, clay-rich beds, and the general inverse relationship between organic carbon and $CaCO_3$. To the extent that upwelling was affected by other variables, particularly local conditions of circulation along the African continental margin and Walvis Ridge, there would have been other periods of increased upwelling and productivity that would not affect the influx of terrigenous clastics. This would explain the additional organic-carbon maxima that are not related to the color cycles, and the shorter periodicities of the organic-carbon cycles.

Acknowledgements

We thank J. R. Herring and L. M. Pratt for their reviews of the manuscript. N. Parduhn and C. Wilson helped with computer graphics in preparing the illustrations.

REFERENCES

Anderson, R. Y. 1964. Varve calibration of stratification. In: Merriam, D. F. (ed.), *Symposium on Cyclic Sedimentation*. State Geological Survey of Kansas Bull. 169, vol. 1, pp. 1–20.

Anderson, R. Y. 1982. A long geodynamic record from the Permian. *J. Geophys. Res.*, **87**, 7285–7294.

Arthur, M. A. 1979. North Atlantic Cretaceous black shales: the record at Site 398 and a brief comparison with other occurrences. In: Sibuet, J.-C., Ryan, W. B. F. et al., *Initial Reports of the Deep Sea Drilling Project*, **47**, part 2. Washington (US Government Printing Office), pp. 719–751.

Arthur, M. A. & Fischer, A. G. 1977. Upper Cretaceous–Paleocene magnetic stratigraphy at Gubbio, Italy: 1. Lithostratigraphy and sedimentology. *Geol. Soc. Am. Bull.*, **88**, 367–389.

Arthur, M. A., Dean, W. E., Bottjer, D. & Scholle, P. A. 1984. Rhythmic bedding in Mesozoic–Cenozoic pelagic carbonate sequences. In: A. Berger, J. Imbrie, J. D. Hays, G. Kukla & B. Saltzman (eds.), *Milankovitch and Climate*. Amsterdam, Riedel, Publ. Co., 191–222.

Broecker, W. S. & Van Donk, J. 1970. Insolation changes, ice volumes and the O^{18} record in deep-sea cores. *Rev. Geophys. Space Phys.*, **8**, 169–198.

Chamley, H., Diester-Haass, L. & Lange, H. 1977. Terrigenous material in east Atlantic sediment cores as an indicator of NW African climates. *'Meteor' Forsch. Ergebnisse*, C, pp. 44–59.

Clifton, E. 1980. Progradational sequence in Miocene shoreline deposits, southeastern Caliente Range, California. *J. Sed. Petrol.*, **51**, 166–184.

Dean, W.E. & Parduhn, N.L. 1984. Inorganic geochemistry of sediments and rocks recovered at DSDP Sites 530 and 532, southern Angola Basin and Walvis Ridge. In: W.W. Hay, J.-C. Sibuet, *et al.*, *Initial Reports of the Deep Sea Drilling Project*, **75**. Washington (US Government Printing Office), 923–958.

Dean, W.E., Gardner, J.V., Jansa, L.F., Cepek, P. & Seibold, E. 1977. Cyclic sedimentation along the continental margin of northwest Africa. In: Y. Lancelot, E. Seibold *et al.*, *Initial Reports of the Deep Sea Drilling Project*, **41**. Washington (US Government Printing Office), pp. 965–989.

Dean, W.E., Gardner, J.V. & Cepek, P. 1981. Tertiary carbonate-dissolution cycles on the Sierra Leone Rise, eastern equatorial Atlantic Ocean. *Mar. Geol.*, **39**, 81–101.

Diester-Haass, L. 1976. Late Quaternary climatic variations in NW Africa deduced from east Atlantic sediment cores. *Quat. Res.*, **6**, 299–314.

Emiliani, C. 1978. The causes of the ice ages. *Earth Planet. Sci. Lett.*, **37**, 349–352.

Fischer, A.G. 1964. The Lofer cyclothems of the Alpine Triassic. In: D.F. Merriam (ed.), *Symposium on Cyclic Sedimentation*. State Geological Survey of Kansas Bull. 169, vol. 1, pp. 107–149.

Gardner, J.V. 1975. Late Pleistocene carbonate dissolution cycles in the eastern equatorial Atlantic. In: W.V. Sliter, A.W.H. Bé & W.H. Berger (eds.), Dissolution of deep-sea carbonates. *Cushman Found. Foram. Res., Spec. Publ.*, **13**, 129–141.

Gardner, J.V. 1982. High-resolution carbonate and organic-carbon stratigraphies for the late Neogene and Quaternary from the western Caribbean and eastern equatorial Pacific. In: W.L. Prell, J.V. Gardner *et al.*, *Initial Reports of the Deep Sea Drilling Project*, **68**. Washington (US Government Printing Office), pp. 347–364.

Gardner, J.V., Dean, W.E. & Wilson, C. 1984. Carbonate and organic-carbon cycles and the history of upwelling at DSDP Site 532, Walvis Ridge, South Atlantic Ocean. In: W.W. Hay, J.-C. Sibuet *et al.*, *Initial Reports of the Deep Sea Drilling Project*, **75**. Washington (US Government Printing Office), 905–921.

Hay, W.W., Sibuet J.C. *et al.* 1982. Sedimentation and accumulation of organic carbon in the Angola Basin and on Walvis Ridge: preliminary results of Deep Sea Drilling Project Leg 75. *Geol. Soc. Am. Bull.*, **93**, 1038–1050.

Hay, W.W., Sibuet, J.-C. *et al.* 1984. *Initial Reports of the Deep Sea Drilling Project*, **75**. Washington (US Government Printing Office).

Hays, J.D., Imbrie, J. & Shackleton, N.J. 1976. Variations in the earth's orbit: pacemaker of the ice ages. *Science*, **194**, 1121–1132.

Hülsemann, J. 1966. On the routine analysis of carbonates in unconsolidated sediments. *J. Sed. Petrol.*, **36**, 622–625.

Kominz, M.A. & Pisias, N.G. 1979. Pleistocene climate: deterministic or stochastic? *Science*, **204**, 171–172.

Matthews, R.K. 1974. *Dynamic Stratigraphy*. Englewood Cliffs, N.J., Prentice Hall, 370 pp.

Matthews, R.K. & Poore, R.Z. 1980. Tertiary $\delta^{18}O$ record and glacio-eustatic sea-level fluctuations. *Geology*, **8**, 501–504.

McCave, I.N. 1979. Depositional features of organic-rich black and green mudstones at DSDP Sites 386 and 387, western North Atlantic. In: B. Tucholke, P. Vogt *et al.*, *Initial Reports of the Deep Sea Drilling Project*, **43**. Washington (US Government Printing Office), pp. 411–416.

Meyers, P.A., Brassell, S.C. & Huc, A.Y. 1984. Geochemistry of organic carbon in South Atlantic sediments from Deep Sea Drilling Project Leg 75. In: W.W. Hay, J.-C. Sibuet, *et al.*, *Initial Reports of the Deep Sea Drilling Project*, **75**. Washington (US Government Printing Office), 967–982.

Milankovitch, M. 1930. Mathematische Klimalehre und astronomische Theorie der Klimaschwankungen. In: W. Koppen & R. Geiger, (eds.), *Handbuch der Klimatologie*, vol. 1, part. A. Berlin, Gebrüder Borntraegar, 176 pp.

Morley, J.J. & Hays, J.D. 1978. Spectral analysis of climatic records from the subtropical South Atlantic (abs.). *Geol. Soc. Am. Abs. Programs*, **10**, 460.

Prell, W.L. 1982. Oxygen and carbon isotope stratigraphy for the Quaternary of Hole 502B. Evidence for two modes of isotopic variability. In: W.L. Prell, J.V. Gardner *et al.*, *Initial Reports of the Deep Sea Drilling Project*, **68**. Washington (US Government Printing Office), 455–464.

Schwartzacher, W. & Fischer, A.G. 1982. Limestone-shale bedding and perturbations of the earth's orbit. In: G. Einsele & A. Seilacher (eds.), *Cyclic and Event Stratification*. New York, Springer-Verlag, pp. 72–95.

Shackleton, N.J. 1967. Oxygen isotope analyses and Pleistocene temperatures reassessed. *Nature*, **215**, 15–17.

7

Pliocene oceanography and climate: an isotope record from the southwestern Angola Basin

H. J. WEISSERT and H. OBERHÄNSLI

Abstract

Pliocene sediments, recovered at DSDP Sites 521, 522, and 523 in the southwestern Angola Basin, preserve a record of South Atlantic paleoceanography and its response to Pliocene climatic change. In the benthic oxygen-isotope curve, stable climatic conditions are recorded between 3.7 m.y. and 3.35 m.y. At the deep sites 522 and 523 we notice a cooling of bottom water at 3.35 m.y., possibly related to the intensification of North Atlantic Deep Water (NADW). In response to the intensification of NADW, production of Antarctic Bottom Water (AABW) was increased and its influence on the southwestern Angola Basin hydrography may be seen in more negative carbon-isotope values at the deep sites 522 and 523 (3.2–2.8 m.y.). At the 2.5 m.y. level a change in the benthic oxygen isotope pattern from a low amplitude to a high amplitude curve signals a major change in the Pliocene climate. Widespread northern and southern hemisphere glaciations are reflected in positive $\delta^{18}O$-spikes at 2.5 m.y., 2.3 m.y., and 1.95 m.y. During times of glaciation, influence of NADW in the deep Angola Basin was weakened and AABW had a more profound impact on bottom water conditions. An increase in carbonate dissolution and more negative benthic $\delta^{13}C$-signals at the deep sites 522 and 523 reflect this oceanographic change.

Introduction

Ice age history, long thought to be limited to the Quaternary, extends with its periodic changing glacial and interglacial climate further and further back into the Tertiary (e.g. Kennett, 1977; Schnitker, 1980; Berger, 1982). While periodic widespread glaciation of Antarctica may have occurred as early as the Eocene (Matthews & Poore, 1980), indication for significant ice build-up on the northern hemisphere continents is only found in sediments of middle Pliocene age (Berggren, 1972; Backman, 1979). The beginning of the northern hemisphere glaciation is one of three distinct climatic and oceanographic events recorded in pelagic sediments of Pliocene age. A period of shrunken glaciers and warm surface and bottom water temperatures, dated as 4.4 m.y., is documented in a silicoflagellate study by Ciesielski & Weaver (1974), where they show evidence for warm Southern Ocean surface temperatures, and in an oxygen-isotope record of DSDP Site 519 from the Central South Atlantic (Weissert et al., 1984). With the exception of this warm event the climatic pattern of the Early Pliocene is dominated by low amplitude changes (Shackleton & Opdyke, 1977; Brunner & Keigwin, 1981; Keigwin, 1982a,b; Weissert et al., 1984). 3.0 m.y. to 3.5 m.y. ago, a major change in oceanic circulation pattern and/or a first step to a more intensely glaciated world shifted the oxygen-isotope record to more positive values (Shackleton & Opdyke, 1977; Keigwin & Thunell, 1979; Keigwin, 1982a,b; Hodell et al., 1983; Leonard et al., 1983; Weissert et al., 1984). A second change in the Pliocene isotope curve occurs at 2.5 m.y. This change coincides with the deposition of the oldest ice rafted debris in the North Atlantic Ocean (Backman, 1979) and may therefore be interpreted as the first signal of widespread northern hemisphere glaciation (Shackleton & Cita, 1979; Shackleton et al., 1984). Thus we are confronted with the question of whether we can assign an age of around 3.2 m.y. or 2.5 m.y. to the beginning of the most recent part of Neogene climatic history.

In the present study we present additional data from DSDP Sites, 521, 522, and 523 which allow us to modify an earlier scenario on Pliocene ocean and climate evolution (Weissert et al., 1984). We will show that the isotope event near 3.3 m.y. is related to changes in global ocean circulation (Keigwin, 1982a,b; Hodell et al., 1983) and that the climate pattern was distinctly modified only at 2.5 m.y. (Shackleton et al., 1984). We use benthic oxygen isotope data as a monitor for past changes in global ice volume and/or bottom water temperature. Additional control on changing bottom water characteristics is given by the benthic carbon-isotope record and by sedimentological data. Global changes in the average carbon-isotope composition of deep water may result from changes in the carbon:phosphorus ratio in sea water (Broecker, 1982). These variations should be equally recorded in locations of varying deep water depth. If, however, we find bathymetric controlled variations in the benthic carbon-isotope record from different localities, we may relate these fluctuations to changes in abyssal circulation (Curry & Lohmann, 1982). In our study we compare the carbon-isotope record of Site 522 with data from a shallower and a deeper locality (Sites 521 and 523) in order to reconstruct Pliocene carbon-isotope gradients in South Atlantic deep water.

Southwestern Angola Basin: stratigraphic and oceanographic framework

Pliocene sediments investigated during this study were cored at DSDP Sites 521, 522, and 523, east of the Mid-Atlantic Ridge at a latitude of 25–30° S (Fig. 1). Today the water depth of these sites ranges between 4125 m (Site 521), 4441 m (Site 522), and 4563 m (Site 523). These values are deeper only by a few tens of metres than the Pliocene depth values calculated according to the back-tracking method of Berger & Winterer (1974). The most widespread sediment deposited throughout the Pliocene in the area studied is a light coloured nannofossil ooze with a $CaCO_3$-content of more than 90%. Only in the upper Pliocene of Sites 522 and 523 do a series of dark nannofossil ooze layers, up to 10 cm thick, alternate with layers of white nannofossil ooze, up to 50 cm thick. The carbonate content in the dark layers varies between 60% and 80% and, in general, dark layers are enriched in the benthic foraminifer *Nuttallides umbonifera*, reflecting greater carbonate undersaturation of bottom water during the formation of these horizons (Weissert *et al.*, 1984). Both the upper and the lower contacts of the dark beds are strongly affected by bioturbation features.

At Sites 521 and 522 undisturbed Pliocene recovery allowed the establishment of a detailed magnetostratigraphy and excellent stratigraphic control (Fig. 2) (Tauxe *et al.*, 1983; Hsü *et al.*, 1984). Although magnetostratigraphic data of the Pliocene interval are missing at Site 523, biostratigraphic information by Percival (1984) and by

Fig. 1. Location of DSDP Sites 521, 522, and 523.

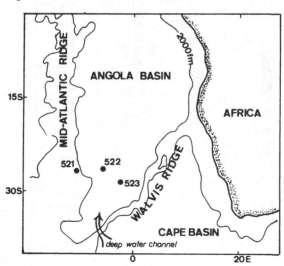

Poore (1984) still allow good calibration of the deepest site with the two shallower ones. Absolute ages assigned to the reversal boundaries and to the biostratigraphic zones, permit an estimate of average Pliocene sedimentation rates. While a value of around 8mm/1000y was calculated for the upper Pliocene light nannofossil ooze, only 2mm of the dark sediment at Sites 522 and 523 were deposited within 1000 years. This estimate is based upon the assumption that the sedimentation rate of the clay minerals remained constant throughout the Late Pliocene (Violanti et al., 1979).

An understanding of the present-day circulation pattern and distribution of water masses in the South Atlantic is essential for reconstructions of Pliocene circulation pattern and water mass changes. Today, nutrient-rich and oxygen-depleted Antarctic Bottom Water (AABW), redefined as Circumpolar Water by Mantyla & Reid (1983),*

* For this discussion we keep the more familiar term 'AABW'.

Fig. 2. Stratigraphic summary of Pliocene sections at DSDP Sites 521, 522, and 523.

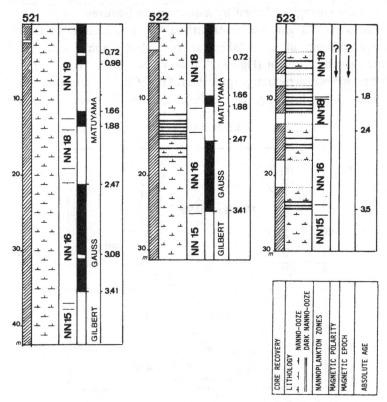

enters the Argentine and Brazil Basin from the south. From the northern Brazil Basin, the AABW escapes through the Romanche fracture zone and enters the Angola Basin. AABW, on its way to the Angola Basin, is modified. It has, for example, only half the silica content of the Brazil Basin deep water, yet this value is still twice as high as the value measured in the North Atlantic Deep Water (NADW) (Mantyla and Reid, 1983). Another, minor entrance for bottom water into the Angola Basin exists in its southwestern corner through a channel in the Walvis Ridge. This entrance, situated near our Sites 522 and 523, allows the inflow of AABW at a depth of more than 4100m, which contributes to the deep water found below 4000m throughout most of the Angola Basin (Connary & Ewing, 1974). Today, Sites 522 and 523 are overlain by AABW and NADW. As a tracer for past fluctuations of AABW flow into the Angola Basin we may use carbon isotopes. The AABW in the central South Atlantic has a carbon-isotope signature which is more negative, by up to 0.5‰, than the one of the overlying NADW (Kroopnick, 1980). Surface water circulation in the area studied is relatively weak since all the sites are located within the South Atlantic gyre system. To the west, warm surface water is transported southward by the Brazil Current, while to the east the cold and less saline water of the Benguela Current controls hydrography.

Oxygen and carbon-isotope stratigraphy

We selected Site 522 for the establishment of a reference isotope stratigraphy for the Pliocene in our area studied. There, the continuously cored pelagic sequence is dated with biostratigraphic and magnetic methods. In addition, we can correlate the isotope stratigraphy with the observed sedimentary cycles, beginning at 2.5 m.y. In the lower part of the section we sampled cores in roughly 100 000 year intervals. The resolution of the record is higher between 2.5m.y. and 1.8m.y. where we sampled individual dark and light nannofossil ooze layers. We chose the benthic species *Nuttallides umbonifera* as a tracer of bottom water history, while *Globigerinoides conglobatus* provided us with information on the surface water evolution. Additional samples from the shallower Site 521 and from the deeper Site 523 allowed us to establish deep water carbon and oxygen-isotope gradients for different time intervals.

The monospecific benthic and planktonic samples chosen for analysis were roasted *in vacuo* for 30 minutes at 400°C, in order to remove possible organic contaminants. In a microline, attached to a Micromass-903 mass-spectrometer, the samples were reacted with 100% H_3PO_4 at

Table 1. *Oxygen and carbon-isotope data from Pliocene sediments of DSDP Sites 521, 522, and 523.*
A. *Site 521 data*

Sample number	Depth (m)	Age (m.y.)	P. wuellerstorfi		N. umbonifera	
			^{18}O	^{13}C	^{18}O	^{13}C
521–4–1, 40–42	12.40	1.72	+3.27	+0.67		
521–4–1, 124–126	13.24	1.85	+3.38	+0.88		
521–4–2, 42–44	13.92	1.92	+3.21	+0.42	+3.39	+0.65
521–4–2, 128–130	14.78	2.01	+3.28	+0.95		
521–4–3, 40–42	15.40	2.08	+3.58	+0.43	+3.57	+0.37
521–4–3, 76–78	15.76	2.12	+3.17	+0.78		
521–5–1, 100–102	17.50	2.17			+2.85	+0.69
521–5–2, 110–112	19.10	2.28	+2.84	+0.64	+3.11	+0.84
521–6–2, 130–132	23.80	2.60	+2.94	+0.72	+3.27	+0.54
521–6–3, 80–82	24.80	2.67			+2.40	+0.86
521–8–2, 34–36	31.33	3.16	+2.46	+0.89	+2.70	+0.80
521–8–2, 133–135	32.83	3.22	+2.22	+0.71		
521–8–3, 34–36	33.34	3.25	+2.45	+0.65	+2.41	+0.30
521–9–1, 135–137	35.85	3.44			+2.72	+0.80
521–9–1, 140–142	35.90	3.45	+2.38	+1.07		
521–9–2, 36–38	36.36	3.48	+2.52	+1.00	+2.23	+0.65
521–9–2, 135–137	37.37	3.54			+2.41	+0.57
521–9–2, 140–142	37.40	3.55	+2.26	+0.96	+2.06	+0.47
521–9–3, 36–38	37.86	3.58	+2.48	+0.94	+2.46	+0.61

B. *Site 522 data*

Sample number	Depth (m)	Age (m.y.)	N. umbonifera		G. conglobatus	
			^{18}O	^{13}C	^{18}O	^{13}C
522–4–1, 31–33	11.11	1.92	+2.82	+0.45	+0.83	+1.91
522–4–1, 52–54	11.32	1.92	+3.78	+0.27	+0.91	+1.58
522–4–1, 90–92	11.70	2.00	+2.66	+0.44	+0.63	+2.31
522–4–1, 118–120	11.98	2.03	+3.18	+0.42	+0.71	+2.07
522–4–1, 132–134	12.12	2.05	+2.34	+0.24	+0.45	+2.15
522–4–2, 54–56	12.84	2.14	+3.28	+0.60	+0.73	+2.22
522–4–2, 76–78	13.06	2.16	+3.34	+0.89	+0.75	+2.04
522–4–2, 82–84	13.12	2.17	+2.87	+0.75	+0.45	+2.14
522–4–2, 104–106	13.34	2.20	+3.84	+0.33	+0.37	+1.76
522–4–2, 131–133	13.61	2.23	+3.49	+0.45	+0.62	+1.77
522–4–2, 146–148	13.76	2.25	+3.71	+0.29	+0.76	+1.70
522–4–3, 14–16	13.94	2.28	+3.56	+0.45	+0.61	+1.79
522–4–3, 37–39	14.17	2.31	+3.47	−0.03	+0.89	+1.68
522–4–3, 56–58	14.36	2.33	+2.57	+0.48	+0.18	+1.95
522–4–3, 98–100	14.80	2.38	+3.00	+0.52	+0.46	+1.82
522–5–1, 50–52	15.70	2.50	+3.25	+0.76	+0.62	+1.82
522–5–1, 83–85	16.03	2.53	+2.48	+0.63	+0.22	+2.16
522–5–1, 100–102	16.20	2.55	+2.93	+0.40	+0.46	+1.60

Table 1. contd.

Sample number	Depth (m)	Age (my)	N. umbonifera		G. conglobatus	
			^{18}O	^{13}C	^{18}O	^{13}C
522–5–1, 138–140	16.58	2.58	+3.45	+0.14		
522–5–2, 50–52	17.20	2.65	+2.75	+0.61	+0.51	+2.11
522–5–2, 100–102	17.70	2.70	+3.00	+0.26	+0.51	+1.83
522–5–2, 140–142	18.10	2.74	+3.13	+0.87	+0.45	+2.12
522–5–3, 20–22	18.40	2.77	+2.74	+0.69		
522–5–3, 42–44	18.60	2.79	+3.23	+0.27	+0.30	+1.80
522–5–3, 80–82	19.00	2.83	+2.74	+0.32	+0.48	+1.74
522–6–1, 50–52	20.10	2.93			+0.62	+2.04
522–6–1, 78–80	20.38	2.97	+3.21	+0.69		
522–6–1, 100–102	20.60	2.98	+2.83	+0.47	+0.50	+2.13
522–6–1, 140–142	21.00	3.02			+0.62	+1.88
522–6–2, 30–32	21.40	3.06	+2.64	+0.48	+0.48	+2.04
522–6–2, 50–52	21.60	3.09	+2.72	+0.05		
522–6–2, 100–102	22.10	3.13			+0.32	+1.87
522–6–2, 140–142	22.50	3.17	+2.76	+0.68		
522–6–3, 20–22	22.80	3.20	+2.80	+0.73	+0.25	+1.95
522–7–1, 50–52	24.50	3.35	+2.90	+0.75	+0.30	+2.35
522–7–1, 100–102	25.00	3.43	+2.52	+0.50	+0.62	+1.63
522–7–1, 140–142	25.40	3.47	+2.45	+1.05	+0.40	+1.90
522–7–2, 50–52	26.00	3.53	+2.52	+0.79	+0.81	+2.40
522–7–2, 100–102	26.50	3.58	+2.59	+0.71	+0.71	+2.27
522–7–2, 140–142	26.90	3.62			+0.46	+2.01
522–7–3, 40–42	27.40	3.67	+2.56	+0.64		
522–7–3, 60–62	27.60	3.69	+2.48	+1.10	+0.41	+1.89

C. *Site 523 data*

Sample number	Depth (m)	Age (m.y.)	P. wuellerstorfi	
			^{18}O	^{13}C
523–3–1, 98–100	9.38	1.76	+3.06	+0.61
523–3–1, 112–114	9.52	1.78	+3.35	+0.11
523–3–1, 123–125	9.63	1.79	+3.13	+0.42
523–3–1, 126–128	9.66	1.80	+3.02	+0.45
523–3–2, 20–22	10.11	1.83	+3.02	+0.73
523–3–2, 45–47	10.35	1.85	+3.69	−0.56
523–3–2, 65–67	10.55	1.87	+2.56	−0.17
523–3–2, 80–82	10.70	1.90	+3.75	−0.10
523–3–2, 110–112	11.00	1.98	+2.67	+0.29
523–3–2, 131–133	11.21	2.01	+2.58	−0.17
523–3–3, 3–5	11.43	2.04	+2.99	+0.18
523–4–1, 119–121	13.89	2.38	+2.73	+0.17
523–4–2, 97–99	15.17	2.56	+2.78	+0.84
523–4–2, 132–134	15.52	2.61	+2.99	+0.70
523–4–3, 23–25	15.93	2.66	+2.62	+0.73

Sample number	Depth (m)	Age (my)	P. wuellerstorfi	
			^{18}O	^{13}C
523–4–3, 80–82	16.50	2.74	+2.83	+0.93
523–5–3, 150–152	17.60	2.89	+2.98	+0.38
523–6–2, 40–42	23.50	3.39	+2.67	+0.74
523–6–2, 105–107	24.15	3.45	+2.74	+0.38
523–6–3, 10–12	24.70	3.50	+2.60	+0.91
523–6–3, 71–73	25.30	3.57	+2.32	+0.76
523–6–3, 85–87	25.45	3.59	+2.42	+0.67

50°C and the evolved gas was measured against Carrara marble as a laboratory standard. The results are expressed in ‰ deviations relative to the PDB-standard. The data are summarized in Table 1 and graphically displayed in Figs. 3–5. Based on the benthic oxygen-isotope record of Site 522 (Fig. 4) we structure the discussion in two parts: (1) The middle Pliocene, 3.7–2.5 m.y. and (2) the Late Pliocene, 2.5–1.8 m.y.

(1) The middle Pliocene: 3.7–2.5 m.y.

Between 3.7 m.y. and 3.4 m.y. the oxygen-isotope record of *N. umbonifera* is remarkably stable with values fluctuating around +2.5‰ (Fig. 4). At 3.35 m.y. the curve shifts by almost 0.5‰ to more positive values, but only after 3.0 m.y. do we recognize distinct fluctuations in the curve with $\delta^{18}O$-depleted values near +2.75‰ and enriched values near +3.25‰. We notice a similar pattern in the curve of the deepest site, 523 (Fig. 5), while at Site 521, we see no indications for a positive shift in the oxygen-isotope record until we reach the 2.5 m.y. level (Figs. 3 and 6). In the planktonic record, we observe no changes of significance throughout the middle Pliocene.

Of importance for our paleoceanographic reconstructions is the evolution of the benthic carbon-isotope record at Site 522 in comparison with the record at Site 521 (Fig. 7). During the late Gilbert Chron and into the early Gauss (3.7–3.2 m.y.) the carbon-isotope values for *N. umbonifera* fluctuate around 0.3‰ to 0.75‰ at both localities. About 3.2 m.y. ago the two curves seem to separate and, between 3.2 m.y. and 2.8 m.y., the $\delta^{13}C$-values measured at Site 522 fall below 0.75‰ while the few available data of Site 521 scatter between +0.75‰ and +1.0‰. Between 2.8 m.y. and 2.5 m.y., the difference between the two curves is less pronounced. A similar pattern develops if we compare the carbon-isotope data of *P. wuellerstorfi* from Sites 523 and 521 (Fig. 8).

The changes we observed in the oxygen-isotope record may be caused by (1) fluctuations in the global ice volume and/or (2) changes in the deep circulation pattern and bottom water temperature. In a first scenario we propose a net increase in global ice volume at 3.35 m.y. (Keigwin & Thunell, 1979). The observed shift in the oxygen-isotope record then should not be limited to the deeper sites studied but also

Fig. 3. Oxygen and carbon-isotope stratigraphy of DSDP Site 521. Absolute age determinations are based on direct interpolation between magnetic polarity reversals. $\delta^{18}O$ and $\delta^{13}C$ values in this figure and Figs. 4–8 are given as ‰.

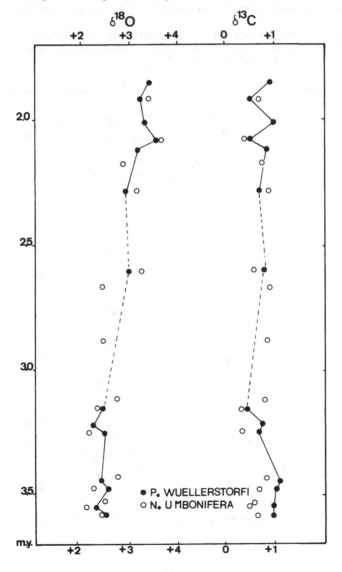

occur in the benthic record of Site 521 and in the planktonic curve. The available data support this hypothesis only if we postulate an increase in temperature of intermediate and surface water, or if we reduce salinity of both water masses by about 1‰.

The data may better be integrated into a second scenario, formulated by Hodell *et al.* (1983). They postulated ocean circulation experienced a major rearrangement near 3.2 m.y. due to the closure of the Isthmus of Panama (Keigwin, 1982b). The middle Pliocene thus would have been the time when present-day deep sea circulation was achieved. Within this hypothesis we envisage the following evolution of the southwestern Angola Basin oceanography: Around 3.5 m.y. ago the blocking of

Fig. 4. Oxygen and carbon-isotope stratigraphy of DSDP Site 522. Absolute age determinations are based on direct interpolation between magnetic polarity reversals.

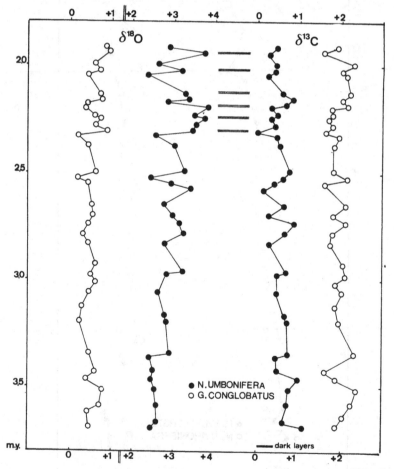

the deep sea connection between the Atlantic and Pacific near Panama triggered significant changes in Atlantic deep water circulation. First, an increased production of nutrient-poor and oxygen-rich NADW is monitored in various isotope records in the Atlantic and Pacific Oceans (Keigwin, 1982b). We correlate the observed positive shift in the benthic oxygen-isotope record at Site 522 with a stronger influence of the NADW in the Angola Basin and with a 2°C cooling of the deep water below 4100 m. Intensification of NADW possibly was an important

Fig. 5. Oxygen and carbon-isotope stratigraphy of DSDP Site 523. Absolute age determinations are based on direct interpolation between nannofossil zone boundaries.

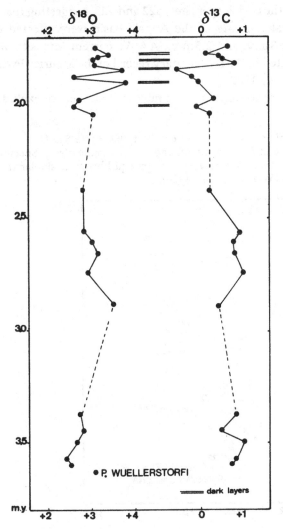

precondition for the formation of an intensified AABW flow. Highly saline NADW could mix with cold Antarctic surface water and produce a mixture which was denser than the two original water masses (Johnson, this volume, chapter 14). Indeed the sedimentary record in the southwest Atlantic Ocean shows an increase in AABW current velocity at about 3.3 m.y. (Ledbetter *et al.*, 1978, Ciesielski *et al.*, 1982). Our present information on the effect of AABW-intensification on the hydrography of the southern Angola Basin is inconclusive. The AABW, nutrient-rich and depleted in oxygen, may be distinguished from NADW by its more negative carbon-isotope signature. In sediments dated as 3.2–2.8 m.y. we observe a trend to ^{13}C- depleted values in benthic samples of the two deeper sites, 522 and 523, suggesting that at least the deeper water masses of the Angola Basin were affected by AABW. Between 2.8 m.y. and 2.5 m.y., AABW current intensity was weakened as indicated in the sedimentary record of the Maurice Ewing Bank (Ciesielski *et al.*, 1982).

Within this picture we drew for the middle Pliocene, we excluded

Fig. 6. Benthic oxygen-isotope record of DSDP Site 521 (*P. wuellerstorfi* and *N. umbonifera*) and 522 (*N. umbonifera*). Species-specific departure from oxygen-isotope equilibrium is similar for both species (Weissert *et al.*, 1984).

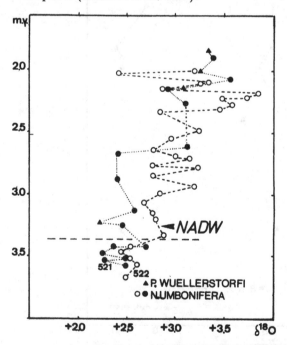

any massive changes in global ice volume. Yet we do not exclude the possibility that at least the low amplitude fluctuations observed in the benthic oxygen-isotope record between 3.0 m.y. and 2.5 m.y. are related to periodic waxing and waning of southern and northern hemisphere glaciers (Shackleton *et al.*, 1984).

(2) The Late Pliocene: 2.5–1.8 my

Near the bottom of the Matuyama Chron at Site 522 at about the 2.5 m.y. level, the benthic low amplitude oxygen-isotope curve of the middle Pliocene is replaced by a curve with high amplitude fluctuations characteristic of the Late Pliocene (Fig. 4). Low values near + 2.5‰ were measured in most samples taken from the light nannofossil ooze. The values from the dark horizons range from + 3.2‰ up to + 3.8‰. The most positive numbers were measured at 2.3 m.y. and 1.95 m.y. High amplitude changes of up to 1.5‰ also are registered in the dark–light nannofossil ooze cycles of Site 523, where again we notice very positive $\delta^{18}O$-values near 2.0 m.y. In the benthic carbon-isotope record we

Fig. 7. Benthic carbon-isotope record of DSDP Sites 521 and 522 (*N. umbonifera*).

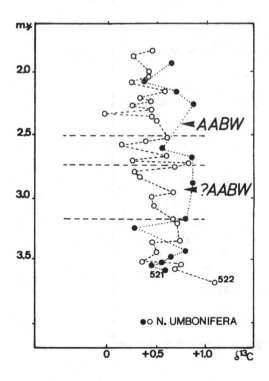

register a trend to more negative δ^{13}C-values at the deepest site, 523, and, even if less pronounced, at Site 522. At Site 521 the difference between middle and Late Pliocene values seems to be of no significant size. At the deep sites the most negative δ^{13}C-values coincide with dark layers. The planktonic oxygen-isotope curve at Site 522 follows in its pattern the benthic curve. Yet, the amplitude of the oxygen-isotope record is smaller than in the benthic record and differences between glacial and interglacial values never exceed 0.75‰. Even less distinct are the fluctuations in the planktonic carbon-isotope record.

The increased amplitude of the benthic δ^{18}O-record at 2.5 m.y. documents the beginning of more widespread northern and southern hemisphere glaciations. Values enriched in O–18 represent periods of extensive glaciations, while O–18-depleted values reflect periods of shrunken glaciers, comparable to the interglacial times of the middle Pliocene. During glacial periods, bottom water in the southern Angola Basin was more corrosive, and enhanced carbonate dissolution led to the formation of dark nannofossil ooze. Deep corrosive water is characterized by its more negative carbon-isotope signature. The strong

Fig. 8. Benthic carbon-isotope record of DSDP Sites 521 and 523 (*P. wuellerstorfi*).

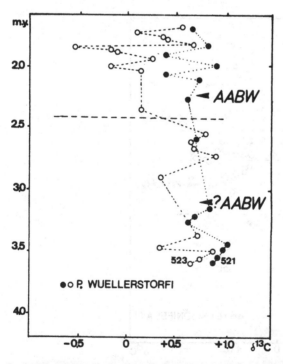

gradient in the carbon-isotope composition between the shallower site, 521, and the deeper sites, 522 and 523, indicates that the corrosive bottom water was admixed with the Angola Basin deep water below a depth of 4100 m (Figs 7 and 8). During glacial periods the intensity of NADW was significantly reduced (Boyle & Keigwin, 1982; Curry & Lohmann, 1982; Corliss, 1983) and AABW could give its stronger imprint on the deep water system of the Angola Basin. Reduction of NADW obviously did not cause the cessation of AABW production. Johnson (this volume, chapter 14) speculates that hypersaline, silica-rich water from the northwest Indian Ocean may have been of additional help in the formation of dense AABW. Evidence for an increase in AABW production during glacial times is in agreement with data from the Southern Ocean. Ciesielski *et al.* (1982) found indications for strong bottom current flow during the latest Gauss and the early Matuyama Chrons. Huang & Watkins (1977) studied sediments in the South Pacific sector of the Southern Ocean. They presented evidence for several pulses of intensified AABW during the early Matuyama. They also noticed high accumulation rates of ice debris at 2.2 my and at 1.9 m.y. These ages coincide with our most positive $\delta^{18}O$-spike ages at Site 522. Widespread southern hemisphere glaciations during early Matuyama previously were postulated in studies by Kennett & Watkins (1976) and by Kennett *et al.* (1979) as well as in a summary of continental data by Mercer (1983). These widespread glaciations between 2.5 m.y. and 1.8 m.y. were not restricted to the southern hemisphere. Backman (1979) revised the stratigraphy of North Atlantic cores containing ice raft debris. He assigned a new age of 2.5 m.y. to the oldest ice raft debris found. On the continents, best evidence for widespread northern hemisphere glaciations is found in loess deposits from China. Heller & Tungsheng (1984) used improved paleomagnetic cleaning methods for the establishment of a more accurate loess magnetostratigraphy. An age of 2.5 m.y. for the oldest loess deposits resulted from these studies. Even if we assign an age of 2.5 m.y. for the first northern hemisphere glaciation and for the beginning of high amplitude climatic changes, the benthic oxygen-isotope data suggest that the most intense Late Pliocene glaciations occurred at 2.3 m.y. and 1.95 m.y. Similar maxima in Late Pliocene benthic records from the Central Pacific and the North Atlantic are noted in studies by Shackleton & Opdyke (1977), Keigwin (1979) and Shackleton *et al.* (1984).

The planktonic data seem at variance with this scenario (Fig. 4). Even though the pattern of the planktonic curve compares well with the benthic pattern, the planktonic record has an amplitude up to 0.5‰

smaller. If indeed bottom water temperature in the southern Angola Basin did not significantly change during the Late Pliocene, we have to postulate either an increase of surface temperature during glacial periods or a decrease in surface salinity due to a change in surface circulation (Leonard *et al.*, 1983). Warm, low latitude surface water would have caused a drastic steepening of the temperature gradient between equatorial and polar regions. An enhanced energy transfer from low to high latitudes in the form of precipitation then facilitated the rapid growth of polar ice sheets (Adam, 1975; Ruddiman & McIntyre, 1979). The northward move of the polar front narrowed the South Atlantic gyre system, and in our area of study highly saline central gyre water may have been replaced by less saline circum-Atlantic water of the Brazil or Benguela Currents. A decrease in surface salinity and no variations in surface water temperatures could best explain the less pronounced signals in the planktonic oxygen-isotope record.

Conclusions

Benthic oxygen and carbon-isotope data from DSDP Sites 521, 522, and 523 in the southwestern Angola Basin record major changes in middle and Late Pliocene paleoceanography and paleoclimate. We chose the benthic oxygen-isotope record of the biostratigraphically and paleomagnetically well dated Site 522 as a reference curve for the reconstruction of glacial history and evolution of bottom water temperatures. The benthic carbon-isotope record of Sites 522 and 523 in comparison with the record of Site 521, situated at a shallower paleodepth, provided additional information on changes in the Pliocene deep water circulation pattern. Between 3.7 m.y. and 3.35 m.y. stable climatic and oceanographic conditions are mirrored in a remarkably constant benthic oxygen-isotope curve. A 0.5‰ positive shift in the oxygen-isotope curve at 3.35 m.y. limited to Sites 522 and 523 indicates a cooling of the deepest Angola Basin water. We relate this cooling to an intensification of NADW caused by the closing of the deep water connection between the Atlantic and Pacific in Central America (Keigwin, 1982b). Another change in deep water circulation is indicated in the evolving carbon-isotope gradient between the more positive Site 521 and the more negative, deeper Sites 522 and 523. We assume, that intensification of AABW near 3.3 m.y. had its influence on bottom water characteristics of the southwestern Angola Basin. The time of strong AABW circulation lasted from 3.3 m.y. to 2.8 m.y. About 2.5 m.y. ago, a change in the oxygen-isotope record from a low amplitude curve to a high amplitude record signals a major shift in the global climate conditions. Widespread glaciations are not restricted to the southern hemisphere but extend to the northern continents (Backman, 1979). We assign ages of 2.3 m.y. and 1.95 m.y. to the most widespread glaciations during the Late Pliocene. Glacial periods led to a rearrangement of South Atlantic deep and surface water circulation. NADW flow was reduced (Curry & Lohmann, 1982) and AABW had a more profound impact on southern Angola Basin hydrography. An increase in carbonate dissolution by the more corrosive AABW is reflected in dark nannofossil ooze layers, and, in the benthic carbon-isotope signature, a shift to more negative values is noticed. A northward shift of the South Atlantic surface water circulation system due to northward migration of the polar front brought the less saline surface water of the Brazil or Benguela Currents to our area of study as reflected in a reduced amplitude of the planktonic oxygen-isotope record.

Acknowledgements

We thank L. Keigwin and D. Bernoulli, who critically reviewed the manuscript. The project was supported by Swiss Science Foundation project 2.898–0.83. This paper is contribution No. 246 of the Laboratory of Experimental Geology, ETH, Zürich.

REFERENCES

Adam, D.P. 1975. Ice ages and the thermal equilibrium of the earth, II. *Quaternary Research*, 5, 161–171.

Backman, J. 1979. Pliocene biostratigraphy of DSDP Sites 111 and 116 from the North Atlantic Ocean and the age of the Northern Hemisphere glaciation. *Stockholm Contributions in Geology*, 32, 115–137.

Berger, W.H. 1982. Deep-sea stratigraphy: Cenozoic climate steps and the search for chemo-climatic feedback. In: G. Einsele & A. Seilacher (eds.), *Cyclic and Event Stratification*, New York, Springer-Verlag, pp. 121–157.

Berger, W.H. & Winterer, E.L. 1974. Plate stratigraphy and the fluctuating carbonate line. In: K. Hsü & H. Jenkyns (eds.), Pelagic sediments on land and under the sea. *Int. Ass. Sedimentol., Spec. Publ.*, 1, pp. 11–48.

Berggren. W.A. 1972. Late Pliocene – Pleistocene glaciation. In: A.S. Laughton, W.A. Berggren et al., *Initial Reports of the Deep Sea Drilling Project*, 12, Washington DC (US Government Printing Office), pp. 953–963.

Boyle, E.A. & Keigwin, L.D. 1982, Deep circulation of the North Atlantic over the last 200 000 years: geochemical evidence. *Science*, 218, 784–787.

Broecker, W.S. 1982. Ocean chemistry during glacial time. *Geochimica et Cosmochimica Acta*, 46, 1689–1705.

Brunner, C.A. & Keigwin, L.D. 1981. Late Neogene biostratigraphy and stable isotope stratigraphy of a drilled core from the Gulf of Mexico. *Marine Micropaleontology*, 6, 397–418.

Connary, S.D. & Ewing, M. 1974. Penetration of Antarctic bottom water from the Cape Basin into the Angola Basin. *Journal Geophysical Research*, 79, 463–469.

Corliss, B.C. 1983. Quaternary circulation of the Antarctic circumpolar current. *Deep Sea Res.*, 30, 47–61.

Ciesielski, P.F. & Weaver, F.M. 1974. Early Pliocene temperature changes in the Antarctic Seas. *Geology*, 2, 511–515.
Ciesielski, P.F., Ledbetter, M.T. & Ellwood, B.B. 1982. The development of Antarctic glaciation and the Neogene paleoenvironment of the Maurice Ewing Bank. *Marine Geology*, 46, 1–51.
Curry, W.B. & Lohmann, G.P. 1982. Carbon isotopic changes in benthic foraminifera from the western South Atlantic: reconstruction of glacial abyssal circulation patterns. *Quaternary Research*, 18, 218–235.
Heller, F. & Tungsheng, L. 1984. Magnetism of Chinese Loess deposits. *Geophysical Journal of the Royal Astronomical Society*, 77, 125–141.
Hodell, D.A., Kennett, J.P. & Leonard, K.A. 1983. Climatically induced changes in vertical water mass structure of the Vema Channel during the Pliocene: Evidence from Deep Sea Drilling Project Holes 516A, 517, and 518. In: P.F. Barker, R.L. Carlson, D.A. Johnson et al., *Initial Reports of the Deep Sea Drilling Project*, 72, Washington, DC (US Government Printing Office), pp. 907–919.
Hsü, K.J., La Brecque, J., Percival, S.F., Wright, R.C., Gombos, A.M., Pisciotto, K., Tucker, P., Peterson, N., McKenzie, J.A., Weissert, H., Karpoff, A.M., Carman, M.F. & Schreiber, E. 1984. Numerical ages of Cenozoic biostratigraphic datum levels: Results of South Atlantic Leg 73 drilling. *Bulletin Geological Society of America*, 95, 863–876.
Huang, T.C. & Watkins, N.D. 1977. Antarctic bottom water velocity: contrasts in the associated sediment record between the Brunhes and Matuyama Epochs in the South Pacific. *Marine Geology*, 23, 113–132.
Keigwin, L. 1979. Late Cenozoic stable isotope stratigraphy and paleoceanography of DSDP sites from the east equatorial and central North Pacific Ocean. *Earth and Planetary Science Letters*, 45, 361–382.
Keigwin, L.D. 1982a. Stable isotope stratigraphy and paleoceanography, of Sites 502 and 503. In: W.L. Prell, J.V. Gardner et al., *Initial Reports of the Deep Sea Drilling Project*, 68, Washington, DC (US Government Printing Office) pp. 445–453.
Keigwin, L. 1982b. Isotopic paleoceanography of the Caribbean and East Pacific: role of Panama uplift in Late Neogene time. *Science*, 217, 350–353.
Keigwin, L.D. & Thunell, R.C. 1979. Middle Pliocene climatic change in the Western Mediterranean from faunal and oxygen isotopic trends. *Nature*, 282, 294–296.
Kennett, J.P. 1977. Cenozoic evolution of Antarctic glaciation, the circum-Antarctic Ocean and their impact on global paleoceanography, *Journal of Geophysical Research*, 82, 3843–3860.
Kennett, J.P. & Watkins, N.D. 1976. Regional deep sea dynamic processes recorded by Late Cenozoic sediments of the southeastern Indian Ocean. *Bulletin Geological Society of America*, 87, 321–339.
Kennett, J.P., Shackleton, N.J., Margolis, S., Goodney, D.E., Dudley, W.C. & Kroopnick, P.N. 1979. Late Cenozoic oxygen and carbon isotopic history and volcanic ash stratigraphy: DSDP site 284, S Pacific. *American Journal of Science*, 279, 52–69.
Kroopnick, P. 1980. The distribution of ^{13}C in the Atlantic Ocean. *Earth and Planetary Science Letters*, 49, 469–484.
Ledbetter, M.T., Williams, D.F. & Ellwood, B.B. 1978. Late Pliocene climate and southwest Atlantic abyssal circulation. *Nature*. 272, 237–239.
Leonard, K.A., Williams, D.F. & Thunell, R.C. 1983. Pliocene paleoclimatic and paleoceanographic history of the South Atlantic Ocean: Stable isotopic records from Leg 72 DSDP Holes 516A and 517. In: P.F. Barker, R.L. Carlson, D.A. Johnson et al., *Initial Reports of the Deep Sea Drilling Project*, 72, Washington, DC (US Government Printing Office), pp. 895–906.
Mantyla, A. & Reid, J.L. 1983. Abyssal characteristics of the world ocean waters. *Deep Sea Research*, 30, 805–833.
Matthews, R.K. & Poore, R.Z. 1980. Tertiary ^{18}O record and glacio-eustatic sea-level fluctuations. *Geology*, 8, 501–504.

Mercer, J.H. 1983. Cenozoic glaciation in the Southern Hemisphere. *Annual Review of Earth and Planetary Science*, 11, 99–132.

Percival, S.F. 1984. Late Cretaceous to Pleistocene calcareous nannofossils from the South Atlantic, Deep Sea Drilling Project Leg 73. In: K.J. Hsü, J.L. La Brecque *et al.*, *Initial Reports of the Deep Sea Drilling Project*, 73, Washington, DC (US Government Printing Office), pp. 391–424.

Poore, R.Z. 1984. Middle Eocene through Quaternary planktonic foraminifers from the southern Angola Basin: Deep Sea Drilling Project Leg 73. In: K.J. Hsü, J.L. La Brecque *et al.*, *Initial Reports of the Deep Sea Drilling Project*, 73, Washington, DC (US Government Printing Office), pp. 429–448.

Ruddiman, W.F. & McIntyre, A. 1979. Warmth of the subpolar North Atlantic Ocean during Northern Hemisphere ice sheet growth. *Science*, 204, 173–175.

Schnitker, D. 1980. Global paleoceanography and its deep water linkage to the Antarctic glaciation. *Earth Science Reviews*, 16, 1–20.

Shackleton, N.J. & Cita, M.B. 1979. Oxygen and carbon isotope stratigraphy of benthnic foraminifera at site 397: Detailed history of climate change during the Late Neogene. In: U. von Rad, W.B.F. Ryan *et al.*, *Initial Reports of the Deep Sea Drilling Project*, 47, Washington, DC (US Government Printing Office) pp. 433–447.

Shackleton, N.J. & Opdyke, N.D. 1977. Oxygen isotope and paleomagnetic evidence for Northern Hemisphere glaciation. *Nature*, 270, 216–219.

Shackleton, N.J. *et al.* 1984. Oxygen isotope calibration of the onset of ice-rafting and history of glaciation in the North Atlantic region. *Nature*, 307, 620–623.

Tauxe, L., Tucker, P., Petersen, N.Y. & La Brecque, J.L. 1983. The magnetostratigraphy of Leg 73 sediments. *Palaeogeography, Palaeoclimatology, Palaeoecology*, 42, 65–90.

Violanti, D., Premoli-Silva, I., Cita, M.B., Kersey, D. & Hsü, K.J. 1979. Quantitative characterization of carbonate dissolution facies of the Atlantic Tertiary sediments. *Rivista Italiana di Paleontologia*, 85, 751–796.

Weissert, H.J., McKenzie, J.A., Wright, R.C., Clark, M., Oberhänsli, H. & Casey, M. 1984. Paleoclimatic record of the Pliocene at Deep Sea Drilling Project Sites 519, 521, 522 and 523 (Central South Atlantic). In: K.J. Hsü, J.L. La Brecque *et al.*, *Initial Reports of the Deep Sea Drilling Project*, 73, Washington DC (US Government Printing Office), pp. 701–715.

8

Paleoceanographic expressions of the Messinian salinity crisis

J. A. McKENZIE and H. OBERHÄNSLI

Abstract

During the Messinian Stage (6.1 to 5.1 Ma) of the latest Miocene, two
significant geologic events occurred: (1) a major build-up of the West
Antarctic ice sheet and (2) the isolation and subsequent desiccation of the
Mediterranean Sea. The timing of these two events is critical to our
understanding of their effect upon the paleoceanography and
paleoclimatology around the Miocene/Pliocene boundary. A combination
of high-resolution isotope-, magneto-, and bio-stratigraphy in pelagic
sediments from the South Atlantic Ocean (DSDP Site 519) allows us to
delineate the paleoceanographic events between 7 and 4 Ma. Site 519 is
presently located beneath the South Atlantic gyre precariously positioned
near the boundary of the Benguela Current. The oxygen-isotope
stratigraphy shows large fluctuations in surface-water temperature and/or
salinity suggesting major current vacillations. During periods of glacial
maxima (cooling), expansions of the polar current would bring Site 519
more under the influence of the colder Benguela Current. Conversely,
during periods of glacial retreat (warming), the warmer waters of the gyre
would dominate. During periods of extreme warming, glacial melt waters
reaching the gyre could produce a melt-water signal in the oxygen-
isotope curve.

Our correlation of isotope stratigraphy from the South Atlantic and
lithostratigraphy from the Mediterranean region enable us to propose the
following scenario:

(1) 6.1 to 5.7 Ma – a distinct cooling phase in the South Atlantic
(Antarctic glaciation) precedes the Mediterranean evaporite deposition
and corresponds to the period of restricted circulation in the
Mediterranean (Tripoli Formation) with little or no return of
Mediterranean water to the Atlantic intermediate waters.

(2) 5.7 to 5.1 Ma – in the South Atlantic two discrete cooling or glacial
events separated by a warming or melt-water event mark the duration of
the Messinian salinity crisis. This two-fold pattern corresponds to the
presence of an upper and a lower evaporite in the Mediterranean which

are separated by a renewed marine incursion of the intra-Messinian transgression.

(3) 5.1 to 4.9 Ma – in the South Atlantic the Miocene/Pliocene boundary is a time of rapid and extreme isotope fluctuations indicating an unstable surface-current pattern related to unstable climatic conditions.

Our isotopic and sedimentologic correlations demonstrate that the latest Miocene expansion of the West Antarctic ice sheet preceded and most likely precipitated the Messinian salinity crisis. Afterwards, extreme fluctuations in the volume of the ice sheet are reflected in lithologic events within the evaporite sequences. At the conclusion of the salinity crisis, the collapse of the Gibraltar dam was the grand finale resulting in global climatic instability, which may have contributed to an overall warming trend culminating with the early Pliocene transgression at 4.3 Ma.

Introduction

The Messinian Stage of the late Miocene is inherently associated with the salinity crisis in the Mediterranean region. The Messinian is the interval of time between the Tortonian Stage and the Miocene/ Pliocene boundary when euryhaline conditions prevailed throughout the Mediterranean (Mayer-Eymar, 1867, Selli, 1960; Cita, 1975a) and was delineated by lithostratigraphic in addition to biostratigraphic criteria. The lithologic changes are indications of major environmental changes of possible global consequence during the Messinian. The end of the salinity crisis, in the earliest Pliocene, is marked by the first reappearance of permanent open marine conditions in the Mediterranean.

The extensive nature of Messinian evaporites was revealed by two Deep Sea Drilling Project (DSDP) cruises into the Mediterranean (Ryan, Hsü *et al.*, 1973 and Hsü, Montadert *et al.*, 1978a), which demonstrated that the evaporites were basin-wide. Drilling showed further that the evaporites from the deeper basins are bounded by deep-water, pelagic sediments. It was concluded that the Mediterranean had been a deep, desiccated basin during the Messinian (Hsü *et al.*, 1978b). Recognizing that an evaporitic event as major as the Messinian salinity crisis must have had global paleoceanographic implications, attempts were made to correlate datum planes between the Mediterranean and other oceans (Ryan *et al.*, 1974) and land sections (van Couvering *et al.*, 1976). A latest Miocene build-up of the West Antarctic ice sheet together with an eustatic sea-level lowering, as evidenced in oxygen-isotope curves from the Southern Ocean (Shackleton & Kennett, 1975a), was proposed as the culprit which finally precipitated the Messinian salinity crisis causing an already restricted Mediterranean to be isolated from the Atlantic Ocean (van Couvering *et al.*, 1976; McKenzie *et al.*, 1979/1980; and Loutit & Keigwin, 1982). This isolation of the Mediterranean by a combination of tectonic and glacial mechanisms must have had corresponding effects upon the paleoceanography of the Atlantic as a source of North Atlantic Deep Water (NADW); the Mediterranean Deep Water (MDW), was eliminated (Bender & Keigwin, 1979). The relatively quick removal of 6% of the ocean's salt through the deposition of an estimated 10^6 km^3 of evaporite minerals in the isolated Mediterranean basins certainly affected the paleosalinity of the ocean if not the paleoclimatology (Ryan, 1973). Likewise, the abrupt end of the Messinian salinity crisis marked by the onset of renewed pelagic sedimentation in the Mediterranean may also have had paleoceanographic and paleoclimatologic consequences.

The purpose of our work was to evaluate the existence of possible paleoceanographic expressions of the Messinian salinity crisis in a deep-sea section from the South Atlantic Ocean. In order to correlate the lithostratigraphic events in the Mediterranean with possible paleoceanographic events during the very short time span of the Messinian (approximately 1 Ma), high-resolution isotope-, magneto- and biostratigraphy on a continuous marine sequence is required. With the advent of the hydraulic piston corer, deep-sea drilling can now provide the necessary material from continuously cored sequences. An initial isotope study (McKenzie *et al.*, 1984) on foraminifera from such a sequence, DSDP Leg 73, Site 519 in the South Atlantic, indicated that this drill hole was potentially excellent for high-resolution stratigraphy, as good magneto- and biostratigraphical controls were obtainable for the latest Miocene/earliest Pliocene. For this study we sampled the interval dated between 7 and 4 Ma in close detail with an average isotope sampling rate of one sample per 50 000 years. This work reports our results: an extended documentation of the marine expressions of the Messinian salinity crisis.

Site description

DSDP Site 519 was located on the edge of a ponded facies on the eastern side of the Mid-Atlantic Ridge (26°8.20′ S, 11°39.97′ W). The site is presently situated under the waters of the South Atlantic gyre at a water depth of 3769 m (Fig. 1). A 151-m thick sequence of pelagic sediments was continuously cored with the hydraulic piston coring apparatus down to the upper Miocene oceanic basement. Recovery was excellent (91.3%). Biostratigraphical dating of the basal sediments gave

a late Miocene (NN 10) age. The upper Miocene–lower Pliocene
sediments analysed in this study were predominantly nannofossil
oozes, which were deposited at water depths between 3250 and 3550 m
as determined from mid-oceanic ridge subsidence curves. Varying
amounts of calcite dissolution of the uppermost Miocene sediments
indicated that the lysocline rose above a paleodepth of 3300 m during
magnetic polarity Epochs 6 and 5. The overlying Pliocene sediments

Fig. 1. Map of the South Atlantic Ocean showing major surface-
current patterns and the location of DSDP sites pertinent to this
study (Sites 513 and 519). Modified after Hsü, La Brecque *et al.*
(1984).

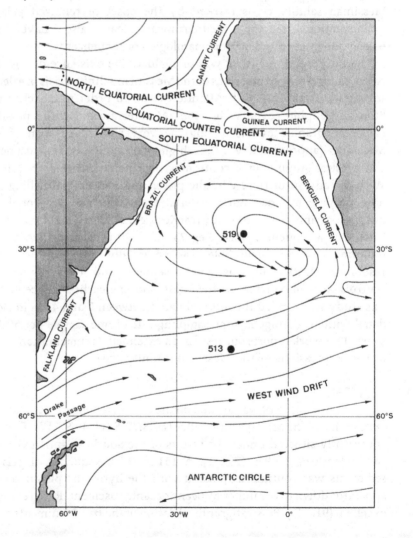

show less dissolution suggesting that the lysocline was depressed below the paleodepth of Site 519 during magnetic polarity Epoch 4 (early Gilbert time). The average sedimentation rate during the late Miocene–early Pliocene was 25 m/Ma (Hsü, La Brecque *et al.*, 1984).

Biostratigraphy and magnetostratigraphy

The correlation of biostratigraphy and magnetostratigraphy for DSDP Site 519 is shown on a time scale based on the magnetostratigraphic ages (Fig. 2). Calcareous nannofossils and foraminiferal datum

Fig. 2. Compilation of magneto- and bio-stratigraphic data from DSDP Site 519 for the late Miocene/early Pliocene (Hsü, La Brecque *et al.*, 1984). FO and LO denote first occurrence and last occurrence, respectively.

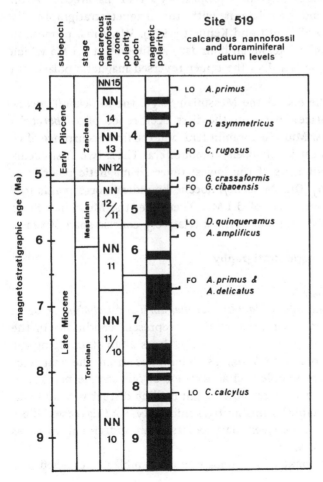

levels and the magnetic reversal pattern were determined by shipboard colleagues (Tauxe *et al.*, 1983, Hsü, La Brecque *et al.*, 1984, p. 48).

Of particular interest to this study is the duration of the Messinian Stage. In the Mediterranean region, the first appearance datum (FAD) of *Globorotalia conomiozea* is considered to be the best stratigraphic marker for the Tortonian/Messinian boundary (d'Onofrio *et al.*, 1975). In New Zealand, the evolutionary FAD of *G. conomiozea* has been found to occur in the upper reversal of Epoch 6, paleomagnetically dated at 6.1 ± 0.1 Ma (Loutit & Kennett, 1979). This FAD is apparently diachronous. A paleomagnetic study in Crete considered the FAD of *G. conomiozea* to occur in Chron 5 at 5.6 Ma (Langereis *et al.*, 1983/1984), while data at Site 519 suggested a FAD not later than Chron 7 (Poore, 1984). Considering this apparent diachronism, we have decided to adopt the magnetostratigraphy proposed by Hsü, La Brecque *et al.* (1984, p. 48) and correlate it with the magnetostratigraphically determined FAD of Loutit and Kennett (1979). We, thus, propose to locate the Tortonian/Messinian boundary at Site 519 at 6.1 Ma which would fall in Zone NN11 in the upper reversed magnetic polarity of Epoch 6.

At Site 519 the end of the Messinian Stage, the Miocene/Pliocene boundary, is placed between the first occurrence of *Globorotalia cibaoensis*, a latest Miocene foraminifera, and the first occurrence of *G. crassaformis*, an earliest Pliocene foraminifera. These two first occurrences are in sediments of the lowest reversed magnetic polarity of Epoch 4 (Gilbert). Our Miocene/Pliocene boundary corresponds to a magnetostratigraphic age of 5.1 Ma. Therefore, the Messinian Stage comprises exactly a million years, a relatively concise span of time.

Stable-isotope stratigraphy

Methodology

The foraminifers *Planulina wuellerstorfi* and *Orbulina universa* were selected for this isotope study to represent conditions in the bottom and surface waters, respectively. Cores 22 to 30 covering the time interval from 4 to 7 Ma, as determined from the magneto-stratigraphy, were sampled on an average of one sample per 50 000 years. The top of each core shows slight drilling disturbance and was considered unsuitable for this stratigraphic study. For this reason, there are discontinuities between data points from individual cores as graphed in Fig. 3, 4, and 5.

Samples were picked for monospecific assemblages, crushed and treated with methanol to remove contaminant particles, and roasted under vacuum at 450 °C for 30 minutes. The microsamples were reacted at 50 °C by using the traditional phosphoric acid method (McCrea, 1950) in a microline reaction system as described by Shackleton *et al.* (1984). The stable-isotope composition of the released carbon dioxide gas was analysed immediately after the completion of the reaction using a V.G. Micromass 903 triple-collecting mass-spectrometer. The results, which are tabulated in Table 1, are reported relative to the PDB international standard using the δ- notation:

$$\delta(‰) = (R_{sample}/R_{standard} - 1) \times 10^3$$

where $R = {}^{13}C/{}^{12}C$ or ${}^{18}O/{}^{16}O$.

Oxygen-isotope stratigraphy

The oxygen-isotope data for both *P. wuellerstorfi* and *O. universa* are plotted versus magnetostratigraphic ages (Figs. 3 and 5). In general, the oxygen-isotope content of both benthic and planktic foraminifera vary simultaneously, but the magnitude of the planktic variations is greater. Apparently, temperature and/or salinity fluctuations, as reflected by the isotopic variations in *O. universa*, are more significant in the surface than in the bottom waters. Of particular note in the planktic curve are the following isotopic events:

(1) a gradual increase in $\delta^{18}O$ from $-0,17‰$ at the base of Core 29 at 6.54 Ma to $+1.08‰$ at 5.93 Ma;

(2) this increase is followed by a gradual decrease to $+0.58‰$ at 5.73 Ma and a return to values around $+1.08‰$ from 5.71 to 5.57 Ma;

(3) a dramatic decrease in $\delta^{18}O$ to a minimum of $-0.39‰$ at the top of Core 29 and the base of Core 28 covering an interval between 5.57 and 5.37 Ma;

(4) in Core 28 between 5.37 and 5.18 Ma there is a major reversal with a progressive oxygen-18 enrichment peaking at a value of $+1.30‰$;

(5) in Core 27 rapid fluctuations ranging from $+1.19$ to $+0.06‰$ are recorded between 5.06 and 4.93 Ma.

A sixth isotopic event, a decrease in $\delta^{18}O$ from $+2.50$ to $+1.92‰$ between 4.58 and 4.34 Ma, was measured only in the benthic foraminifera.

The preservation of a high-resolution isotopic record in sediments with a relatively low sedimentation rate (25 m/Ma) is noteworthy. Presumably, bioturbation of slowly accumulating sediments should dampen the isotope signal (Shackleton, 1977), but this is apparently not

Table 1. *Stable Isotope data for DSDP Site 519*

Sample	Core depth (m)	Age (Ma)	Orbulina universa $\delta^{18}O/\delta^{13}C$ (PDB‰)	Planulina wuellerstorfi $\delta^{18}O/\delta^{13}C$ (PDB‰)
519–22–2, 63–65 cm	90.2	4.18		+2.21/+0.84
22–2, 83–85	90.4	4.18	+0.92/+2.06	+2.56/+0.78
23–2, 115–117	95.2	4.34		+1.92/+0.60
23–3, 26–28	95.8	4.36	+0.68/+2.08	+2.29/+0.73
24–1, 34–36	97.2	4.39	+0.69/+2.34	+2.01/+0.61
24–1, 55–57	97.4	4.40		+2.17/+0.82
24–2, 130–132	99.7	4.45		+2.34/+1.16
25–1, 14–16	101.4	4.49	+0.60/+2.26	+2.30/+1.22
25–1, 60–62	101.9	4.50	+0.63/+2.09	+2.24/+0.81
25–1, 107–109	102.4	4.52	+0.57/+2.18	
25–2, 49–51	103.3	4.56	+0.65/+2.45	+2.35/+1.10
25–2, 92–94	103.7	4.58	+0.65/+2.47	+2.50/+1.00
26–1, 38–40	106.1	4.72	+0.76/+2.19	
26–1, 60–62	106.3	4.72	+0.46/+2.28	
26–1, 122–124	106.9	4.75	+0.65/+2.32	+2.40/+1.00
26–2, 49–51	107.7	4.80	+0.53/+2.10	
27–1, 9–11	110.2	4.93		+2.48/+0.90
27–1, 24–26	110.4	4.94	+0.81/+1.94	+2.59/+1.15
27–1, 65–67	110.8	4.96	+1.17/+2.15	+2.59/+1.39
27–1, 82–84	110.95	4.965		+2.73/+1.15
27–1, 87–89	11.10	4.97	+0.39/+2.46	
27–1, 99–101	111.1	4.976	+0.06/+2.32	+2.48/+0.92
27–1, 109–111	111.2	4.98	+0.77/+2.75	+2.57/+1.13
27–1, 120–122	111.3	4.987	+0.43/+2.67	
27–1, 135–137	111.4	4.99	+0.44/+2.70	
27–2, 15–17	111.8	5.01	+0.08/+2.07	
27–2, 34–36	111.9	5.02	+1.19/+1.90	+3.01/+1.11
27–2, 88–90	112.5	5.05	+0.74/+2.02	+2.52/+1.04
27–2, 98–100	112.6	5.056	+0.97/+2.06	+2.34/+0.94
27–2, 107–109	112.7	5.06	+0.43/+1.39	+2.73/+0.79
28–1, 50–52	115.0	5.18	+1.26/+1.83	
28–1, 75–77	115.2	5.19	+1.30/+1.80	
28–1, 107–109	115.6	5.21	+0.84/+2.14	+2.72/+1.09
28–1, 109–111	115.6	5.21		+2.64/+1.11
28–2, 3–5	116.0	5.24	+0.60/+2.15	
28–2, 39–41	116.4	5.26	+0.42/+1.92	+2.58/+0.66
28–2, 79–81	116.8	5.29	+0.77/+2.19	+2.61/+1.05

Table 1 (contd.)

Sample	Core depth (m)	Age (Ma)	Orbulina universa $\delta^{18}O/\delta^{13}C$ (PDB‰)	Planulina wuellerstorfi $\delta^{18}O/\delta^{13}C$ (PDB‰)
28–2, 99–101	117.0	5.32	+0.73/+2.29	+2.72/+1.11
28–2, 109–111	117.1	5.33	+0.62/+2.36	+2.45/+0.65
28–2, 119–121	117.2	5.34		+2.63/+1.07
28–2, 126–128	117.3	5.36	+0.56/+1.94	+2.43/+0.54
28–2, 133–135	117.4	5.37	+0.44/+2.12	
28–2, 139–141	117.4	5.37	−0.11/+1.83	+2.42/+0.73
29–1, 30–32	118.7	5.56	−0.39/+2.04	+2.47/+0.52
29–1, 40–42	118.9	5.57	+0.96/+2.41	
29–1, 60–62	119.1	5.60	+0.64/+1.76	+2.76/+0.80
29–1, 80–82	119.3	5.63	+1.08/+2.18	
29–1, 100–102	119.5	5.65	+0.98/+2.18	+2.58/+1.03
29–1, 120–122	119.7	5.68	+1.06/+2.24	
29–1, 140–142	119.9	5.71	+1.08/+2.23	
29–2, 10–12	120.1	5.73	+0.53/+1.86	+2.57/+0.73
29–2, 30–32	120.3	5.76	+0.63/+2.22	+2.35/+0.91
29–2, 50–52	120.5	5.83	+0.91/+2.33	
29–2, 69–71	120.7	5.93	+1.08/+2.15	+2.84/+0.77
29–2, 90–92	120.9	6.04		+2.42/+1.14
29–2, 98–100	121.0	6.09	+0.96/+2.62	+2.49/+1.02
29–2, 115–117	121.2	6.19	+0.79/+2.07	+2.60/+1.04
29–2, 125–127	121.3	6.24	+0.81/+2.29	+2.17/+0.87
29–2, 141–143	121.4	6.29	+0.81/+2.72	+2.65/+1.28
29–3, 10–12	121.6	6.40	+0.57/+2.36	+2.34/+1.00
29–3, 23–25	121.7	6.45	+0.52/+2.66	+2.32/+1.28
29–3, 41–43	121.9	6.54	−0.17/+2.17	+2.36/+1.10
30–1, 41–43	123.7	6.78	+0.65/+2.25	
30–1, 90–92	123.8	6.79	+0.71/+2.49	+2.51/+0.95
30–1, 100–102	123.9	6.80	+0.46/+2.26	
30–1, 110–112	124.0	6.82	+0.64/+2.35	
30–1, 115–117	124.1	6.83		+2.44/+1.24
30–1, 130–132	124.2	6.84	+0.43/+2.11	+2.20/+1.12
30–2, 5–7	124.4	6.86	+0.47/+2.11	+1.97/+0.92
30–2, 30–32	124.7	6.91	+0.55/+1.90	+2.75/+1.22
30–2, 90–92	125.3	6.99	+1.02/+2.61	+2.59/+1.58
30–2, 138–140	125.8	7.05	+0.68/+2.58	+2.43/+1.38
30–3, 56–58	126.5	7.15		+2.21/+1.74

an effective mechanism for eliminating the isotope record at Site 519. The position of Site 519 under the South Atlantic gyre, a zone of low productivity and, hence, low organic input to the bottom waters, could account for decreased benthic activity and inefficient homogenization of the sediments. Crowley & Matthews (1983) recognized this phenomenon in their isotope study of the planktic record in a late Quaternary core from beneath the low-productivity region of the North Atlantic subtropical gyre. They concluded that the low degree of record modification in sediments with a low sedimentation rate (12 m/Ma) was a result of low input of organic matter to the sea floor which could support only small populations of burrowing organisms. This finding is

Fig. 3. Oxygen-isotope data for planktic (*Orbulina universa*) and benthic (*Planulina wuellerstorfi*) foraminifers plotted against time as determined by magnetostratigraphy (Hsü, La Brecque *et al.*, 1984). The right-hand column shows the semiquantitative variation of shell diameter of *O. universa*, which is interpreted as a function of surface-water salinity. See text for discussion.

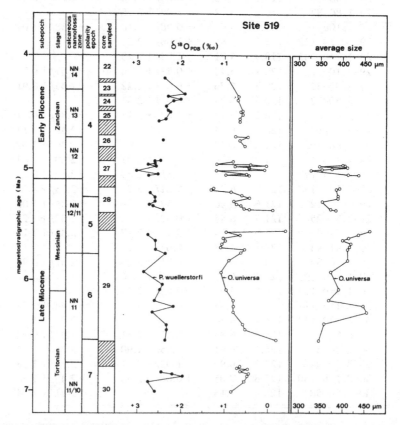

significant because cores from specific areas of interest with low sedimentation rates (e.g. underlying gyre systems) are now available for high-resolution studies.

Paleoceanography

At Site 519 the high-resolution, oxygen-isotope stratigraphy for planktic foraminifera representing a relatively short period of 1.6 Ma between 6.5 and 4.9 Ma displays rapid fluctuations of over 1‰ (Fig. 3). The magnitude of these same fluctuations for the benthic foraminifera is significantly less. We propose that for this 1.6-Ma period the low-amplitude changes in the benthic signal reflect global ice-volume

Fig. 4. Carbon-isotope data for planktic (*Orbulina universa*) and benthic (*Planulina wuellerstorfi*) foraminifers plotted against time as determined by magnetostratigraphy (Hsü, La Brecque *et al.*, 1984). The right-hand column shows variations in the surface-to-bottom water carbon-13 gradient, which fluctuates around an average value of 1.23‰. See text for discussion.

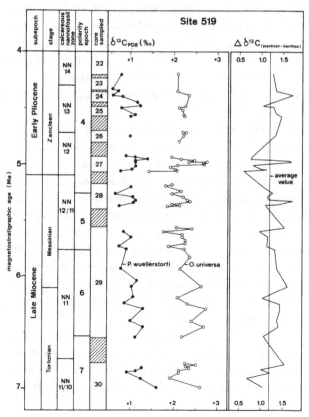

changes; the more positive $\delta^{18}O$ values represent glacial expansion and vice versa. Furthermore, the planktic signal contains this ice-volume effect superimposed upon an isotopic signal related to current fluctuations in the surface waters of the South Atlantic.

The modern surface waters overlying Site 519 are within the subtropical gyre of the South Atlantic (Fig. 1). An intensification of the Benguela Current on the eastern boundary of the gyre during periods of glacial expansion in the southern polar regions could bring Site 19 more under the influence of these colder waters flowing northward from the Southern Ocean because of the northward migration of the Antarctic Convergence and West Wind Drift. Therefore, periods of glacial expansion (isotopic events 1, 2, and 4) are expressed as increases in the oxygen-18 content of planktic foraminifera caused by larger ice-volumes and colder surface-water temperatures.

Isotopic event 3, the dramatic decrease in $\delta^{18}O$ by about 1.5‰, probably represents a sudden deglaciation when large volumes of isotopically light melt water were introduced into the Benguela Current between 5.60 and 5.37 Ma. This isotopic signal is then predominantly a reflection of paleosalinity rather than paleotemperature. The swift oxygen-18 fluctuations in the planktic values occurring between 5.06 and 4.93 Ma (isotopic event 5) could be interpreted as a result of extreme instability in the current pattern with rapid alternations between 'normal' gyre salinity and low-salinity melt water introduced via the Benguela Current.

This paleosalinity interpretation of the oxygen-isotope stratigraphy is supported by variations in the average size of *O. universa* (Fig. 3). Measurements of shell diameters for 100 to 400 *O. universa* specimens present in each of 31 samples from this study show a strong correlation between their size and the oxygen-18 content of their shells. When the oxygen-18 values decreases, there is a tendency for the shell diameter to increase, i.e. with decreased salinity the size of the planktic foraminifera increases. Hecht *el al.* (1976) previously described this relationship between size and salinity for *O. universa* samples taken from plankton tows and sediment core tops in the Indian Ocean. They found a direct correlation between salinity and shell diameter and porosity. Large, thin-walled specimens were found in the less-saline equatorial water masses, while small, thick-walled specimens came from the denser central water masses. We conclude that during the late Miocene/early Pliocene a similar phenomenon governed the relationship between the size of *O. universa* and the salinity of their habitat at Site 519.

Both the morphologic and isotopic variations in *O. universa* indicate salinity changes. Although it is difficult to separate all the factors influencing the oxygen-isotope stratigraphy at Site 519, there were apparently periods of decreased salinity (decreased oxygen-18) which we associate with interglacials and glacial melting. At other times the oxygen-18 content of the planktic foraminifera increased due to an increased continental ice volume (also recorded in benthic foraminifera) and an intensification of the Benguela Current allowing colder water to infringe upon the denser waters of the subtropical gyre. We associate these latter periods with increased glaciation, most likely an expansion of the West Antarctic ice sheet.

Isotopic event 6 between 4.58 and 4.34 Ma is different from the others in that it is more strongly registered in the benthic foraminifera. The planktic $\delta^{18}O$ value remains relatively constant throughout this time interval. This constancy could signify a trend towards greater influence of dense gyre water on the surface waters at Site 519. Superimposed upon this trend is the gradual decrease in $\delta^{18}O$ which is recorded in the benthic fauna. This decrease represents a major transgression during the early Pliocene. The onset of this transgression (glacial melting or global warming) possibly was first registered by surface-water dwellers during isotopic event 5 between 5.06 and 4.93 Ma. Afterwards there were no further sudden pulses of melt water and the more saline water mass of the South Atlantic gyre recovered its dominancy at Site 519. The global ice-volume continued to decrease until 4.3 Ma, as documented by the benthic curve, but this tendency towards more negative $\delta^{18}O$ values for the global oceans is masked in the surface waters at Site 519 by a trend towards a stabilization of the isotopically heavier waters of the subtropical gyre. The same transgression has been recognized by various authors on the basis of their studies elsewhere (e.g. Hayes, Frakes *et al.*, 1975; Ludwig, Krasheninikov *et al.*, 1983).

Carbon-isotope stratigraphy

The carbon-isotope data for both *P. wuellerstorfi* and *O. universa* are plotted against magnetostratographic ages in Fig. 4. The difference between the $\delta^{13}C$ values for the planktic and benthic foraminifera from the same sample ($\Delta\delta^{13}C$) is also included in the graph. This latter curve compares changes in the surface-to-bottom water $\delta^{13}C$ profile. Between 7.0 and 4.93 Ma the $\delta^{13}C$ values for both *P. wuellerstorfi* and *O. universa* vary approximately simultaneously, but the magnitude of the *O. universa* fluctuations are invariably greater. This is reminiscent of the $\delta^{18}O$ curves for the same time interval. Analogous to the $\delta^{18}O$

variations these planktic $\delta^{13}C$ variations could be explained as paleoceanographic changes regulated by glacial–interglacial periods.

Kroopnick (1980) extensively documented the distribution of carbon-13 in the modern Atlantic Ocean showing that individual water masses have distinct $\delta^{13}C$ signals which can be used as tracers. Perhaps, the low-amplitude benthic variations reflect more or less intensification of the Antarctic Bottom Water (AABW) while the larger planktic fluctuations result from encroachment of the Benguela Current upon the central water masses. Surface waters of the Southern Ocean tend to be up to 0.5‰ more negative in carbon-13 than those of the South Atlantic subtropical gyre, and AABW is approximately 0.5‰ more negative than the southward flowing North Atlantic Deep Water (NADW) (Kroopnick, 1980). In general, the curves in Fig. 4 show an upward trend towards a more negative $\delta^{13}C$ value of both *P. wuellerstorfi* and *O. universa* throughout Cores 29 and 28, i.e. colder climatic conditions in the southern high latitudes resulted in an intensification of the AABW and Benguela Current. For *O. universa*, the most negative $\delta^{13}C$ values of the top of Core 28 and base of Core 27 correspond to the maximum latest Miocene glaciation, as indicated by the $\delta^{18}O$ data (Fig. 3).

In Core 27 between 5.06 and 4.93 Ma the unstable surface-water conditions of oxygen-isotopic event 5 are curiously associated with a very distinct carbon-isotope peak. In a period of 80 000 years, the $\delta^{13}C$ value increases by over 1‰ returning, within 40 000 years to the pre-peak value. As this peak is a surface-water phenomenon, one explanation could be that a period of heightened productivity in the earliest Pliocene resulted in carbon-13 enrichment of the surface waters. Afterwards a steady decrease in the $\delta^{13}C$ value of the bottom waters reaching a minimum of 4.34 Ma is recorded by *P. wuellerstorfi*. This carbon-13 minimum correlates with the oxygen-18 minimum, which corresponds to the peak of the early Pliocene transgression, and is not recorded in the planktic record. A tentative speculation could postulate that the period of heightened productivity at 5.0 Ma was not only a local phenomenon at Site 519 but occurred world-wide. The extra organic matter produced at 5 Ma could have been gradually oxidized and returned to the oceans over the following 700 000 years decreasing the $\delta^{13}C$ value of the oceanic carbon reservoir. Obviously, more high-resolution stratigraphy in early Pliocene cores from other locations is required to determine whether carbon peaks and the associated oxygen-isotopic events 5 and 6 observed at Site 519 have a global significance.

The occurrence of a carbon-isotope shift towards lighter $\delta^{13}C$ values by about 0.8‰ has been reported for numerous deep-sea sites and has been placed between 6.10 and 5.90 Ma (Haq *et al.*, 1980; Vincent *et al.*, 1980). Bender & Keigwin (1979) proposed that this shift was related to a geologically instantaneous change in the turnover rate of oceanic circulation, possibly associated with the cessation of the inflow of Mediterranean Deep Water into the Atlantic. The carbon-isotope data from Site 519 (Fig. 4) does not indicate the presence of a shift between 6.10 and 5.90 Ma but does show instead a progressive transition towards slightly more negative $\delta^{13}C$ values in both benthic and planktic samples throughout the latest Miocene. Thus, the late Miocene carbon-isotope shift is manifested as a transition at Site 519.

Late Miocene Antarctic glaciation

From their oxygen-isotope curves of deep-sea sediments cored at drill sites in the Southern Ocean, Shackleton & Kennett (1975a,b) surmised a rapid build-up of a permanent ice cap in East Antarctica between 14 and 10 Ma which was followed in the latest Miocene by an even sharper deterioration of climatic conditions. They suggested that this latter period of severe continental glaciation may have exceeded all other Neogene glaciations in its intensity. Evidence of late Miocene glaciation of the Ross Sea region was revealed by seismic investigation during Leg 28 deep-sea drilling (Hayes, Frakes *et al.*, 1975). Biostratigraphic studies of carbonate versus siliceous ooze distributions in sediments from DSDP sites in the Southern Ocean and South Atlantic (summarized in Wise, 1981 and Wise *et al.*, this volume, chapter 15) also indicate a sudden northward shift of the Antarctic Convergence in the latest Miocene. In addition to this biogenic transition, the presence of ice-rafted debris and erosional episodes related to intensification of bottom-water currents could, in paleomagnetically dated piston cores from the Maurice Ewing Bank, indicate the formation of a floating but partially grounded West Antarctic ice sheet in the latest Miocene and the production of the first true Antarctic Bottom Water (AABW) (Ciesielski *et al.*, 1982). Furthermore, these marine interpretations are supported by land geologists working on glacial deposits in the southern hemisphere, as reviewed by Mercer (1983). In particular, Rutford *et al.*, (1972) proposed the establishment of an ice sheet in West Antarctica by around 7 Ma, while latest Miocene or earliest Pliocene glaciation in southern Argentina indicates climatic conditions more extreme than those prevailing today (Mercer & Sutter, 1982).

A recent detailed correlation of ice-rafted debris and erosional episodes with paleomagnetics at DSDP Site 513 in the subantarctic region of the Atlantic Ocean (Fig. 1) implies that a grounded West Antarctic ice sheet was formed by Chron C-9 or approximately 8.7 Ma (Ciesielski & Weaver, 1983). The authors also note that this ice sheet was probably highly unstable and frequently became ungrounded during the latest Miocene. During these periods the West Antarctic ice sheet became a floating ice shelf and may have produced a greater volume of AABW which could erode or inhibit sedimentation in the Southern Ocean. The recovery of continuously deposited sediments from the earliest Chron C-6 ($\approx$ 6.5 Ma) to the latest Chron C-5 ($\approx$ 5.35 Ma), with the exception of a brief erosional episode in late Chron C-5, suggests that the West Antarctic ice sheet was grounded during this period. The grounding line of the ice sheet on the continental shelf is dependent upon sea-level depth, whereby if sea level drops the ice sheet will be grounded further seaward and vice versa (Hollin, 1962). Therefore, the data from Site 513 imply that during earliest Chron C-6 and until latest Chron C-5 there was increased glaciation and, hence, lowered sea level. This latest Miocene glaciation was interrupted by at least one warming period when ungrounding or destruction of the West Antarctic ice sheet occurred, as manifested by the disconformity in uppermost Chron C-5. Also, two distinct occurrences of ice-rafted debris record these two latest Miocene periods of glacial enhancement, one in the middle of Chron C-5 and one in the later normal of Chron C-5 (C-5N1) to earliest Gilbert Chron (Bornhold, 1983 and Ciesielski & Weaver, 1983).

The fluctuations in the oxygen-isotope date from Site 519 between 6.7 and 5.1 Ma indicate repeated phases of glacial enhancement during the latest Miocene and are interpreted as cooling and warmings in Fig. 5. In particular, the extreme fluctuations in the data for the planktic foraminifera could directly reflect the instability of the West Antarctic ice sheet. During coolings isotopically light water is stored in continental ice producing a progressive increase in the $\delta^{18}O$ value of marine waters. At the same time sea level drops, and the West Antarctic ice sheet becomes more firmly grounded. During warmings the melting of Antarctic ice introduces tremendous quantities of isotopically light water into the surface currents flowing west and northwards into the southeast Atlantic Ocean. The sea level rises, the West Antarctic ice sheet becomes ungrounded and greater amounts of AABW are produced.

These rapid injections of melt water are seen as negative spikes in the planktic isotope curve (Fig. 5). In the latest Miocene, at least four such warming events are recorded with a particularly large one in the middle of polarity Epoch 5. Although it is difficult to exactly correlate the isotopic changes at Site 519 with the data from Site 513, the warming and cooling patterns at the two sites are apparently similar. We conclude that the negative oxygen-isotope excursions represent warming periods and are most likely melt-water spikes, as might be expected if large-scale melting of an unstable West Antarctic ice sheet occurred (P.F. Ciesielski, personal communication). On the other hand, the positive oxygen isotope events represent major cooling which result in eustatic sea-level lowering. It is undoubtedly not a coincidence that

Fig. 5. Oxygen-isotope data for planktic (*Orbulina universa*) and benthic (*Planulina wuellerstorfi*) foraminifers plotted against time as determined by magnetostratigraphy (Hsü, La Brecque *et al.*, 1984). Positive excursions in the planktic values are interpreted as cooling periods with ice accumulation in Antarctica, while negative excursions are representative of warming periods with significant melting of Antarctic ice. The isotope data are correlated with time-equivalent lithostratigraphic events from the Mediterranean region. The MPL foraminifera zones are after Cita (1975b).

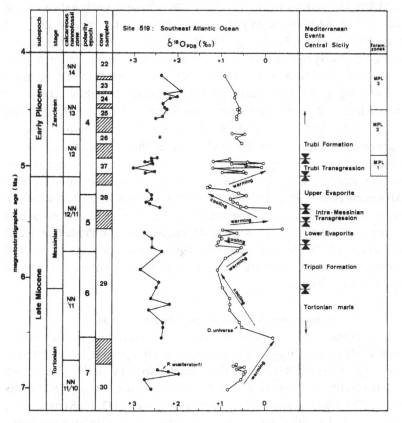

oxygen-isotope curves for the latest Miocene/earliest Pliocene of the Blind River section, New Zealand, contain a warming and cooling pattern which resembles that of Fig. 5 (Loutit & Kennett, 1979).

Early Pliocene transgression

Global sea-level curves based on seismic stratigraphy show a relatively low stand in the late Miocene followed by a high stand in the early Pliocene (Vail *et al.*, 1977), and the study of marine sediments from the Southern Ocean indicates a warm interval during the early Pliocene which separated latest Miocene and mid-Pliocene coolings (Hay & Opdyke, 1967). In the Southern Ocean warm subantarctic assemblages are found in Antarctic and high subantarctic sediments which are dated as early Gilbert (5.26 to 4.7 Ma) (Keany, 1978). An early Pliocene amelioration of the climate between 4.3 and 3.95 Ma led to the diminution of eroding currents in the southwest Atlantic (Ciesielski *et al.*, 1982) and, perhaps, water temperatures were even too warm for the existence of a West Antarctic ice sheet (Ciesielski & Weaver, 1974).

At Site 519 in the southeast Atlantic, this climatic amelioration is first recorded in the planktic oxygen-isotope curve of the earliest Pliocene between 5.06 and 4.93 Ma (Fig. 5). The extreme and rapid fluctuations in the curve may indicate sudden pulses of melt-water injection into the surface currents as the volume of the West Antarctic ice sheet adjusted to the changing climatic conditions. After about 4.93 Ma the $\delta^{18}O$ value of the surface waters was apparently no longer influenced by the melt waters but a continued rise in sea level until 4.34 Ma is indicated by a progressive decrease in the $\delta^{18}O$ value of the benthic foraminifera. After 4.34 Ma the oxygen-isotope curve becomes enriched in the heavier isotope as colder climatic conditions once again prevailed (McKenzie *et al.,*. 1984; Weissert & Oberhänsli, this volume, chapter 7). In the southwest Atlantic the accumulation rate of ice-rafted debris increases and major eposides are more frequent after 4.35 Ma (Ciesielski & Weaver, 1983). This permanent change in the deposition of ice-rafted debris suggests an increased stability of the West Antarctic ice sheet into the late Neogene.

Messinian salinity crisis in the Mediterranean

Numerous attempts have been made to correlate global eustatic events with the timing of the lithological changes in the Mediterranean region which pertain to the events surrounding the Messinian salinity crisis (e.g. Ryan *et al.*, 1974; van Couvering *et al.*, 1976; Adams *et al.*, 1977; Cita & Ryan, 1979; Loutit & Keigwin, 1982; McKenzie *et al.*, 1984). With the high-resolution stratigraphy obtainable from the deep-sea sediments cored at Site 519, we propose yet another correlation attempt. In Fig. 5 we correlate Messinian lithologic events from the Mediterranean, as seen in the Sicilian sedimentary sequence (Decima & Wezel, 1973), with the paleoceanographic record from the South Atlantic. As mentioned above, the oxygen-isotopic signal in the benthic foraminifera is assumed to represent ice-volume changes presumably on Antarctica, while the planktic signal reflects the ice-volume changes amplified by changes in the surface-current patterns. These changes are both induced by relatively warmer or cooler climatic conditions in the high southern latitudes. The planktic date even suggest that warmer conditions could produce a significant injection of melt water from the West Antarctic ice sheet into the northward flowing surface currents.

We have divided our oxygen-isotope events into time periods based on magnetostratigraphic ages and compared these events to their Mediterranean counterparts. This isotope and lithostratigraphic correlation allows us to propose the following scenario:

(1) 6.54 to 5.93 Ma – a cooling period in the South Atlantic spans the Tortonian/Messinian boundary. In Sicily, this boundary is marked by a change from Tortonian marls to the Tripoli Formation, a sequence of alternating diatomites and claystones (McKenzie *et al.*, 1979/1980). These sediments indicate that restrictive conditions predominated in the deeper Mediterranean basins during this period. Gersonde (1980) placed the base of the Tripoli Formation in the upper reversal of polarity Epoch 6 based on diatom and paleomagnetic correlations. The first appearance of *Globorotalia conomiozea* at the base of the Tripoli Formation (d'Onofrio *et al.*, 1975), which corresponds to a datum level of 6.1 Ma (Loutit & Kennett, 1979), approximately places the beginning of diatom deposition at the Tortonian/Messinian boundary. Therefore, during this period a steady build-up or consolidation of the West Antarctic ice sheet could have lowered the sea level enough to cause restrictive circulation in the Mediterranean and, perhaps, reduce the amount of dense Mediterranean water flowing into the Atlantic. A change in the Mediterranean circulation pattern most likely resulted in an increased supply of nutrients to the surface waters promoting diatom blooms. The cyclic sedimentation within the Tripoli Formation suggests multiple (maximum 34) sea-level fluctuations (McKenzie *et al.*, 1979/1980). These shorter-term fluctuations are not seen in the South Atlantic record, only the higher frequency variations.

(2) 5.93 to 5.57 Ma – Gersonde (1980) placed the top of the Tripoli Formation in the lower normal polarity of magnetic Epoch 5. From his extensive sedimentological study, Bennet (1980) estimated that 500 000 years is a reasonable duration for the deposition of the entire formation.

Combining both sets of data allows us to place the top of the formation at
5.7 Ma during a warming period, as indicated by the oxygen-isotope
curve. Gersonde's analysis of the faunal diversity showed a drastic
decrease in diversity in the uppermost part of the Tripoli Formation
with a short-lived increase right before the onset of evaporite
deposition. This short-lived reversal in an increasingly stressed
environment complements the isotope data. From 5.7 to 5.57 Ma there
was a period of intensive cooling. We consider this relatively short
period of about 130 000 years to represent the time of deposition of the
Lower Evaporite in the Mediterranean (Decima & Wezel, 1973; Hsü et
al., 1978b). The Lower Evaporite contains thick halite deposits often
associated with potassium salts in the deeper parts of the basins. The
deposition of the Main Salt unit, which is up to or more than a thousand
metres thick, requires a continued flux of Atlantic sea water into the
Mediterranean without refluxing. We propose that the Antarctic
glaciation between 5.7 and 5.57 Ma was sufficient to lower the sea level
enough to cause a circulation imbalance allowing for the deposition of
the Lower Evaporite complex with a continued influx of marine water.
This would require that 6% of the ocean's salt, as estimated by Ryan
(1973), was removed during a short interval of 130 000 years. Ryan
predicted that the subsequent decrease in surface salinity could have
led to a significant climatic deterioration, but this is not what we
observe in the South Atlantic isotope record.

(3) 5.57 to 5.37 Ma – a dramatic change from glacial to interglacial
conditions is recorded in the oxygen-isotope curve for this time
interval. We interpret this to be a result of a melting of the West
Antarctic ice sheet which could have produced an elevated sea level.
The intra-Messinian event in the Mediterranean (Hsü et al., 1978b)
begins with its total isolation from the Atlantic and is manifested in the
sediments as an unconformity with erosion and recycling of salts. A
marine transgression over the intra-Messinian unconformity resulted in
marine deposition at the base of the Upper Evaporite. We correlated this
intra-Messinian transgression with the South Atlantic melt-water spike
at ≈ 5.5 Ma in the Site 519 isotope record.

(4) 5.37 to 5.18 Ma – a second, severer glacial event occurs in the
isotope record of the latest Miocene (Fig. 5). The magnitude of this sea-
level drop must have been large enough to cut-off once again the
Mediterranean from its source of marine water. The Upper Evaporite
complex comprises predominately gypsum beds alternating with marls
(Decima & Wezel, 1973). The fact that the late-stage evaporative
minerals (halite etc.) are rare in the Upper Evaporite indicates that the
marine influx differed from that during the deposition of the Lower

Evaporite. The isotope data from Site 519 could suggest that the Atlantic-Mediterranean connection was even more restricted during the deposition of the Upper Evaporite. The gypsum and halite sediments of the Upper Evaporite could have resulted from an internal redistribution by dissolution-redeposition processes under a hydrological regime confined to the Mediterranean region. Near the end of the Messinian an influx of water from the Parathetys led to the deposition of lacustrine sediments (Lago Mare) in the eastern Mediterranean while restricted marine sedimentation continued in the western Mediterranean (Hsü *et al.*, 1978b).

(5) 5.18 to 4.93 Ma – this period containing the Miocene/Pliocene boundary shows extreme and rapid isotopic fluctuations, which we interpret as an instability in the South Atlantic current patterns possibly related to a climatic instability. In Fig. 5 we correlate this period to the earliest Mediterranean Pliocene foraminiferal zone (MPL–1), which extends from the Miocene/Pliocene boundary to about 300 000 years afterwards (Cita, 1975b). The MPL-1 zone is characterized by an acme of *Sphaeroidinellopsis* and an increase in the bulk carbonate content of the sediments. This zone, the Trubi Transgression, marks the end of the Messinian salinity crisis and the resumption of marine sedimentation in the Mediterranean with the opening of the Gibraltar Straits. We propose that the flooding of the Mediterranean caused a disequilibrium in the Atlantic circulation pattern, as demonstrated by the wide excursions seen in the surface waters at Site 519. The re-establishment of an antiestuarine circulation pattern in the Mediterranean with the injection of Mediterranean Deep Water (MDW) into the North Atlantic Deep Water (NADW) surely influenced or altered the early Pliocene circulation in the Atlantic Ocean. The stabilization of the surface-water conditions at Site 519 after 4.93 Ma, as indicated by a $\delta^{18}O$ value lying midway between the glacial and melt-water extremes, may testify to an equable climate which allowed a stable gyre system to dominate. During this stable interval from 4.93 to 4.34 Ma, progressive changes in both the $\delta^{13}C$ and $\delta^{18}O$ values of the bottom water indicate increased input of carbon-12 into the AABW as well as a gradual sea-level rise. A coupling of the termination of the salinity crisis with the early Pliocene climatic amelioration and global transgression is only conjecture at this point.

Conclusions

The latest Miocene/earliest Pliocene was a period of significant climatic changes recognized as major regressions and transgressions. High-resolution stratigraphy in continuously cored deep-sea sections allowed us to study the paleoceanographic expressions of these climatic changes and to correlate these specific expressions with recognized events elsewhere. In particular, we attempted to correlate events from the Mediterranean region, which occurred during the two-million-year period straddling the Miocene/Pliocene boundary, with paleoceanographic changes registered in deep-sea sediments from the South Atlantic, DSDP Site 519. Using high-resolution stratigraphy, it is possible to evaluate the exact timing of events in order to delineate 'causes and effects'. In this study, we demonstrated that the Messinian salinity crisis in the Mediterranean was intrinsically related to southern hemisphere glaciations. The timing of events indicated that the isolation or flooding of the Mediterranean was a consequence of the waxing and waning of the West Antarctic ice sheet. On the other hand, events in the Mediterranean, such as the deposition of 6% of the ocean's salt in the Lower Evaporite or the breakthrough of the Gibraltar Straits in the very earliest Pliocene, enhanced or possibly even stimulated climatic fluctuations and, thus, influenced paleoceanography. As in Plio-Pleistocene sequences, detailed high-resolution isotope stratigraphy promises to be a powerful method of defining climatic evolution prior to 3 Ma.

Acknowledgements

We would like to thank the Deep Sea Drilling Project for fulfilling our numerous sample requests. One of us (JAM) would especially like to acknowledge the many colleagues who, over the past ten years, have initiated her into the very exciting problems surrounding the Messinian salinity crisis. In particular, she recognizes the inspiration of M. B. Cita, A. Decima, K. J. Hsü, and B. C. Schreiber. We thank P. F. Ciesielski for reviewing the manuscript. This is contribution No. 244 of the Laboratory of Experimental Geology, ETH, Zürich.

REFERENCES

Adams, C.G., Benson, R., Kidd, R.B., Ryan, W.B.F. & Wright, R.C. 1977. The Messinian salinity crisis and evidence of late Miocene eustatic changes in the world ocean. *Nature*, **269**, 383–386.

Bender, M.L. & Keigwin, L.D. 1979. Speculations about the upper Miocene change in abyssal Pacific dissolved bicarbonate $\delta^{13}C$. *Earth and Planetary Science Letters*, **45**, 383–393.

Bennet, G. 1980. The sedimentology, diagenesis and palaeo-oceanography of diatomites from the Miocene of Sicily. Ph.D. thesis, University of Durham, England, 223 pp.

Bornhold, B.D. 1983. Ice-rafted debris in sediments from Leg 71, southwest Atlantic Ocean. In: W.J. Ludwig, V.A. Krasheninnikov *et al.*, *Initial Reports of* the *Deep Sea Drilling Project*, **71**, part 1, Washington DC, US Government Printing Office, pp. 307–316.

Ciesielski, P.F. & Weaver, F.M. 1974. Early Pliocene temperature changes in the Antarctic Seas. *Geology*, **2**, 511–515.

Ciesielski, P.F. & Weaver, F.M. 1983. Neogene and Quaternary paleoenvironmental history of Deep Sea Drilling Project Leg 71 sediments, southwest Atlantic Ocean. In: W.J. Ludwig, V.A. Krasheninnikov *et al.*, *Initial Reports of the Deep Sea Drilling Project*, **71**, part 1, Washington DC, US Government Printing Office, pp. 461–477.

Ciesielski, P.F., Ledbetter, M.T. & Ellwood, B.B. 1982. The development of Antarctic glaciation and the Neogene paleoenvironment of the Maurice Ewing Bank. *Marine Geology*, **46**, 1–51.

Cita, M.B. 1975a. The Miocene/Pliocene boundary, history and definition. In: T. Saito, & L.H. Burcke (eds.), Late Neogene epoch boundaries. *Micropaleontology Special Publication*, No. 1. New York, Micropaleontology Press, American Museum of Natural History, pp. 1–30.

Cita, M.B. 1975b. Studi sul Pliocene e sugli strati di passaggio dal Miocene al Pliocene, VIII. Planktonic foraminiferal biozonation of the Mediterranean Pliocene deep sea record. A revision. *Rivista Italiana di Paleontologia e Stratigrafia*, **81**, 527–544.

Cita, M.B. & Ryan, W.B.F. 1979. Late Neogene environmental evolution. In: U. von Rad, W.B.F. Ryan *et al.*, *Initial Reports of the Deep Sea Drilling Project*, **47**, part 1, Washington DC, US Government Printing Office, pp. 447–459.

Crowley, T.J. & Matthews, R.K. 1983. Isotope–plankton comparisons in a late Quaternary core with a stable temperature history. *Geology*, **11**, 275–278.

Decima, A. & Wezel, F.C. 1973. Late Miocene evaporites of the Central Sicilian Basin, Italy. In: W.B.F. Ryan, K.J. Hsü *et al.*, *Initial Reports of the Deep Sea Drilling Project*, **13**, part 2, Washington DC, US Government Printing Office, pp. 1234–1239.

d'Onofrio, S., Giannelli, L., Iaccarino, S., Morlotti, E., Romeo, M., Salvatorini, G., Sampò, M. & Sprovieri, R. 1975. Planktonic foraminifera of the upper Miocene from some Italian sections and the problem of the lower boundary of the Messinian. *Bolletino della Società Paleontologica Italiana*, **14**, 177–196.

Gersonde, R. 1980. Paläoökologische und biostratigraphische Auswertung von Diatomeenassoziationen aus dem Messinium des Caltanissetta-Beckens (Sizilien) und einiger Vergleichsprofile in SO-Spanien, NW-Algerien und auf Kreta. Ph.D. thesis, Christian-Albrechts-Universität, Kiel, FRG, 393 pp.

Haq, B.U., Worsley, T.R., Burckle, L.H., Douglas, R.G., Keigwin, L.D., Jr, Opdyke, N.D., Savin, S.M., Sommer, M.A., II, Vinent, E. & Woodruff, F. 1980. Late Miocene marine carbon-isotopic shift and synchroneity of some phytoplanktonic biostratigraphic events. *Geology*, **8**, 427–431.

Hays, D.E., Frakes, L.A. *et al.* 1975. *Initial Reports of the Deep Sea Drilling Project*, **28**, Washington DC, US Government Printing Office, 1017 pp.

Hays, J.D. & Opdyke, N.D. 1967. Antarctic radiolaria, magnetic reversals, and climatic changes. *Science*, **158**, 1001–1011.

Hecht, A.D., Bé, A.W.H. & Lott, L. 1976. Ecologic and paleoclimatic implications of morphologic variations of *Orbulina universa* in the Indian Ocean. *Science*, **194**, 422–424.

Hollin, J.T. 1962. On the glacial history of Antarctica. *Journal of Glaciology*, **4**, 173–195.

Hsü, K.J., Montadert, L. *et al.* 1978a. *Initial Reports of the Deep Sea Drilling Project*, **42**, Part 1, Washington DC, US Government Printing Office, 1249 pp.

Hsü, K.J., Montadert, L., Bernoulli, D., Cita, M.B., Erikson, A., Garrison, R.E., Kidd, R.B., Mélières, F., Müller, C. & Wright, R. 1978b. History of the Mediterranean salinity crisis. In: K.J. Hsü, L. Montadert *et al.*, *Initial Reports of the Deep Sea Drilling Project*, **42**, part 1, Washington DC, US Government Printing Office, pp. 1053–1078.

Hsü, K.J., La Brecque, J.L. *et al.* 1984. *Initial Reports of the Deep Sea Drilling Project*, **73**, Washington DC, US Government Printing Office, 798 pp.

Keany, J. 1978. Paleoclimatic trends in early and middle Pliocene deep-sea sediments of the Antarctic. *Marine Micropaleontology*, **3**, 35–49.

Kroopnick, P. 1980. The distribution of ^{13}C in the Atlantic Ocean. *Earth and Planetary Science Letters*, **49**, 469–484.

Langereis, C. G., Zachariasse, W. J. & Zijderveld, J. D. A. 1983/1984. Late Miocene magnetobiostratigraphy of Crete. *Marine Micropaleontology*, **8**, 261–281.

Loutit, T. S. & Keigwin, L. D., Jr, 1982. Stable isotopic evidence for the latest Miocene sea-level fall in the Mediterranean region. *Nature*, **300**, 163–166.

Loutit, T. S. & Kennett, J. P. 1979. Application of carbon isotope stratigraphy to late Miocene shallow marine sediments, New Zealand. *Science*, **204**, 1196–1199.

Ludwig, W. J. Krasheninnikov, V. A., et al. 1983. *Initial Reports of the Deep Sea Drilling Project*, **71**, Washington DC, US Government Printing Office, 1187 pp.

Mayer-Eymar, K. 1867. *Catalogue systématique et descriptif des fossiles des terrains tertiaires qui se trouvent au Musée Fédéral de Zürich.* Zürich, vol. 1, 37 pp, vol. 2, 65 pp.

McCrea, J. M. 1950. The isotopic chemistry of carbonates and a paleotemperature scale. *Journal of Physical Chemistry*, **18**, 849–857.

McKenzie, J. A., Jenkyns, H. C. & Bennet, G. G. 1979/1980. Stable isotope study of the cyclic diatomite-claystones from the Tripoli Formation, Sicily: a prelude to the Messinian salinity crisis. *Palaeogeography, Palaeoclimatology, Palaeoecology*, **29**, 125–141.

McKenzie, J. A., Weissert, H., Poore, R. Z., Wright, R. C., Percival, S. F., Jr, Oberhänsli, H. & Casey, M. 1984. Paleoceanographic implications of stable-isotope data from upper Miocene–lower Pliocene sediments from the Southeast Atlantic (Deep Sea Drilling Project Site 519). In: K. J. Hsü, J. L. La Brecque et al., *Initial Reports of the Deep Sea Drilling Project*, **73**, Washington DC, US Government Printing Office, pp. 717–724.

Mercer, J. H. 1983. Cenozoic glaciation in the southern hemisphere. *Annual Review of Earth and Planetary Sciences*, **11**, 99–132.

Mercer, J. H. & Sutter, J. F. 1982. Late Miocene–earliest Pliocene glaciation in southern Argentina: implications for global ice sheet history. *Palaeogeography, Palaeoclimatology, Palaeoecology*, **38**, 185–206.

Poore, R. Z. 1984. Middle Eocene to Quaternary planktic foraminifers. In: K. J. Hsü, J. L. La Brecque et al., *Initial Reports of the Deep Sea Drilling Project*, **73**, Washington DC, US Government Printing Office, pp. 429–448.

Rutford, R. H., Craddock, C., White, C. M. & Armstrong, R. L. 1972. Tertiary glaciation in the Jones Mountains. In: R. J. Adie (ed.), *Antarctic Geology and Geophysics*. Oslo, Universitetsforlaget, pp. 239–250.

Ryan, W. B. F. 1973. Geodynamic implications of the Messinian crisis of salinity. In: C. W. Drooger (ed.), *Messinian Events in the Mediterranean*. Amsterdam, North-Holland Publishing Company, pp. 26–38.

Ryan, W. B. F., Hsü, K. J. et al. 1973. *Initial Report of the Deep Sea Drilling Project*, **13**, Washington DC, US Government Printing Office, 1447 pp.

Ryan, W. B. F., Cita, M. B, Rawson, M. D., Burckle, L. H. & Saito, T. 1974. A paleomagnetic assignment of Neogene stage boundaries and the development of isochronous datum planes between the Mediterranean, the Pacific and Indian Oceans in order to investigate the response of the world ocean to the Mediterranean 'salinity crisis'. *Rivista Italiana di Paleontologia e Stratigraphia*, **80**, 631–688.

Rutford, R. H., Craddock, C., White, C. M. & Armstrong, R. L. 1972. Tertiary glaciation in the Jones Mountains. In: R. J. Adie (ed.), *Antarctic Geology and Geophysics*. Oslo, Universitetsforlaget, pp. 239–250.

Selli, R. 1960. The Mayer-Eymar Messinian 1867. Proposal for a neostratotype. *Twenty-first International Geological Congress, Copenhagen*, part 28, pp. 311–333.

Shackleton, N. J. 1977. The oxygen isotope stratigraphic record of the late Pleistocene. *Royal Society of London Philosophical Transactions*, **B280**, 169–182.

Shackleton, N. J. & Kennett, J. P. 1975a. Late Cenozoic oxygen and carbon isotopic changes at DSDP Site 284: implications for glacial history of the northern hemisphere and Antarctica. In: J. P. Kennett, R. E. Houtz et al., *Initial Reports of the Deep Sea Drilling Project*, **29**, Washington DC, US Government Printing Office, pp. 801–807.

Shackleton, N.J. & Kennett, J.P. 1975b. Paleotemperature history of the Cenozoic and the initiation of Antarctic glaciation: oxygen and carbonisotope analyses in DSDP Sites 277, 279, and 281. In: J.P. Kennett, R.E. Houtz *et al.*, *Initial Reports of the Deep Sea Drilling Project*, **29**, Washington DC, US Government Printing Office, pp. 743–755.

Shackleton, N.J, Hall, M.A. & Boersma, A. 1984. Oxygen and carbon isotope data from Leg 74 foraminifers. In: Moore, T.C., Rabinowitz, P.D. *et al.*, *Initial Reports of the Deep Sea Drilling Project*, **74**, Washington DC, US Government Printing Office, pp. 599–612.

Tauxe, L., Tucker, P., Petersen, N.P. & La Brecque, J.L. 1983. The magnetostratigraphy of Leg 73 sediments. *Palaeogeography, Palaeoclimatology, Palaeoecology*, **42**, 65–90.

Vail, P.R., Mitchum, R.M., Jr & Thompson, S., III, 1977. Seismic stratigraphy and global changes of sea level, Part 4: Global cycles of relative changes of sea level. In: C.E. Payton (ed.), Seismic stratigraphy-applications to hydrocarbon exploration. *Am. Ass. Petrol. Geol. Mem.*, **26**, 83–97.

van Couvering, J.A., Berggren, W.A., Drake, R.E., Aguirre, E. & Curtis, G.H. 1976. The terminal Miocene event. *Marine Micropaleontology*, **1**, 263–286.

Vincent, E., Killingley, J.S. & Berger, W.H. 1980. The magnetic epoch-6 carbon shift: A change in the ocean's $^{13}C/^{12}C$ ratio 6.2 million years ago. *Marine Micropaleontology*, **5**, 185–203.

Wise, S.W., Jr, 1981. Deep sea drilling in the Antarctic: Focus on late Miocene glaciation and applications of smear-slide biostratigraphy. In: J.E. Warme, R.G. Douglas & E.L. Winterer (eds.), The Deep Sea Drilling Project: A decade of progress. *Society of Economic Paleontologists and Mineralogists, Special Publication* No. 32, pp. 471–487.

9

The Paleogene oxygen and carbon isotope history of Sites 522, 523, and 524 from the central South Atlantic

H. OBERHÄNSLI and M. TOUMARKINE

Abstract

From DSDP Leg 73, short time-series of the Late Paleocene–earliest Eocene (Zones NP 4 to NP 10; Site 524), Middle Eocene (Zones NP 15 to NP 17; Site 523) and the Eocene/Oligocene transition (upper part of Zone NP 20 to lower part of Zone NP 21; Site 522) have been studied in detail. For the stable isotope investigation we have chosen monospecific and monogeneric planktic and benthic foraminifera samples.

The most obvious feature of the Paleocene stable isotope record is a major change in the carbon-13 record of benthic and planktic foraminifera from depleted, to enriched, to more depleted values. It occurred from Zones NP 6 to NP 9. The period of enriched δ^{13}C values (planktics: $+4.3\permil$, benthics: $+2.5\permil$) can probably be calibrated with the uppermost part of the magnetic Chron C 25R and the lowermost part of C 24R. This carbon-13 event is most probably due to global change in the carbon-13 budget perhaps due to a significant world-wide increase in biological productivity and/or changes in the storage rates of organic matter in shelf areas or marginal seas. The Late Paleocene δ^{18}O record of benthics increases from Zone NP 5 to the top of NP 8 (approximately 62–60.5 Ma) and decreases from Zones NP 9–NP 12 (approximately 60–55 Ma), suggesting a probable cooling of 2–3 °C, followed by a subsequent warming of at least 3–4 °C in the bottom water environments. The climatic changes are less pronounced in the surface water record.

We have deduced other important climatic changes that occurred during the later Middle Eocene ($\approx$ 45–41 Ma) and at the Eocene/Oligocene boundary (36.9–36.6 Ma, Chrons C 13R3–C 13N2). Both events indicate possible coolings of 3–5 °C in surface and bottom water environments.

The Middle Eocene climatic deterioration was accompanied by a significant enrichment (about 1‰) in the δ^{13}C record of benthics. It may indicate a bottom water circulation event, which probably was related to a critical opening stage of the Norwegian–Greenland Sea.

During the Eocene/Oligocene cooling, the δ^{13}C values of both benthic and planktic foraminifera increased by 1‰. This δ^{13} carbon event probably represents a global change in the carbon-13 budget, and may be attributed to an increase in the terrestrial biomass.

Introduction

Stable isotope stratigraphy is a valuable tool for tracing climatic history (Emiliani, 1954; Savin *et al.*, 1975; Shackleton & Kennett, 1975; Kennett & Shackleton, 1976; Savin, 1977) during the Tertiary. In broad outline, the climatic evolution deduced from stable isotope data from different DSDP sites displays a world-wide conformity. During the Paleogene, climatic evolution shows important cooling events in the mid and high latitudes of the Pacific and Atlantic Oceans, which occurred in a step-like manner rather than continuously (Savin *et al.*, 1975; Shackleton & Kennett, 1975; Keigwin, 1980; Corliss, 1981; Corliss & Keigwin, 1983; Miller, 1983; Keigwin & Keller, 1984; Snyder *et al.*, 1984, Miller & Thomas, 1985).

In the Atlantic Ocean the extent of a single deterioration can be correlated between the South and North Atlantic sites, although the timing of the different cooling steps is generally not very accurate because of the low resolution of the isotopic record. Boersma & Shackleton (1977a) and Vergnaud-Grazzini *et al.* (1978) showed that surface temperatures at mid-latitude sites of the northern and southern hemisphere of the Atlantic Ocean decreased from the Late Paleocene/ Early Eocene to the Early Oligocene by 5–10°C. Bottom water temperatures at mid-latitudes decreased by 6–8°C within this time interval. At most of the investigated Atlantic sites, surface, and also bottom water temperatures at some sites, decreased by a further 4°C during the Late Oligocene (NP 24/NP 25; Boersma & Shackleton, 1977a,b; Biolzi, 1983; Keller, 1983; Keigwin & Keller, 1984). Boersma & Shackleton (1977b) noted a temperature increase at the termination of the Oligocene.

Recently, a controversy over the interpretation of Paleogene oxygen isotope stratigraphy has arisen. Basically two different models can be applied to explain the earlier Tertiary oxygen isotope pattern: (1) no Antarctic ice-sheets existed prior to the Middle Miocene (Savin *et al.*, 1975; Shackleton & Kennett, 1975; Savin, 1977), and (2) Antarctic ice-caps existed at least since the Early Oligocene (Matthews & Poore, 1980; Miller & Fairbanks, 1983; Hsü *et al.*, 1984a; Keigwin & Keller, 1984; Poore & Matthews, 1984; Miller & Thomas, 1985). Poore & Matthews (1984) and Hsü *et al.* (1984b) assumed that the volume of continental ice during the Early Oligocene was only slightly smaller than that of the Late Pleistocene glacial maximum. For the Late Eocene time they postulated an ice volume comparable to that of the present day. This latter model implies that we should consider an ice-volume effect of at least 0.5–1.0‰ when interpreting the $\delta^{18}O$ values of the Oligocene and probably even the Eocene (Miller & Fairbanks, 1983; Keigwin & Keller, 1984; Miller & Thomas, 1985; and Poore & Matthews, 1984).

The tectonic history of the Atlantic Ocean reveals a major reorganization during the earlier Tertiary. The Norwegian–Greenland Sea started opening during the latest Paleocene (63–60 Ma; Sclater *et al.*, 1977a) and became significantly wider during the Eocene. By the beginning of the Tertiary the South Atlantic was a wide and deep ocean which, besides the Mid-Atlantic Ridge, was topographically structured by the Rio Grande–Walvis Ridge complex. Paleobathymetric reconstructions of the South Atlantic Ocean by Sclater *et al.* (1977a) show that the Vema Channel became deep enough during the Early Tertiary to allow the passage of bottom waters into the northern basins of the South Atlantic, although the Walvis Ridge and the Rio Grande Rise acted as efficient barriers to bottom water exchange. During the Paleocene and Eocene both morphological highs subsided, and by the Oligocene the Walvis Ridge reached a depth exceeding 3000 m. The deeper eastern South Atlantic was probably also open to the north, so that deep water exchange with the North Atlantic occurred after the earlier Tertiary (Sclater *et al.*, 1977a; Parker *et al.*, 1984).

Climatic changes often coincide with circulation changes, thus suggesting a cause and effect relationship. Moreover, reorganization of current systems seems to be linked intimately to changes in geometry of ocean basins. Circulation patterns in the oceans changed fundamentally during the Paleogene: the circum-equatorial circulation, with low vertical and latitudinal temperature gradients, was successively replaced by a circulation pattern dominated by circum-Antarctic currents (Talwani *et al.*, 1976; Berggren & Hollister, 1977; Kennett, 1977, 1982). By the Eocene/Oligocene boundary the cold water masses, formed at high latitudes, spread over the ocean floor, and vertical temperature gradients in low and mid-latitude oceans increased considerably. The main purpose of this paper is to focus on the threshold events in the climatic and paleoceanographic evolution of the eastern South Atlantic during the Paleogene. Key events occurred within the following time intervals: (1) the Late Paleocene, (2) the Early/Middle Eocene transition, (3) the later Middle Eocene, and (4) the Eocene/Oligocene boundary. The discussion is based on stable isotope data of monospecific and monogeneric benthic and planktic foraminifera from DSDP Leg 73, Sites 522, 523, and 524. The generally continuous recovering at the sites, situated on the southern flank of the Walvis Ridge and in the southern Angola Basin (Fig. 1), provides an excellent paleomagnetic record (La Brecque *et al.*, 1983; Poore *et al.*, 1983; Tauxe *et al.*, 1983; Hsü *et al*,

1984c) and allows a precise age correlation with the stable isotope events.

Remarks on materials and methods

The sites studied were drilled along a transect between latitudes of 25° and 30° south in the central South Atlantic (Fig. 1). Site 524 is situated on the southern slope of the Walvis Ridge. Sites 522 and 523 are located in the southern Angola Abyssal Plain.

At Site 524 the Paleocene sediments are generally marly and clayey. Intercalations of reworked shallower biogenic material were observed in the lower to upper Paleocene section. In the uppermost Paleocene section the detritic influence diminishes suddenly and pelagic sediments prevailed. In the upper Paleocene to lower Eocene section, chert layers indicate brief intervals of extensive siliceous planktic fertility. Back-tracking (Berger & Winter, 1974; Sclater et al., 1977b) of the depositional depth at this site was questionable (Hsü et al., 1984b).

Fig. 1. Location map of Leg 73 sites studied from the eastern South Atlantic.

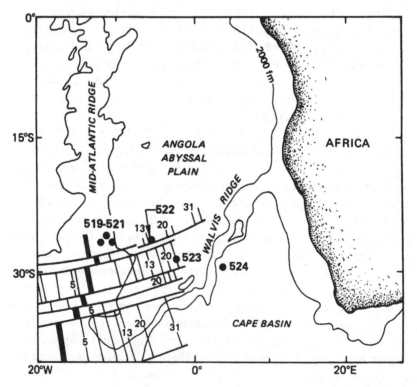

At Site 523 a 100-m thick sequence of nannofossil oozes, with several hemipelagic intercalations, was deposited during the Middle Eocene to Early Oligocene (NP 15–NP 22). This site was situated at a depth of about 2600 m during the early Middle Eocene and subsided to 3400 m by the Eocene/Oligocene boundary.

At Site 522 nannofossil oozes were deposited during the Late Eocene and Oligocene. This site subsided from a depth of 2600 m at the end of the Eocene to 3600 m at the end of the Oligocene.

For investigations of the stable isotopic record, monospecific and monogeneric planktic and benthic foraminifera samples were selected. The isotopic results* of Sites 524, 523, and 522 are plotted in Figs. 2–5. Isotopic results, which have not yet been published are listed in Table 1. The other data are given in Oberhänsli *et al.* (1984). Ages are reported according to the time scale of Hsü *et al.*, 1984c).

At Site 524 the upper Paleocene to basal lower Eocene section has been studied isotopically (Fig. 2). The preservation of the planktic foraminifera is generally very poor during this time interval (samples 524-9-2, 125–127 cm to 524-4-2, 12–13 cm). Samples below 524-5-5, 89–91 cm contain either no planktic foraminifera or only fragments. Thus the isotopic record of the surface inhabitants is rather scarce during the Late Paleocene (Fig. 2). Changes in surface water environments are documented by isotopic values of planktic species such as *Morozovella* spp., *Subbotina* spp., *Acarinina* spp., and *Chiloguembelina* spp. The bottom water changes are documented by isotopic data of the benthic genera *Oridorsalis*, *Nuttallides*, and *Gavelinella*.

Benthic and planktic foraminifera from Site 523 provide a stable isotopic record of the surface and bottom water conditions during the Middle Eocene (Fig. 3). Stable isotope ratios of *Globigerinatheka mexicana* and *Subbotina* spp. as well as of *Acarinina* spp. characterize surface water environments. Changes in bottom water environments are recorded in the isotopic data of *Oridorsalis umbonatus*. From Site 522 samples, a detailed isotopic study across the Eocene/Oligocene boundary has been made (Fig. 4). For the Eocene/Oligocene isotopic record, *Globigerina venezuelana* s.l. – a shallow to intermediate dweller (Douglas

---

* Before isotopic processing the samples were cleaned ultrasonically, crushed, and afterwards heated in a vacuum at 400 °C for 30 minutes. The carbon dioxide was released by reaction with 100% orthophosphoric acid at 50 °C. The isotopic composition of the gas was measured by mass-spectrometers, Micromass 903C at the ETH Zürich. The isotopic ratios are reported in the δ-notation as per mil deviation from the PDB standard. For oxygen and carbon isotopic ratios the analytical precision is better than ±0.10 per mil.

Table 1. Oxygen and carbon isotope data of monospecific and monogeneric planktic and benthic foraminifera samples of Holes 524 and 524B. The stable isotope data are reported as per mil deviation from the PDB Standard.

	Depth below sea floor (m)	Morozovella spp. $\delta^{18}O/\delta^{13}C$ (‰) PDB	Subbotina spp. $\delta^{18}O/\delta^{13}C$ (‰) PDB	Acarinina spp. $\delta^{18}O/\delta^{13}C$ (‰) PDB	Other planktic foraminifera $\delta^{18}O/\delta^{13}C$ (‰) PDB	Nuttallides truempyi $\delta^{18}O/\delta^{13}C$ (‰) PDB	Other benthic foraminifera $\delta^{18}O/\delta^{13}C$ (‰) PDB
524B-3-2, 45–47 cm	9.5	−1.26. +2.37[a]	−0.74/+1.11	−1.22/+2.02[g]	−1.18/+0.54[j] −0.88/+0.69[k]	−0.95/−0.32	
524B-5-2, 135–137 cm	18.9	−1.33/+2.30[a]	−0.57/+1.11[e]	−1.08/+2.30[h]		−0.73/+0.37	−0.35/−0.06[m]
524B-7-3, 12–14 cm	28.1	−1.53/+3.31[b]	−0.66/+1.71			−0.53/+1.00	
524-2-1, 65–66 cm	10.2	−1.92/+1.74[a]	−0.64/+1.04	−1.45/+1.77[h]			−0.85/+0.48[n]
534-3-1, 96–97 cm	29.5	−1.19/+2.37[a]	−1.09/+1.43	−1.00/+2.43[h]		−0.53/+0.70	
524-3-3, 75–76 cm	32.3	−1.27/+2.78[a]	−0.59/+1.40	−0.92/+2.19[h]			
524-4-1, 83–84 cm	48.3		−0.81/+2.67	−1.04/+3.28[h]		−0.01/+2.64	
524-4-2, 12–13 cm	49.1						
524-5-1, 145–146 cm	58.5				−0.59/+3.44[l]	−0.08/+2.38	
524-5-2, 62–64 cm	59.1					−0.15/+2.35	

Sample	Depth (m)					
524-5-3, 130–132 cm	61.3				−0.22/+2.22	
524-5-4, 100–102 cm	62.5				−0.02/+2.42	
524-5-5, 89–91 cm	63.8	−0.86/+4.28[c]	−0.47/+3.00	−0.74/+3.88[i]	+0.38/+2.40	
524-5-5, 103–105 cm	64.0			−0.60/+3.69[l]		
524-6-1, 11–12 cm	66.6				+0.13/+2.40	
524-6-2, 131–133 cm	69.3				+0.34/+1.67	
524-7-1, 133–135 cm	77.3				+0.04/+1.57	
524-8-1, 90–92 cm	86.4				0.00/+1.35	
524-8-2, 58–60 cm	87.6				+0.16/+1.89	
524-8-3, 118–120 cm	89.7				+0.20/+1.75	
524-8-4, 23–24 cm	90.2				0.00/+1.58	
524-8-5, 16–18 cm	91.7				−0.20/+1.52	
524-9-1, 63–65 cm	95.6				+0.04/+1.42	
524-9-2, 125–127 cm	97.8					
524-9-3, 25–27 cm	98.3	−1.14/+2.59[c]	−1.14/+1.92[e]		−0.46/+1.08[o]	
524-11-2, 31–33 cm	117.3	+0.57/+1.70[d]	−0.17/+2.06[f]			

[a] Morozovella subbotinae
[b] Morozovella marginodentata
[c] Morozovella velascoensis
[d] Morozovella pseudobulloides
[e] Subbotina triangularis
[f] Subbotina triloculinoides
[g] Acarinina soldadoensis angulosa
[h] Acarinina soldadoensis soldadoensis
[i] Acarinina mckannai
[j] Chiloguembelina subcylindrica
[k] Chiloguembelina wilcoxensis
[l] mixed planktic assemblage
[m] Oridorsalis umbonatus
[n] Bolivina spp.
[o] Gavelinella beccariformis

132   H. Oberhänsli and M. Toumarkine

Fig. 2. Site 524: stable isotope record of monospecific and monogeneric benthic and planktic foraminifera samples from the Late Paleocene to the earliest Eocene. Faunal zonation and magnetostratigraphy after Poore *et al.* (1983) and Tauxe *et al.* (1983).

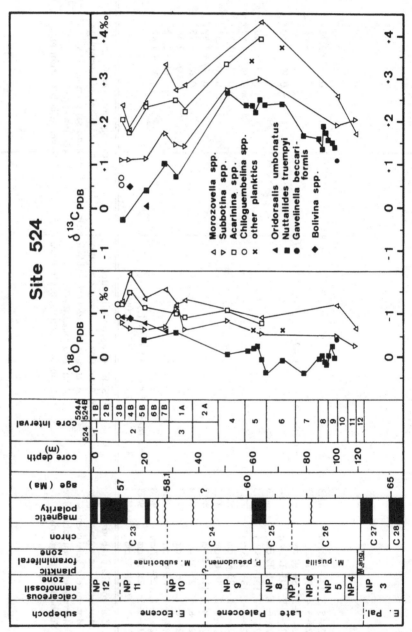

& Savin, 1978; Biolzi, 1983), *Catapsydrax dissimilis* – a deeper dwelling planktic form (Biolzi, 1983), and the benthic *Stilostomella* spp. were chosen, because of their abundance and their continuous occurrence.

Late Paleocene–Early Eocene (65–58 Ma; Site 524)

The oxygen and carbon isotope history of the bottom water is well documented from the top of Zone NP 5 through the base of Zone NP 10 (Fig. 2). The $\delta^{18}O$ values of benthic foraminifera increase slightly (0.5‰) from the upper part of Zone NP 5 to the top of Zone NP 8. The $\delta^{18}O$ values of the planktic *Morozovella* spp. appear to show a similar trend. From Zones NP 9 to NP 10 the $\delta^{18}O$ values of benthic and planktic foraminifera species decrease by 1.00‰ and 0.50‰ respectively. Within this short time interval, the vertical gradient between $\delta^{18}O$ values of benthics and planktics has diminished considerably and has vanished completely by around 58 Ma.

The $\delta^{13}C$ values of benthic and planktic foraminifera indicate pronounced changes within the earlier Paleogene interval (Fig. 2). The Late Paleocene carbon isotope record of Site 524 reveals a prominent positive excursion. This change seems to be more pronounced in the surface waters (1.00–1.50‰) than in the bottom water record (0.70‰). *Morozovella* spp. and *Acarinina* spp. have maximum $\delta^{13}C$ values of + 4.3‰ and + 3.9‰ respectively (524-5-5, 89–91 cm). Benthics show maximum $\delta^{13}C$ values of + 2.5‰. Through the late NP 9 and early NP 10 Zones, the $\delta^{13}C$ values of both planktic and benthic foraminifera decrease by 1.5–2.0‰ and 2.5‰ respectively.

The oxygen isotope increase from Zone NP 5 to the top of Zone NP 8, if uniquely interpreted as resulting from a temperature change, would indicate a cooling of 2–3 °C in bottom waters. The decrease in the $\delta^{18}O$ record from NP 10 would then correspond to a temperature increase of 5 °C in bottom waters and 2–3 °C in surface waters.

The oxygen isotope change from Zones NP 9 to NP 11 coincides with faunal and floral migrations in the North Atlantic, when low-latitude planktic assemblages migrated into higher latitudes (Haq et al., 1977). According to Haq et al. (1977) and Haq (1980), this warming culminated in the Early Eocene, which is considered to be the warmest interval of the entire Tertiary. From the Southern Ocean and adjacent continents, as well as from the Pacific Ocean, faunal, floral, and isotopic studies within this time interval (Paleocene/Eocene transition) indicate a similar climatic amelioration to that observed in the Atlantic (Savin et al., 1975; Shackleton & Kennett, 1975; Kemp, 1978, 1981; Kennett, 1978).

Calibration with the magnetostratigraphy indicates that the carbon isotope increase probably occurs from the upper C 25R to the lower C 24R (La Brecque *et al.*, 1983; Tauxe *et al.*, 1983). However, comparison between the polarity pattern in Site 524, cores 4 through 11, and the standard magnetic anomaly pattern was complicated by a sampling or drilling gap within core 6. Thus, the magnetic normal interval in core 5 should be either C 26N or C 25N (La Brecque *et al.*, 1983; Poore *et al.*, 1983; Tauxe *et al.*, 1983). Poore *et al.* (1983) concluded, after comparing the results of Leg 73 with those of Leg 74, that in Site 524 Subchron C 26N occurs in the unrecovered interval of core 6. Similar positive $\delta^{13}C$ values have been reported from planktic foraminifera and calcareous nannofossil samples in other Atlantic sites, Pacific and Indian Ocean bulk samples, and most are correlated with Zone P 4 (Site 305: Douglas & Savin, 1975; Hole 398D: Arthur *et al.*, 1979; DSDP Leg 74 Sites: Shackleton & Hall, 1984; Shackleton *et al.*, 1984; Sites 217 and 237: Oberhänsli, 1985).

The strongly increasing and later decreasing $\delta^{13}C$ values of benthic and planktic foraminifera during the Late Paleocene (Fig. 2) are probably due to a global change in the ^{13}C budget. This global change, in turn, could be related to changes in the biological productivity in oceans and/or on continents. The organic matter, strongly enriched in the light carbon isotope (^{12}C), is the most important factor in the carbon balance. Thus formation and storage rates of organic matter are crucial for the $\delta^{13}C_{\Sigma CO_2}$ values in the ocean.

An increase of productivity in surface water *in situ*, which would have depleted the dissolved carbonate species of the surface water in carbon-12, is rather unlikely. In the sedimentary record of Site 524, besides several cherty or diatom-rich horizons, no indications of increased productivity have been found. The sedimentation rate did not increase noticeably at Site 524 during the Late Paleocene, which may be due in part to the rather shallow lysocline. Preservation of planktic foraminifera in cores 4 to 8 is generally very poor (except in sample 524-5-5, 89–91 cm, where planktics reveal the most enriched $\delta^{13}C$ values).

If productivity increased only in the surface water, the $\delta^{13}C$ values of planktics would increase but those of benthics would decrease (Kroopnick, 1980). Thus this model can not explain satisfactorily the $\delta^{13}C$ increase in bottom and surface waters. Alternative interpretations are given by Shackleton (1977) and Broecker (1982). According to these authors the $\delta^{13}C$ record may reflect changes in the global terrestrial biomass (Shackleton 1977) and/or that organic matter was preferentially stored in shelve areas and marginal seas (Broecker 1982).

This is a scenario which could explain the changes of the carbon isotope ratios in planktic and benthic foraminifera during the Late Paleocene. But mechanisms, which induced and started the observed world-wide geochemical (?) changes, documented in the transient carbon-13 enrichment, cannot be enumerated yet. Isotopic comparison, however, can be drawn with the Mid-Cretaceous anoxic events, although the black shale facies, typical for the Mid-Cretaceous, is not known during the Late Paleocene. Further studies are required to shed light on the Late Paleocene carbon isotope event.

Early–Middle Eocene (approximately 50 Ma)

The Early/Middle Eocene boundary was not cored in Leg 73 sites. Nevertheless, comparing the isotope data from the Early Eocene of Site 524 with Middle Eocene data from Site 523 (Fig. 2 and 3), we note striking changes which occurred within this unrecovered interval. The $\delta^{18}O$ values of *Nuttallides truempyi* and *Oridorsalis umbonatus* increased by 1–1.5‰ from the Early Eocene (Fig. 2; 524B-3-2, 45–47 cm) to the Middle Eocene (Fig. 3; 523-49-2, 35–37 cm and 523-47-2, 108–110 cm). Within the same interval the $\delta^{18}O$ values of *Morozovella* spp. and *Acarinina* spp., both surface water inhabitants, increased by 1.0‰ and 0.5‰ respectively, and those of *Subbotina* spp., an intermediate dweller, by 0.7‰. These oxygen isotope changes observed between Sites 524 and 523 may partly be due to the fact that the sites were drilled at different depths in different basins, but may also be due to a cooling during the time period not recovered in Leg 73 cores.

Coeval oxygen isotope increases, although of fairly different extents (0.5 to 2.0‰), are known from the North Atlantic, the North Pacific, the Southern Ocean, and the Indian Ocean (Douglas & Savin, 1975; Shackleton & Kennett, 1975; Arthur *et al.*, 1979; Letolle *et al.*, 1979; Oberhänsli, 1985). This $\delta^{18}O$ increase has been observed in the surface waters on a world-wide scale. For the bottom waters, due to the sparse benthic isotope record, the increase has only been noted elsewhere in the Northern Pacific and the Southern Ocean (Douglas & Savin, 1975; Shackleton & Kennett, 1975). The increase is most pronounced in the Southern Ocean and Indian Ocean, where oxygen isotope ratios during the early Middle Eocene are 1.5–2.0‰ more positive than during the Early Eocene. In the Indian Ocean, the $\delta^{18}O$ increase occurred within 1–2 Ma from NP 13 to NP 14 (Oberhänsli, 1985). Thus the increase seems to be due to a sudden cooling event, breaking the Early Eocene climatic optimum, rather than to a gradual decline of temperature. Paleontological data may partly corroborate this interpretation (Haq *et al.*, 1977;

Kemp, 1978, 1981; McGowran, 1978; Berggren, 1982; Benson *et al.*, this volume, chapter 16), although it seems that the Indopacific realm was more intensely affected by this climatic deterioration than the Atlantic Ocean.

Fig. 3. Site 523: stable isotope record of monospecific and monogeneric benthic and planktic foraminifera samples from the Middle Eocene to the very Early Oligocene. Percentages of *Nuttallides* spp. are reported from Leg 73, Site Reports (Hsü *et al.*, 1984b). Faunal zonation and magnetostratigraphy after Poore *et al.* (1983) and Tauxe *et al.* (1983).

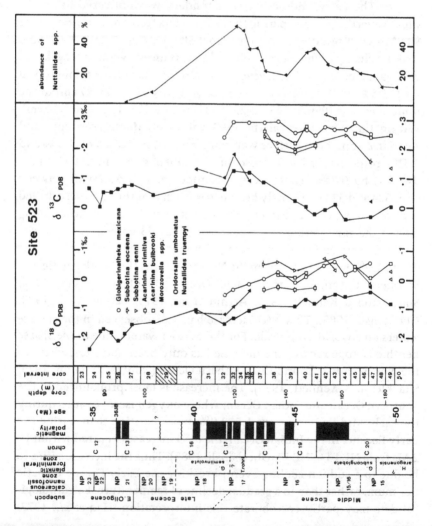

Middle–Early Late Eocene (49–39 Ma; Site 523)

The Middle Eocene oxygen isotope record (Fig. 3) is character-ized by a steady increase in ^{18}O for planktic and benthic foraminifera. The oxygen isotope ratios of planktics and benthics increase by approximately 1‰ from 46 to 41 Ma and approximately 0.75‰ from 45 to 41 Ma respectively. Parallel to the $\delta^{18}O$ increase at Site 523, the $\delta^{13}C$ values of benthics increase by 1.3‰ from Zone NP 16 to the base of Zone NP 17, whereas the carbon isotope ratios of planktics remain stable at a fairly high average level (+ 2.75‰ for *Globigerinatheka mexicana*; Fig. 3).

The gradual oxygen isotope increase (about 1‰) that occurred during the later Middle Eocene (Fig. 3; Zones NP 16–NP 17) is evident in both the benthic and planktic records at Site 523. This increase is not restricted to the eastern South Atlantic. Similar oxygen isotope changes are known from the western South Atlantic (Boersma & Shackleton, 1977a; Muza *et al.*, 1983) as well as from the Pacific (Douglas & Savin, 1973; 1975; Shackleton & Kennett, 1975). Recently, Keller (1983) published isotope data from DSDP Sites 363 and 219 from the late Middle to Late Eocene where $\delta^{18}O$ values of benthic foraminifera increase significantly in both the Atlantic and the Indian Ocean sites within Zone NP 17 (approximately P 14). This positive shift is of about the same extent at all studied sites, although the changes apparent are not isochronous. However, precise dating of the isotopic event is often difficult because of biostratigraphic uncertainties and insufficient isotopic data.

Possible causes for such a widespread and extensive oxygen isotope change could be (1) a temperature decline (Shackleton & Kennett, 1975; Corliss, 1981), or (2) a change in the evaporation/precipitation pattern in several main source areas of surface and bottom water masses (Oberhänsli *et al.*, 1984), and thus a reorganization of the current pattern, or (3) ice formation.

Paleontological studies revealed drastic changes from the Middle to Late Eocene in surface and bottom waters. The geographical distribu-tion pattern of calcareous nannofossils and planktic foraminifera assemblages from the North Atlantic indicates an environmental change within Zone NP 16 (Haq *et al.*, 1977), which may be due to a cooling of high-latitude surface waters. Moreover, modifications in the planktic foraminiferal assemblages at the Middle/Late Eocene boundary (P 14/ P 15) are well known. Thus, according to these authors at least two climatic deteriorations occurred in surface water: the NP 16 and P 14/P 15 events, which would correspond to 46–43 Ma and 40 Ma respectively.

In bottom water environments, benthic foraminiferal assemblages gradually changed throughout the Middle Eocene to the Early Oligocene (Corliss, 1981). Tjalsma & Lohmann (1982) noted that the replacement of the *Nuttallides truempyi* assemblage may have been diachronous in the Southern Atlantic and Caribbean. In shallower sites (< 3 km) the *N. truempyi* assemblage, typical for the Early/Middle Eocene, was replaced in the early Middle Eocene, whereas in deeper sites (> 3 km) the replacement of this assemblage by a Late Eocene benthic assemblage with *Globocassidulina subglobosa*, *Gyroidinoides* spp., *Cibicidoides ungerianus*, and *Oridorsalis umbonatus* occurred during the earliest Late Eocene (40 Ma). The latter corresponds with the observation made at Site 523, which had subsided below 3000 m depth at the Middle/Late Eocene transition, where the abundance of *N. truempyi* decreases drastically just above the Middle/Late Eocene boundary (approximately 40 Ma). This coincides with the observations reported by Miller *et al.* (1984). From benthic foraminiferal studies in Site 523, Parker *et al.* (1984) noted that the Middle Eocene seems to be a time of faunal instability, due probably to faunal migrations into the basin. According to these authors the faunal change was not a progressive process, but a rather sporadic phenomenon. The sporadic migrations of a typical Late Eocene assemblage into the Angola Basin during the Middle Eocene seem to herald the oceanic turnover at the Eocene/Oligocene boundary.

The progressive oxygen isotopic changes suggest slow but continuous climatic changes during the Middle Eocene. In addition, the fragmentation rate of planktic foraminifera increases at the base of Zone NP 17 (Site Reports, Leg 73: Hsü *et al.*, 1984b). The suddenly increasing dissolution rate implies that by this time the site had subsided to the critical depth of about 3200 m (Hsü *et al.*, 1984a) and the site was at the lysocline level. The increasing abundance of *N. truempyi* (Fig. 3), which is positively correlated with the corrosiveness of bottom waters (Tjalsma & Lohmann, 1982), supports this interpretation.

From benthic foraminiferal and isotopic studies in the North Atlantic, Miller (1983) found a negative correlation between $\delta^{13}C$ values of benthics and the abundance of *Nuttallides* spp. The lowest $\delta^{13}C$ values (early Middle Eocene and Middle Oligocene) are associated with the highest percentages of *N. truempyi* and *N. umbonifera* respectively. Conversely, this author noted that percentages of *Nuttallides* spp. are lowest when $\delta^{13}C$ values are highest. This author concludes that *Nuttallides* appears and prospers in response to a corrosive and sluggish bottom water circulation.

A different mechanism must be found to explain the positive correlation between maximum occurrence of *N. truempyi* and positive $\delta^{13}C$ values, as observed in the Southern Atlantic Site 523. If the $\delta^{18}O$ enrichment in the bottom water record during the late Middle Eocene may at least partly be attributed to a temperature drop, then the bottom water circulation was activated during this time interval. As a consequence, the old bottom waters, with a negative $\delta^{13}C$ signal, were replaced by cooler and younger bottom waters with a relatively heavy $\delta^{13}C$ signal.

From benthic foraminiferal studies (Parker *et al.*, 1984) evidence for a different, probably northern origin of bottom waters in the Angola Basin exists from the late Middle Eocene, although the beginning of a southern circulation event ('proto-AABW') may not be excluded. The bottom water circulation pattern in the eastern South Atlantic could be linked to the progressive opening of the Norwegian–Greenland Sea (considered as source area of 'proto-NADW') which had started around 63 to 60 Ma (Kennett, 1982). Miller & Tucholke (1983), however, suggested, that the North Atlantic deep water circulation began in the Late Eocene to Early Oligocene. According to the authors the widely distributed seismic reflector R4, indicating an unconformity, shows a change in the abyssal current system.

Paleontological data indicate that a significant cooling occurred at 40 Ma. According to Corliss (1981) and Benson *et al.* (this volume, chapter 16) this climatic deterioration was probably a threshold event which initiated the thermohaline bottom water circulation and which then continued to the Eocene/Oligocene boundary. Furthermore, it was probably crucial for the establishment of the psychrosphere. At present the isotope data from Site 523 (Fig. 3) do not conclusively indicate whether the cooling shown by the $\delta^{18}O$ increase between 45 and 41 Ma, or an additional cooling at 40 Ma, which we did not observe in our isotopic record, was the causal event for the faunal changes at 40 Ma.

Eocene/Oligocene boundary (37.7–36.3 Ma; Sites 522 and 523)

The stable isotope history across the Eocene/Oligocene boundary is documented at Site 522 (Fig. 4) and at Site 523 (Fig. 3). The latest Eocene (Zone NP 20 to lowermost Zone NP 21; Fig. 4) is characterized by relatively stable $\delta^{18}O$ values (average values: *G. venezuelana* s.l. $+0.6\%$, *C. dissimilis* $+1.0\%$, *Stilostomella* $+1.25\%$). The vertical gradient ($\delta^{18}O$) was not very pronounced at this time. Later, in Zone NP 21 (approximately 36.95–36.8 Ma; samples 522-36-2, 68–70 cm to 522-34-3, 43–45 cm), the $\delta^{18}O$ values of *G. venezuelana* s.l., *C. dissimilis*, and

Stilostomella spp. increase in a step-like manner by 0.9‰, 1.2‰, and 1.3‰ respectively. The maximum increase is recorded above the Eocene/Oligocene boundary. This increase occurred within 10^5 years. Calibration with the magnetostratigraphic scale (Fig. 4) shows that the maximum increase may be placed within the magnetic interval from C 13R3 to C 13N2. The sudden $\delta^{18}O$ increase across the Eocene/Oligocene boundary was accompanied by increasing $\delta^{13}C$ values for benthics and planktics (approximately 1.0‰ and 0.75‰ respectively), although the increase in $\delta^{13}C$ values precedes the $\delta^{18}O$ increase. Within the detailed isotopic record of Site 522, especially in the benthic record, extensive

Fig. 4. Site 522: detailed stable isotope record of monospecific planktic and monogeneric benthic foraminifera samples across the Eocene/Oligocene boundary. Faunal stratigraphy and magentostratigraphy after Poore *et al.* (1983) and Tauxe *et al.* (1983).

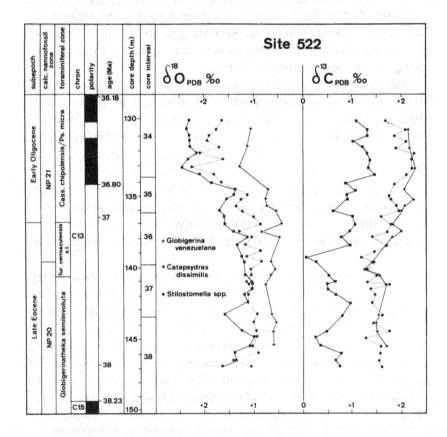

$\delta^{13}C$ fluctuations (up to 1.0‰) of relatively short duration (less than 10^5 years) occur.

From other Atlantic and Pacific DSDP sites similar oxygen isotope increases in bulk samples and planktic and benthic samples were observed at the Eocene/Oligocene boundary (Douglas & Savin, 1973, 1975; Savin *et al.*, 1975; Shackleton & Kennett, 1975; Kennett & Shackleton, 1976; Keigwin, 1980; Miller & Curry, 1982; Keigwin & Keller, 1984; Miller & Thomas, 1984; Miller *et al.*, 1984; Snyder *et al.*, 1984). At the southwestern Pacific Site 277, the oxygen isotope increase occurred, as in the site studied, within 10^5 years (Shackleton & Kennett, 1975). According to Savin *et al.* (1975), Shackleton & Kennett (1975), and Kennett & Shackleton (1976) this 1.0‰ to 1.3‰ oxygen isotope increase in bottom water environments and in mid- and high-latitude surface water environments indicates a 4–5 °C temperature drop at high latitudes. This cooling, which probably lowered seawater temperatures to the proximity of the freezing point, led to the first sea-ice formation in the Ross Sea (Kennett 1982). Thus, this cooling caused the formation of cold bottom waters at high latitudes, which, by this time, had spread extensively over the deeper oceans. As a consequence the psychrospheric oceans became permanently established (Benson, 1975; Shackleton & Kennett, 1975). These authors have suggested that no significant ice-sheets had formed by this time, although local glaciation since the later Eocene is probably confirmed (Hayes & Frakes, 1975). The ice volume on Eastern Antarctica therefore remained relatively small until the Middle Miocene and had no significant influence on the global $\delta^{18}O_{water}$ signal.

An alternative interpretation of the $\delta^{18}O$ record at the Eocene/Oligocene is given by Matthews & Poore (1980), Miller & Fairbanks (1983), Hsü *et al.* (1984a), Keigwin & Keller (1984), Poore & Matthews (1984) and Miller & Thomas (1984). These authors state that the build-up event of the ice-cap in Eastern Antarctica occurred at 40 to 38 Ma rather than at 15 Ma. Thus the world-wide increase in $\delta^{18}O$ values of benthic and mid- and high-latitude planktic foraminifera would be due to the ice-effect rather than to a cooling. This would imply that, since the latest Eocene, the $\delta^{18}O_{seawater}$ was globally enriched in ^{18}O by at least 1‰ (0.5‰: Miller & Fairbanks, 1983; Keigwin & Keller, 1984; Miller & Thomas, 1985) depending on the extent of the ice-sheets. At present we do not have independent evidence, such as further ice-rafted sediments, to support this hypothesis. Selected drill sites may shed further light on this problem in the near future.

From paleontological data, however, a significant cooling is inferred from the Late Eocene to the Early Oligocene record. Sancetta (1979) reported three latitudinal planktic faunal and floral provinces (tropical, transitional, and temperate) from the Pacific and the Southern Oceans during the Middle and Late Eocene (late Zone NP 15–early Zone NP 16 and Zone NP 17 respectively). During the Early Oligocene (Zone NP 22– early NP 23) Sancetta distinguished five faunal and floral provinces (tropical, subtropical, transitional, subantarctic, and antarctic) in the southern hemisphere. This diversification of the zoo- and phyto-geographic pattern during the later Paleogene doubtless points to a steepening of the latitudinal thermal gradients. Causes may be seen in a high-latitude cooling accompanied by a reorganization of the current system which prevented heat transport from lower latitudes towards higher latitudes. Thus, the suggestion of Sancetta (1979) that by the time of the Eocene/Oligocene transition a precursor of the Antarctic Convergence developed, or became established, may be very meaningful.

Benthic foraminifera and ostracod studies from the Atlantic and the Pacific surprisingly enough do not show major faunal changes related to the isotopic bottom water event at the Eocene/Oligocene boundary (Benson, 1975; Corliss, 1981; Miller, 1983; Miller et al., 1984; Snyder et al., 1984; Benson, et al., this volume, chapter 16). Corliss (1981) and Miller (1983) suggested that the benthic foraminiferal species must have had a fairly wide environmental tolerance at the Eocene/Oligocene transition, and stenotopic species evolved only some time after the development of the psychrosphere. According to Corliss, this assumption could imply that the oxygen-18 enrichment reflects a decrease in the bottom water temperature of less than 3 °C rather than the build-up of the ice-cap on the Antarctic continent.

The $\delta^{13}C$ increase in the benthic record may be linked to the cooling event which activated bottom water circulation. The Angola Basin was filled with young bottom waters which initially had a relatively positive $\delta^{13}C$ signal. The source area of the bottom waters which invaded the eastern South Atlantic during the Eocene/Oligocene transition might have been the Norwegian–Greenland Sea (Miller & Tucholke, 1983; Parker et al., 1984) and/or the Southern Ocean (Kennett, 1978).

An alternative interpretation of the $\delta^{13}C$ increase is given by Hsü et al. (1984a). These authors attributed the enrichment in ^{13}C in the surface and bottom waters to an increase of the biomass on the continents. Due to the lowering of the sea level by about 100 m (30 to 40 m, respectively, according to Keigwin, 1980 and Keigwin & Keller, 1984), as a consequence of the Antarctic glaciation at the Eocene/Oligocene boundary, coastal plains became significantly enlarged, and plant growth was favoured. However, this trend is contrary to the one observed in the glacial–interglacial cycles of the Pleistocene (Shackleton, 1977). This author reported, that the ^{13}C values of benthics are lower during glacial time than during interglacial time.

Conclusions

In the eastern South Atlantic, the evolution of paleoceanography seems to be strongly influenced by the world-wide climatic history and major tectonic events. Most of the events recorded in the isotopic history of Leg 73 sites may be compared with similar isotopic events observed in other Atlantic, Pacific, and Indian Ocean sites. Thus, local influences may only have smoothed or strengthened the extent of a single isotopic shift or cycle. For the Paleogene paleoceanographic and climatic evolution, the isotopic history of several time intervals such as the Late Paleocene, the Early/Middle Eocene transition, the Middle/ early Late Eocene, and across the Eocene/Oligocene boundary are highly significant. These isotopic events, surprisingly enough follow or coincide with a major reorganization of the ocean/continent geometry.

From the oxygen isotope record, if uniquely interpreted as temperature change, we deduce the following climatic events:

The Late Paleocene/earliest Eocene climatic history shows a distinct cooling (Zones NP 4–NP 8) in the bottom water record (less pronounced in the surface water record) and a later warming (Zones NP 10–NP 12), which again is most pronounced in the bottom water record.

At the Early/Middle Eocene transition a cooling event (at approximately 50 Ma) took place. We suggest that the separation of Antarctica and Australia had a significantly controlling effect on the climatic evolution during the earlier Plaeogene, although the effect seems to be smoothed in the South Atlantic.

The enrichment of $\delta^{18}O$ in the benthic and planktic record provides evidence for a next climatic deterioration by about 4°C in surface and bottom water environments during the late Middle Eocene.

The interpretation of the increase in oxygen-18 at the Eocene/ Oligocene boundary is still controversial. A high-latitude cooling or a build-up of the continental ice-sheets could account for this increase. Paleontological data indicate that a temperature change was at least partly responsible for the oxygen-18 enrichment. Further studies will prove whether, or to what extent, a continental glaciation contributed to this increase.

The carbon isotope record may be interpreted as (1) global changes in the ^{13}C budget or as (2) change in the abyssal circulation pattern:

During the latest Paleocene the carbon isotope increase (Zones NP 8–NP9) noted in the surface and bottom water record is probably a result of a global change in the ^{13}C budget. Causes may be seen in a higher productivity of plants on continents and/or a higher storage rate of organic matter in shelf sediments or marginal basin sediments.

During the later Middle Eocene cooling, the increasing $\delta^{13}C$ values of benthics indicate a reorganization of the bottom water circulation. This may be due to the basin evolution in the eastern South Atlantic (progressive subsidence) or to the fact that the progressively opening Norwegian–Greenland Sea may have reached a critical width so that by 46–45 Ma this area became a significant source area for bottom waters, although an Antarctic origin of the bottom water in the eastern South Atlantic cannot be excluded.

At the Eocene/Oligocene boundary the $\delta^{13}C$ increase of benthic and planktic foraminifera probably indicates a global change in the ^{13}C budget. The geochemical event began slightly earlier than the climatic deterioration.

Acknowledgements

We would like to thank Ken Hsü for inviting us to contribute this chapter to the *South Atlantic Paleoceanography* volume. Constructive discussions with Peter Hochuli and Helmut Weissert were very stimulating. The paper benefited from the review of Kenneth Miller and Roger Fairbanks. We thank Albert Uhr for assistance with diagrams.

REFERENCES

Arthur, M.A., Scholle, P.A. & Hasson, P. 1979. Stable isotopes of oxygen and carbon in carbonates from Sites 398 and 116 of the Deep Sea Drilling Project. *Init. Rep. Deep Sea Drill. Proj.*, **47** (2), 477–491.

Benson, R.H. 1975. The origin of the psychrosphere as recorded in changes of deep sea ostracode assemblages. *Lethaia*, **8**, 69–83.

Berger, W.H. & Winterer, E.L. 1974. Plate stratigraphy and the fluctuating carbonate line. *Intern. Assoc. Sed., Spec. Publ.*, **1**, 11–48.

Berggren, W.A. 1982. Role of ocean gateways in climatic change. In: *Studies in Geophysics, Climate in Earth History*, National Academy Press, Washington, DC, pp. 118–125.

Berggren, W.A. & Hollister C.D. 1977. Plate tectonics and paleocirculation–commotion in the ocean. *Tectonophys.*, **38**, 11–48.

Biolzi, M. 1983. Stable isotope study of Oligocene–Miocene sediments from DSDP Site 354, Equatorial Atlantic. *Mar. Micropaleont.*, **8**, 121–139.

Boersma, A. & Shackleton, N.J. 1977a. Tertiary oxygen and carbon isotope stratigraphy, Site 357 (mid latitude South Atlantic). *Init. Rep. Deep Sea Drill. Proj.*, **39**, 911–924.

Boersma, A. & Shackleton N.J. 1977b. Oxygen and carbon isotope record through the Oligocene, DSDP Site 366, Equatorial Atlantic. *Init. Rep. Deep Sea Drill. Proj.*, **41**, 957–962.

Broecker, W.S. 1982. Glacial to interglacial changes in ocean chemistry. *Prog. Oceanogr.*, **11**, 151–197.

Corliss, B.H. 1981. Deep-sea benthonic foraminiferal faunal turnover near the Eocene/Oligocene boundary. *Mar. Micropaleont.*, **6**, 367–384.

Corliss, B.H. & Keigwin, L.D., Jr, 1983. The Eocene/Oligocene event in the deep sea. *First International Conference on Paleoceanography, Zürich, 1983*, (abstract), p. 16.

Douglas, R.G. & Savin, S.M. 1973. Oxygen and carbon isotope analyses of Cretaceous and Tertiary foraminifera from the Central North Pacific. *Init. Rep. Deep Sea Drill. Proj.*, **17**, 591–605.

Douglas, R.G. & Savin, S.M. 1975. Oxygen and carbon isotope analyses of Tertiary and Cretaceous microfossils from Shatsky Rise and other sites in the North Pacific Ocean. *Init. Rep. Deep Sea Drill. Proj.*, **32**, 509–520.

Douglas, R.C. & Savin, S.M. 1978. Oxygen isotopic evidence for the depth stratification of Tertiary and Cretaceous planktic foraminifera. *Mar. Micropaleont.*, **3**, 175–196.

Emiliani, C. 1954. Depth habitats of some species of pelagic foraminifera as indicated by oxygen isotope ratios. *Am. J. Sci.*, **252**, 149–158.

Haq, B.U. 1980. Paleogene paleoceanography: Early Cenozoic oceans revisited. *Oceanol. Acta*, Special Publication, 71–82.

Haq, B.U., Premoli-Silva, I. & Lohmann, G.P. 1977. Calcareous plankton paleobiogeographic evidence for major climatic fluctuations in the Early Cenozoic Atlantic Ocean. *J. Geophys. Res.*, **82**, 3861–3876.

Hayes, D.E. & Frakes, L.A. 1975. General synthesis Deep Sea Drilling Project, Leg 28. *Init. Rep. Deep Sea Drill. Proj.*, **28**, 919–942.

Hsü, K.J., McKenzie, J.A., Oberhänsli, H., Weissert, H. & Wright, R.C. 1984a. South Atlantic Cenozoic paleoceanography. *Init. Rep. Deep Sea Drill. Proj.*, **73**, 771–785.

Hsü, K.J., La Brecque, J.L. *et al.* 1984b. *Init. Rep. Deep Sea Drill Proj.*, **73**, Washington, DC, 798 pp.

Hsü, K.J., Percival, S.F., Jr, Wright, R.C. & Petersen, N.P. 1984c. Numerical ages of magnetostratigraphic stratigraphically calibrated biostratigraphic zones. *Init. Rep. Deep Sea Drill. Proj.*, **73**, 623–636.

Keigwin, L.D., Jr, 1980. Oxygen and carbon isotope analyses from Eocene/Oligocene boundary at DSDP Site 277. *Nature*, **287**, 722–728.

Keigwin, L. & Keller, G. 1984. Middle Oligocene cooling from Equatorial Pacific DSDP Site 77B. *Geology*, **12** (1), 16–19.

Keller, G. 1983. Paleoclimatic analyses of Middle Eocene through Oligocene planktic foraminiferal faunas. *Palaeogeogr. Palaeoclimat. Palaeoecol.*, **43**, 73–94.

Kemp, E.M. 1978. Tertiary climatic evolution and vegetation history in the Southeast Indian Ocean region. *Palaeogeogr. Palaeoclimat. Palaeoecol.*, **24**, 169–208.

Kemp, E.M. 1981. Tertiary palaeogeography and the evolution of Australian climate. In: A. Keast, (ed.), *Ecological Biogeography of Australia*, Dr W. Junk Publishers, The Hague, pp. 33–49.

Kennett, J.P. 1977. Cenozoic evolution of Antarctic glaciation, the Circumantarctic Ocean, and their impact on global paleoceanography. *J. Geophys. Res.*, **82** (27), 3843–3860.
Kennett, J.P. 1978. The development of planktonic biogeography in the Southern Ocean during the Cenozoic. *Mar. Micropaleont.*, **3**, 301–345.
Kennett, J.P. 1982. *Marine Geology.* Prentice Hall Inc., New Jersey, 752 pp.
Kennett, J.P. & Shackleton, N.J. 1976. Oxygen isotopic evidence for the development of the psychrosphere 39 Myr ago. *Nature*, **260**, 513–515.
Kroopnick, P. 1980. The distribution of ^{13}C in the Atlantic Ocean. *Earth Planet. Sci. Lett.*, **49**, 469–484.
La Brecque, J.L., Hsü, K.J., Carman, M.F., Jr, Karpoff, A.-M., McKenzie, J.A., Percival, S.F., Jr, Petersen, N.P., Pisciotto, K.A., Schreiber, E., Tauxe, L., Tucker, P., Weissert, H.J. & Wright, R. 1983. DSDP Leg 73: Contributions to Paleogene stratigraphy in nomenclature, chronology and sedimentation rates. *Palaeogeogr. Palaeoclimat. Palaeoecol.*, **42**, 91–125.
Letolle, R., Vergnaud-Grazzini, C. & Pierre, C. 1979. Oxygen and carbon isotopes from bulk carbonates and foraminiferal shells at DSDP Sites 400A, 401, 402, 403, and 406. *Init. Rep. Deep Sea Drill. Proj.*, **48**, 741–755.
Matthews, R.K. & Poore, R.Z. 1980. Tertiary $\delta^{18}O$ record and glacioeustatic sea level fluctuations. *Geology*, **8**, 501–504.
McGowran, B. 1978. Stratigraphic record of Early Tertiary oceanic and continental events in the Indian Ocean region. *Mar. Geol.*, **26**, 1–39.
Miller, K.G. 1983. Eocene – Oligocene paleoceanography of the deep Bay of Biscay: benthic foraminiferal evidence. *Mar. Micropaleont.*, **7**, 403–440.
Miller, K.G. & Curry, W.B. 1982. Eocene to Oligocene benthic foraminiferal isotopic records in the Bay of Biscay. *Nature*, **96**, 347–350.
Miller, K.G. & Fairbanks, R.G. 1983. Evidence for Oligocene – Middle Miocene abyssal circulation changes in the western North-Atlantic. *Nature*, **306**, 250–253.
Miller, K.G. & Thomas, E. 1985. Late Eocene to Oligocene benthic foraminiferal isotopic record, Site 574 equatorial Pacific. In: L. Mayer, F. Theyer *et al.*, *Init. Rep. Deep Sea Drill. Proj.*, **85**, 549–589.
Miller, K.G. & Tucholke, B.E. 1983. Development of Cenozoic abyssal circulation south of the Greenland–Scotland ridge. In: M. Bott, S. Saxov, M. Talwani & J. Thiede (eds), *Structure and Development of the Greenland–Scotland Ridge*, Plenum Publishing Corporation, New York.
Miller, K.G., Curry, W.B. & Ostermann, D.R. 1984. Late Paleogene benthic foraminiferal paleoceanography of the Goban Spur Region, DSDP Leg 80. *Init. Rep. Deep Sea Drill. Proj.*, **80**, (in press).
Muza, J.P., Williams, D.F. & Wise, S.W. 1983. Paleogene oxygen isotope record for deep sea drilling sites 511 and 512, subantarctic south Atlantic ocean: paleotemperatures, paleoceanographic changes, and the Eocene/Oligocene boundary event. *Init. Rep. Deep Sea Drill. Proj.*, **71**, 409–422.
Oberhänsli, H. 1985. Latest Cretaceous – Early Neogene oxygen and carbon isotopic history at DSDP sites in the Indian Ocean. *Mar. Micropaleont*, (in press).
Oberhänsli, H., McKenzie, J.A., Toumarkine, M. & Weissert, H. 1984. A paleoclimatic and paleoceanographic record of the Paleogene in the Central South Atlantic (Leg 73, Sites 522, 523, 524). *Init. Rep. Deep Sea Drill. Proj.*, **73**, 737–748.
Parker, W.C., Clark, M.W., Wright, R.C. & Clark, R.K. 1984. Population dynamics, Paleogene abyssal benthic foraminifers, Eastern South Atlantic. *Init. Rep. Deep Sea Drill. Proj.*, **73**, 481–486.
Poore, R.Z. & Matthews, R.K. 1984. Late Eocene–Oligocene oxygen and carbon isotope record from the South Atlantic Ocean DSDP Site 522. *Init. Rep. Deep Sea Drill. Proj.*, **73**, 725–736.
Poore, R.Z., Tauxe, L., Percival, S.F., Jr, La Brecque, J.L., Wright, R., Petersen, N.P., Smith, C.L., Tucker, P. & Hsü, K.J. 1983. Late Cretaceous–Cenozoic magnetostratigraphic and biostratigraphic correlations of the South Atlantic Ocean: DSDP Leg 73. *Palaeogeogr. Palaeoclimat. Palaeoecol.*, **42**, 127–149.

Rabussier-Lointier, D. 1980. Variations de composition isotopique de l'oxygène et du carbone en milieu marin et coupures stratigraphiques du Cénozoique. Thèse doct. 3e cycle. Univ. Pierre-et-Marie-Curie, Paris.

Sancetta, C. 1979. Paleogene Pacific microfossils and paleoceanography. *Mar. Micropaleont.*, **4**, 363–398.

Savin, S.M. 1977. The history of the earth surface temperature during the past 100 million years. *Ann. Rev. Earth Planet. Sci.*, **5**, 319–355.

Savin, S.M., Douglas, R.G. & Stehli, F.G. 1975. Tertiary marine paleotemperatures. *Bull. Geol. Soc. Am.*, **86**, 1499–1510.

Sclater, J.G., Hellinger, S. & Tapscott, C. 1977a. The paleobathymetry of the Atlantic Ocean from the Jurassic to the Present. *J. Geol.*, **85** (5), 509–522.

Sclater, J.G., Abbott, D. & Thiede, J. 1977b. Paleobathymetry and sediments of the Indian Ocean. In: J.R. Heirtzler (eds.), *Indian Ocean Geology and Biostratigraphy*, American Geophysical Union, Washington, pp. 25–59.

Shackleton, N.J. 1977. Carbon-13 in Uvigerina: Tropical rainforest history and the Equatorial Pacific carbonate dissolution cycles. In: N.R. Anderson & A. Malahoff (eds.), The fate of fossil fuel CO_2 in the ocean, *Mar. Sci.*, **6**, 401–427.

Shackleton, N.J. & Hall, M.A. 1984. Carbon isotope data from Leg 74 sediments. *Init. Rep. Deep Sea Drill. Proj.*, **74**, 613–620.

Shackleton, N.J. & Kennett, J.P. 1975. Paleotemperature history of the Cenozoic and the initiation of Antarctic glaciation: Oxygen and carbon isotope analyses in DSDP Sites 277, 279 and 281. *Init. Rep. Deep Sea Drill. Proj.*, **29**, 743–755.

Shackleton, N.J., Hall, M.A. & Boersma, A. 1984. Oxygen and carbon isotope data from Leg 74 foraminifers. *Init. Rep. Deep Sea Drill. Proj.*, **74**, 599–612.

Snyder, S.W., Mueller, C. & Miller, K.G. 1984. Eocene–Oligocene boundary: Biostratigraphic recognition and gradual paleoceanographic change at DSDP Site 549. *Geology*, **12**, 112–115.

Talwani, M., Udinstev, G. et al. 1976. Tectonic synthesis. *Init. Rep. Deep Sea Drill. Proj.*, **38**, 1213–1240.

Tauxe, L., Tucker, P., Petersen, N.P. & La Brecque, J.L. 1983. The magnetostratigraphy of Leg 73 sediments. *Palaeogeogr. Palaeoclimat. Palaeoecol.*, **42**, 65–90.

Tjalsma, R.C. & Lohmann, G.P. 1982. Paleocene–Eocene bathyal and abyssal benthic foraminifera from the Atlantic Ocean. *Micropaleontol., Spec. Publ.*, **4**, 90p.

Vergnaud-Grazzini, C. & Rabussier-Lointier, D. 1980. Compositions isotopiques de l'oxygène et du carbone des foraminifères tertiaires en Atlantique équatorial (site 366 du DSDP). *Rev. Géol. dynamique Géogr. physique*, **22** (1), 63–74.

Vergnaud-Grazzini, C., Pierre, C. & Letolle, R. 1978. Paleoenvironment of the North-East Atlantic during the Cenozoic: Oxygen and carbon isotope analyses at DSDP Sites 398, 400A and 401. *Oceanol. Acta*, **1** (3), 381–390.

10

History of calcite dissolution of the South Atlantic Ocean

K. J. HSÜ and R. WRIGHT

Abstract

Six sites were drilled in the South Atlantic during the 1980 DSDP Leg 73 cruise. Continuous cores, taken mainly by hydraulic piston coring, were obtained for precise biostratigraphy and magnetostratigraphy. The samples provided a record of the history of calcite dissolution in the South Atlantic.

Three parameters for dissolution were examined, namely the insoluble residue (IR) content, the relative abundance of benthic foraminifers (BF) in microfaunas, and the relative abundance of the fragmented foraminiferal tests (FFT). Our results indicate periodic variations in the level of the calcite-compensation depth (CCD), with lower stands during the Oligocene and Plio-Quaternary, and higher stands during the Eocene and Miocene Epochs. Short-term fluctuations during the Plio-Quaternary have also been detected, and they are correlated to glacial and interglacial stages.

Correlation of the calcite-dissolution record to other paleoceanographic data indicates that the rise and fall of CCD was influenced by many factors, including climate, ocean temperatures, ocean circulation, nutrient supply, and regional variations in productivity. The long-term variations, with periodicity of about 10^7 years, are related mainly to productivity of calcareous plankton in open oceans, but second-order fluctuations, with periodicity of 10^4 or 10^5 years, may be interpreted in terms of differential solution, resulting from changing chemistry of bottom waters during glacial and interglacial stages.

Introduction

The occurrence of Miocene red clays in the South Atlantic was one of the surprising discoveries of the DSDP Leg 3 cruise. The inexperienced shipboard staff, including the first author of this report, were not familiar with marine geochemistry. Assuming a simple relation of calcite-dissolution to paleodepth, the red clays were interpreted as deposits on subsided ridge-crest at times when seafloor-spreading slowed down or stopped altogether (Hsü & Andrews, 1970). Shortly after the publication of the cruise report, it was pointed out that the calcite-compensation depth (CCD) could not have remained constant through geologic time (Hay, 1970). With the accumulation of deep-sea data, it also became clear that the rate of seafloor-spreading should have been more or less linear since the late Mesozoic (La Brecque *et al.*, 1977). The facies variations of the Atlantic pelagic sediments thus became the basis of deciphering the ups and downs of the CCD, and the dominance of the Middle Miocene red clays was cited as evidence of a very high level of CCD during that epoch.

With the observation that the newly formed ridge-crest is almost everywhere 2600 deep, and that the seafloor subsides as an oceanic lithospheric plate moves away from a spreading center (Sclater *et al.*, 1971), the paleodepth at any part of the ocean could be computed on the basis of the back-tracking method. The wealth of deep-sea data permitted reconstructions of the fluctuation of the 'carbonate line' on the ocean bottom (cf. snowline in the mountains) during the Cenozoic (e.g., Berger & Winterer, 1974; van Andel *et al*, 1977). The Atlantic records suggested a deep CCD at a depth of about 4.5–5 km during the Oligocene, but rising to an anomalously shallow depth (less than 3.5 km) during the Middle Miocene before plunging down to greater than 4.5 km depth again in the Early Pliocene (see van Andel *et al.*, 1977). The cause of the anomalous CCD, however, has remained a mystery.

A transect of holes to study the Middle Miocene CCD-crisis of the Atlantic was proposed at the start of the International Program for Ocean Drilling (IPOD) in 1975 by the first author. Realising that the high CCD at Middle Miocene time should have dissolved away calcareous fossils at deeper sites, thereby eliminating both stratigraphic and paleoceanographic information, drill-holes were proposed on the Mid-Atlantic Ridge, so positioned as to obtain Miocene sediments deposited at relatively shallow paleodepths (< 3000 m). The selection of the exact locations required a reliable record of magnetic anomalies; consequently the region at 30 °S latitude was selected, an area where the magnetic signatures are clear, and where the weather is suitable for drilling operations throughout the year. Since the west flank of the ridge was drilled during Leg 3, the new transect was designed for the east flank to test the symmetry (with respect to the ridge-axis) of paleoceanographic processes.

While the planning progressed, the hydraulic piston-coring (HPC) device was invented and successfully tested in 1979. This new device opened the way for precision stratigraphy. With undisturbed and nearly complete recovery, we could devise plans to study short-term

(10^4–10^5 years) variations, as well as to obtain a long continuous record. Eventually our Mid-Atlantic Ridge transect was scheduled to be drilled during Leg 73, one of the five legs to investigate South Atlantic paleoenvironments in 1980.

With five years of lead-time for planning, and the help of the HPC-device to obtain several excellent suites of cores, we had a successful cruise. The preliminary results and conclusions of Leg 73 drilling have been published in the *Initial Reports of the Deep Sea Drilling Project* (Hsü, La Brecque, *et al.*, 1984). This article includes an up-to-date revision of information on calcite dissolution, and presents a working hypothesis to explain the changes of calcite compensation level during the Cenozoic in the South Atlantic.

Regional framework and methodology

Glomar Challenger left Santos Brazil on 13 April, and docked in Cape Town, South Africa, on 1 June 1980. During the 49 days at sea, 13 holes were drilled at six sites and more than 1000 m of cores were recovered (Fig. 1, Table 1).

Fig. 1. Location map showing positions of Leg 73 sites, marine magnetic anomalies, and the 2000 fathom bathymetric contour.

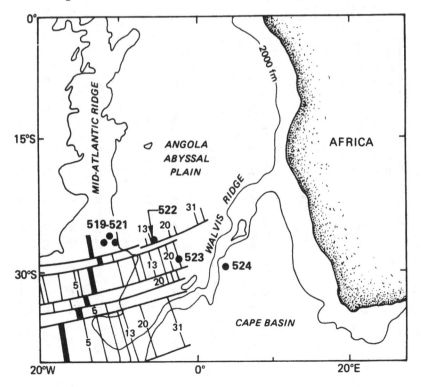

Table 1. Leg 73 coring summary

Hole	Latitude	Longitude	Water depth (m)	Hole depth (m)	Number of cores	Meters cored	Meters recovered	Recovery (%)	Oldest sample	Magnetic epochs	Seafloor anomaly	Age (my)
519	26°08.20'S	11°39.97'W	3778.5	151.6	37	151.6	138.4	91.3	Upper Miocene	10	5^R	10.1
519A	26°08.20'S	11°39.97'W	3778.5	180.0	7	84.6	61.4	73.1	Basalt	10	5^R	10.1
520	25°31.40'S	11°11.14'W	4217.0	458.5	31	246.5	69.4	28.1	Middle Miocene–Langhian	—	$5B^R$	14.3
520A	25°31.40'S	11°11.14'W	4217.0	18.5	1	2.4	2.4	100	Quaternary	—	—	—
521	26°04.43'S	10°15.87'W	4141.0	84.0	21	84.0	75.4	89.7	Middle Miocene–Langhian	16	$5C^N$	15.9
521A	26°04.54'S	10°15.59'W	4141.6	71.1	17	71.1	64.3	90.4	Middle Miocene–Langhian	16	$5C^N$	15.9
522	26°06.84'S	05°06.78'W	4456.6	148.7	39	148.7	137.7	92	Upper Eocene	C-16	16^N	38.4
522A	26°06.84'S	05°06.78'W	4456.6	156.0	31	106.0	97.9	89.9	Upper Eocene	C-16	16^N	38.4
522B	26°06.84'S	05°06.78'W	4456.6	170.4	6	40.5	25.3	62.6	Upper Eocene	C-16	16^N	38.4
523	28°33.13'S	02°15.08'W	4572.0	193.5	51	182.5	149.2	81.9	Middle Eocene	$C-20^n$	20^R	51.0
524	29°29.05'S	03°30.74'E	4806.0	348.5	35	306.5	199.4	65	Upper Cretaceous	$C-31^r$	?	71
524A	29°29.05'S	03°30.74'E	4805.0	47.5	2	19.0	9.9	52	Upper Cretaceous	$C-31^r$	?	—
524B	29°29.07'S	03°30.74'E	4804.5	29.5	7	29.5	20.7	7	Upper Cretaceous	$C-31^r$	?	—

The Neogene crust on both sides of the Mid-Atlantic Ridge constitutes the so-called 'rough-basement'. Block faulting resulted in a topography of local basins surrounded by submarine hills. We noted during the planning sessions that the sedimentary cover on the hills is very thin. The basin sequences are thicker, but may include many turbidite beds. After drilling the first two sites (519 and 520) during the first part of Leg 73, we concluded that the thin sedimentary sequences on submarine hills may be condensed, but they include no other hiatus than solution unconformities. The basinal turbidite sequences on the other hand, cannot give clear stratigraphic signals, because they are too thick to be sampled by hydraulic piston-coring. Consequently sites 521, 522, and 523 were all spudded on submarine hills. Site 524 was located in the Cape Basin and cored by rotary drill, but the sequence was sufficiently well consolidated to yield good recovery of samples for magnetostratigraphy.

Holes 519 to 523 were positioned on magnetic anomalies 5 (R), 5B (R), 5C (N), 16 (N), and 20 (R), respectively, and drilled into crust of Miocene and Eocene ages. The sites were so chosen, because those epochs were times when the CCD was relatively elevated; we had to drill into sediments deposited near the crest of the Mid-Atlantic Ridge to obtain calcareous and fossiliferous sediments. Hole 524 was drilled into volcanics at the foot of Walvis Ridge – an aseismic ridge that stood above CCD during late Cretaceous and Paleocene. The cores from Leg 73 thus provided a fairly complete record of Cenozoic changes in CCD.

All sites are sufficiently distant from land so that the input of detritus for continents was minimal. The Mid-Atlantic Ridge sediments encountered by our drilling were mainly pelagic, but the Cape Basin sequence at Site 524 included pelagic as well as volcanoclastic turbidites. From a paleoceanographic perspective these middle-latitude sites are located on the periphery of the South Atlantic gyre, weakly influenced by the Benguela and by Brazil Currents (Fig. 2).

Our methodology was straightforward: magnetostratigraphy and biostratigraphy provided precise dating of samples, with the synchroneity of the relative ages accurate within 10^5 or 10^6 years. The sampling density depended upon sedimentation rate and upon the nature of the problems involved. We were able to study significant paleoceanographic changes with a duration of 10^3 or 10^4 years, and to establish a record of long-term variations for the Cenozoic. Analyses of bulk compositions, grain size, microfossil preservation, and benthic percentages give quantitative expressions of calcite dissolution. Stable-isotope data served as a basis for interpreting changes in temperatures,

chemistry, biologic productivity, and circulation patterns of past
oceans. Paleoenvironmental studies of microfaunas and nannofloras
provided further information on paleoceanography. The basic information is summarized graphically for each site (Figs. 3–5 and 7). For the
sake of brevity in expression we have adopted the terms *alytic*, *eolytic*,
oligolytic, *mesolytic*, *pleistolytic*, and *hololytic*, signifying, respectively,

Fig. 2. Distribution of the Pliocene diatoms and the Oligocene
Braarudosphaera chalk in the South Atlantic and their relation to
the present-day current systems.

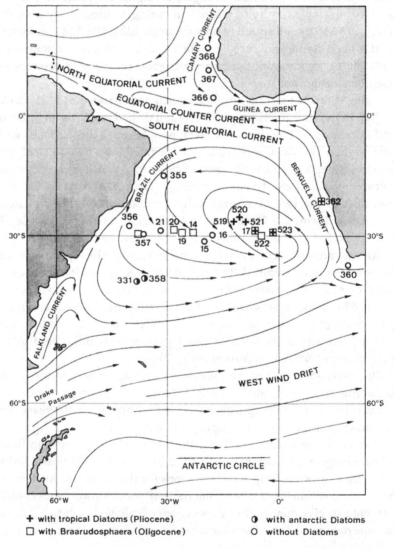

+ with tropical Diatoms (Pliocene)   ◑ with antarctic Diatoms
☐ with Braarudosphaera (Oligocene)   ○ without Diatoms

not dissolved, beginning to be dissolved, slightly dissolved, moderately dissolved, mostly dissolved, and all dissolved; these definitions were given by Violanti *et al.* (1979).

The current practices of magnetostratigraphic nomenclature use two sets of terms. Epochs such as Brunhes, Matuyama, Gauss, Gilbert, 5, 6, etc. have been recognized back to the end of the Paleogene (Theyer & Hammond, 1974). After the successful correlation of the magneto-stratigraphy of Cenozoic sediments to the linear anomalies on the seafloor, we proposed that magnetostratigraphic chrons be established, and the chron numbers, with prefix C, should correspond to seafloor anomaly numbers (Hsü, La Brecque *et al.*, 1984). Chron C-5(N), for example, is the epoch corresponding to the time when the strip of seafloor with Cenozoic Magnetic Anomaly 5 (normally magnetized) was formed. Chron C-5(N) is thus a synonym of the 'long-normal magnetostratigraphic epoch', which has been variously considered Epoch 9 (by Dreyfus & Ryan, 1972) or Epoch 11 (by Foster & Opdyke, 1970). In this article we shall retain the nomenclature of late Neogene magnetostratigraphic epochs, because they are well known and are commonly used at the present time. We shall, however, apply the chron-nomenclature to designate older Neogene and Paleogene magnetostratigraphic units.

Site 519

A 151-m thick sequence was continuously cored to basement at this site by using the hydraulic piston-corer (HPC). The nearly complete recovery and absence of drilling disturbances enabled us to work out the precise bio- and magnetostratigraphy. Except for one thick slump-deposit at 15–30 m subbottom and sequence is pelagic. The Plio-Quaternary sediments are chalk oozes, and the Upper Miocene ones are marly.

Relevant data on calcite dissolution include the percentages of insoluble residues (IR), of benthic foraminifers (BF) relative to all foraminifers, and of fragmented planktonic foraminiferal tests (FFT). Their variations with time are shown by Fig. 3.

The oldest sediments of magnetostratigraphic Epoch 10 or Chron C-5(R), (10 m.y.), show considerable dissolution, containing 34% IR, 10% BF, and 78% FFT. The overlying sediments of Epochs 9–7 or Chrons C-5(N), C-4, and C-4' (8–10 m.y.) have undergone less dissolution, even though the ridge-crust subsided rapidly from 2600 to 3200 m depth during the deposition of those early Late Miocene sediments. Assuming that the mesolytic sediments of Epoch 10 were deposited near the top of

Fig. 3. Paleoceanographic data, Hole 19. Magnetostratigraphic epochs (not chrons) are given; they are B₁ Brunhes (B), Matuyama (M), Gauss (Ga) and Gilbert (Gi); magnetic events are Jaramillo (J), Olduvai (O), Kaena (K), Mammoth (M), Cochiti (Co), Nunivak (N), C₁ and C₂. Arrows mark the datum of first increase of *N. umboniferus*.

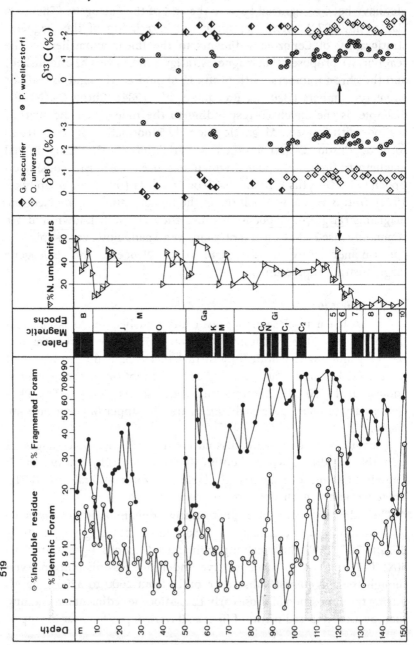

lysocline, the CCD must have been depressed more rapidly than subsidence during the subsequent 2 m.y. Minimum dissolution (6% IR) was reached at the time of Epoch 7/8 boundary or Chron C-4A, (8 m.y.), before the lysocline rose again. The peak dissolution occurred during Epoch 6 or Chron C-3A(R) at about 6 m.y. according to the IR data (32% IR), or during Epoch 5 or Chron C-3A(N) at about 5.5 m.y. according to BF data. The lysocline probably rose to a level above the paleodepth of 3300 m at that time. The lysocline was rapidly depressed again, reaching another minimum during early Gilbert or Chron C-3(R) time (4.5 m.y.) so that a sediment deposited during C_1 Event has a minimum IR of 4.6%. The lysocline remained depressed during the next 1.5 m.y., except for one short dissolution event during the Cochiti Event. During the late Gauss and the Brunhes, or Chrons C-2A and C-1, subordinate peaks of IR may either indicate slightly increased dissolution (as suggested by high FFT) or substantial increase of terrigenous input (as suggested by the low BF percentages).

The extent of calcite dissolution is reflected in the sedimentation rates as shown by the time–depth plot for this site (Fig. 8). The average accumulation rates for bulk sediment and for IR are also tabulated for various intervals (Table 2). The rates of IR-deposition were 0.7–1.0 m/ m.y. for the Late Miocene, but 1.5–2.7 m/m.y. for the Plio-Quaternary, confirming the general observation that the terrigenous influx to the Atlantic was unusually high during the glacial times. The lower Upper Miocene sediments deposited during Epochs 9–7 have about the same IR percentages as the Plio-Quaternary, but they have experienced, on the whole, twice as much dissolution, as indicated by the rates of bulk-sedimentation (7.0–10.4 m/m.y. versus 15–26.7 m/m.y.). this more advanced degree of dissolution is also shown by the FFT data (Fig. 3). The upper Upper Miocene sediments deposited during Epochs 6–5 have about three times as much IR as the Plio-Quaternary average (25% versus approximately 8%), because they have undergone a five-times more severe dissolution, as suggested by their bulk-sedimentation rates (4.2 m/m.y. versus approximately 20 m/m.y.).

A comparison with the dissolution studies of DSDP Site 16 yields interesting results. That site on the west flank of the ridge, was drilled on Anomaly 5(N) and terminated in basalt below Tortonian sediments slightly (< 1 m.y.) younger than the oldest sediments at Site 519 (see Hsü & Andrews, 1970). The Plio-Quaternary ooze (Albatross) is a chalk ooze deposited at a rate of 18 m/m.y., with a terrigenous influx of 1.7 m/ m.y.; rates similar to those of correlative sediments at Site 519. A lithologic change occurs in the lowest Pliocene sediments at Site 16 (as at Site 519). The Messinian ooze (Blake) is a remarkably white chalk ooze, containing, on average, as much insoluble residue (9.3%) as the Plio-Quaternary ooze at that site, even though the Messinian has undergone more severe dissolution, as can be seen from the comparative sedimentation rates (12 m/m.y. versus 18 m/m.y.). The bulk compositions of the two different oozes are coincidentally identical because of a proportionally slower rate of terrigenous influx 1.1 m/m.y.) in the more dissolved Messinian sediments. The latest Late Miocene rate of IR accumulation is about the same at Sites 16 and 519, but a comparison of the IR and the bulk-sedimentation rates indicates that the ooze of this age on the east side of the Mid-Atlantic Ridge has undergone three times as much dissolution as that on the west side although the difference in paleodepth of the two sites was less than 300 m. The older Upper Miocene ooze (Challenger) at Site 16 was deposited at 10 m/m.y. while the IR was accumulated at 1.0 m/m.y. – rates similar to those of the

Table 2. *Sedimentation rates at Site 519.*

Datum	Depth (m)	Age (m.y.)	% IR	Sedimentation rate (m/m.y.)	
				Interval (Bulk)	IR
			12	15.4	1.9
Jaramillo, top	14	0.91			
			–	(slump block)	–
Jaramillo, bottom		0.98			
			10	15	1.5
Kaena, top	62	2.92			
			7.5	22.6	1.7
Gauss/Gilbert	73	3.41			
			7.5	26.7	2.0
Cochiti, top	85	3.86			
			12	21.4	2.6
G_1, top	95.5	4.35			
			7	38.2	2.7
C_2, top	102	4.52			
			17	6.1	1.0
Gilbert/Epoch 5	116.5	5.26			
			25	4.2	1.0
Epoch 6/Epoch 7	121.9	6.54			
			9	7.6	0.7
Epoch 7/Epoch 8	132.1	7.88			
			9.5	10.4	1.0
Epoch 8/Epoch 9	139.2	8.56			
			12	7.0	0.8
Epoch 9/Epoch 10	149.1	10.00			

synchronous sediments at Site 519 – indicating the same degree of dissolution at both sites during the time interval of Epochs 7–9 (6.5–8.5 m.y.).

We have made an estimate of the percentage of the benthic foraminifer species *Nuttallides umboniferus*. This species is found today in moderate amounts at shallower abyssal depths where the bottom water is the North Atlantic Deep Water (NADW), but the species becomes dominant in areas underlying Antarctic Bottom Water (AABW). Although we realize that many factors influence the ecology of *N. umboniferus*, large variations in its abundance may be indications of changing bottom circulations. *N. umboniferus* first occurred in significant numbers at Site 519 during Epoch (6 m.y.) when the site had subsided to 3400 m below sealevel, presumably to a depth below the top of the AABW. Another remarkable increase in the population of *N. umboniferus* took place about 3.5 m.y. ago (just below the Gilbert–Gauss boundary), signifying an intensification of the AABW, more or less synchronous with the mid-Pliocene expansion of the Antarctic Ice Sheet (see Shackleton & Opdyke, 1977). A detailed investigation indicates synchronous oscillations of the *N. umboniferus* population and the $\delta^{18}O$ values in cyclically deposited late Pliocene sediments. This fact suggests pulsations of AABW in response to alternating glacial and interglacial stages (Weissert *et al.*, 1984). Yet another increase of *N. umboniferus*, with a corresponding oxygen-shift, took place at about 2.5 m.y. (Gauss–Matuyama transition), indicating a further intensification of the AABW, corresponding to another expansion of polar ice caps (see Shackleton & Opdyke, 1977).

A detailed examination of the cyclically deposited Pliocene ooze at this and other sites revealed that the darker oozes contain more IR, higher percentages of BF, more FFT and more *N. umboniferus*, and are characterized by a more positive $\delta^{18}O$ than lighter oozes. Although the high IR could be interpreted as having resulted from a higher influx of terrigenous detritus, the other parameters clearly indicate a higher degree of dissolution of the marl oozes deposited during glacial stages, when the AABW was also more active. The periodicity of the Pliocene cycles has been estimated to be about 10^5 years (Weissert *et al.*, 1984).

A dissolution event – its beginning indicated by an abrupt increase of benthic foraminiferal percentages at 122 m subbottom, and its end by an abrupt decrease at 111 m subbottom – took place during the Late Miocene. McKenzie and Oberhänsli's study of carbon and oxygen isotopes (this volume, chapter 8) indicates a reduction of plankton-productivity at the beginning of the dissolution event, and an increase at the end. The coincidence in timing seems to suggest that the rise in CCD may well be related to an episode of decreased productivity.

Site 520

Hole 520 penetrated a Neogene sequence 450 m thick and was bottomed in basalt below Middle Miocene (Langhian) sediments. The Plio-Quaternary and Upper Miocene sediments include thick resedimented deposits, but a Middle Miocene pelagic sequence was encountered at a depth below 400 m subbottom.

The Plio-Quaternary chalk oozes and turbidites show few effects of dissolution. The Upper and Middle Miocene sediments are, however, marly or diatomaceous. Selected analyses of the Miocene samples indicate variable IR content, ranging from about 20 to 80%. The more calcareous oozes are mainly resedimented, laid down too rapidly to have undergone much dissolution. The $CaCO_3$-poor sediments are either biogenic (diatom oozes, diatomites) or terrigenous (dark brown marls, red clays). Practically all samples examined contain identifiable nannofossils, although the preservation is poor in sediments from Middle Miocene nannofossil zones. One sample in core 29, late Middle Miocene (12–14 m.y.) in age, is a hololytic red clay devoid of calcium carbonate, indicating that the CCD rose above the paleodepth of about 3600 m during the peak of this dissolution-crisis. Another red clay (2% $CaCO_3$) is interlayered in diatomites of the NN 11 Zone, signifying another maximum height of CCD during the time interval between 8 and 6 m.y. ago. The average rate of pelagic sedimentation during the Middle Miocene here (6.2 m/m.y.) falls within the ranges of the Late Miocene rates at Site 519 (4.2–10.4 m/m.y.). *Nuttallides umboniferus* fauna first became dominant during the time represented by the interval from 438 to 398 m subbottom, or between 8 and 14 m.y. ago. The first appearance was earlier than that at Site 519, probably because this site, with its older crust, subsided to a water depth of 3400 m at an earlier time (13.5 m.y. versus 6 m.y.). Intense calcite dissolution seems to have preceded the arrival of the fauna; nannofossils are poorly preserved even in the oldest sediment of this site.

One anomalous fact is the presence of thick Upper Miocene sediments of the *Discoaster quinqueramus* (NN 11) Zone. Sediments of this age are either very thin or entirely missing at other Mid-Atlantic Ridge sites drilled during Leg 73. The very high sedimentation rate here resulted from resedimentation and from diatom preservation in a basinal environment on the eastern flank of the Mid-Atlantic Ridge.

The discovery of diatomites at an open ocean site distant from coastal, polar, and equatorial high-productivity belts was a surprise. A diatom flora is present in the Pliocene as well as the Late Miocene sediments. The assemblage is composed predominantly of strongly silicified, solution-resistant species, *Ethmodiscus rex* being the most abundant diatom species. The diversity of the overall assemblage is not great; only about 30 species were observed. The flora is comparable to the subtropical or tropical assemblage from the Walvis Ridge (Site 362) and is distinct from the Antarctic assemblage reported in the Argentine Basin (Gombos, 1984).

The Pliocene *Ethmodiscus* oozes, like those found at Site 17, are intercalated in nannofossil oozes that are bioturbated and contain normal foraminiferal assemblages. The laminated diatomites of the Upper Miocene show, however, no evidence of bioturbation, and they do not contain a benthic microfauna. The organic carbon content of the diatomites and associated sediments ranges from 0.14 to 0.36% (Herbin & Deroo, 1984). This content is about an order of magnitude less than that in the sapropels of the South Atlantic. Much, if not all, of the organic material from siliceous or calcareous organisms seems to have been oxidized before reaching the ocean bottom. However, the organic content is still an order of magnitude greater than that in most of the oceanic oozes sampled during the leg (it was 0.026–0.52% at Site 519, for example). The lack of bioturbation and benthic fossils suggests an oxygen-poor, if not anoxic, condition; thus the organic matter that did settle on the basin floor was largely preserved. The organic carbon content is thus still considerably greater than that of oozes deposited on the well-oxygenated ocean bottom.

Gombos (1984) believed that the occurrence of the diatomites at this site was the result of favourable preservation. The *Ethmodiscus* assemblage is the type that results from differential dissolution; the valves of less resistant species were presumably dissolved as they sank downward through the deep water column. The more resistant skeletons managed to settle on a basin floor where the CO_2-rich, stagnant bottom water mass had a relatively low pH and was thus favorable for the dissolution of $CaCO_3$ and preservation of siliceous fossils.

Diatom assemblages are also present, however, in the Plio-Quaternary oozes deposited on a well-oxygenated bottom at several southeastern Atlantic sites (17, 362, 519). The presence thus could not solely be attributed to preservation in a basinal euxinic environment. The record at Site 362 indicates unusually high plankton fertility during the late Neogene in the south-eastern Atlantic (Bolli, Ryan *et al.*, 1978). Both siliceous and calcareous plankton have been well preserved at this relatively shallow Walvis Ridge site, and the sedimentation rate during NN 11 time was unusually high, like that at Site 520. Calcarous plankton were, however, poorly or not preserved in the basinal environment where the bottom water must have been very corrosive.

Marine diatoms of Late Miocene or Pliocene age are absent at Sites 366 to 369 near the West African Coast (Schrader, 1978). It seems that the Pliocene–Miocene diatom blooms took place in the waters of the Benguela Current system, while the Canary Current did not yield sufficient nutrients for much diatom production. The distribution of the Benguela Current is shown in Fig. 2.

We note that not only Site 362, but also Sites 17, 519, and 520, lie under the Benguela Current. The sites on the west side of the ocean basin lie beneath the Brazil Current. The late Neogene sediments there have no indigenous diatom flora.

Site 521

The two holes drilled at this site penetrated similar sedimentary sequences. In Hole 521 we cored 84 m of foraminifer–nannofossil ooze, marly nannofossil ooze, and nannofossil clay. The adjacent section of Hole 521A was similar but only 71.1 m thick.

The Plio-Quaternary section consists of homogeneous, pale brown foraminifer–nannofossil ooze with carbonate content commonly ranging between 80 and 95%. The lowest Pliocene and Miocene sediments include dark brown nannofossil marls and clays, with $CaCO_3$-content varying between 10 and 90%.

The degree of dissolution has been manifested by the insoluble residue content (IR), by the abundance of benthic foraminiferas (BF), and by the fragmentation of foraminiferal tests (FFT). The three sets of criteria yield, on the whole, concordant data. Two major dissolution events have been recognized (Fig. 4).

The basal sediments at this site, deposited at 2600–3400 m paleodepth, are mainly oligolytic, containing 15–30% IR. Two occurrences of BF maxima (up to 30%) during the Middle Miocene may represent brief shifts of CCD. Table 3 shows that the average sedimentation rate during NN 5 time was about 7 m/m.y., comparable to the Late Miocene rates (7.0–10.4 m/m.y.) at Site 519 when it sank to a similar paleodepth. The rate of the Middle Miocene IR accumulation at 1.4 m/m.y. is, however, appreciably greater (versus 0.7–1.0 m/m.y.). Whether the discrepancy is real or a consequence of inaccurate chronology cannot be ascertained.

Fig. 4. Paleoceanographic data, Hole 521. Polarity and magnetic epochs as in Fig. 3 Dashed, solid, and heavy arrows denote, respectively, the second significant increase of *N. umboniferus*, times of major shift in CCD, and the datum level of major oxygen shift.

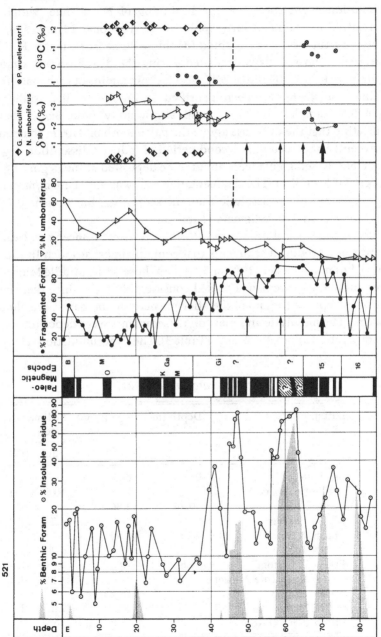

The IR content decreases systematically from 36% to 11% shortly after the *Orbulina* Datum (Epochs 15/16 boundary) or at about the same time as a positive shift in oxygen-18 anomaly (from +1.7 to +2.7%). This interval of a dissolution minimum lasted less than 1 m.y. The main dissolution event started at the beginning of NN 6 time, which is correlative to magnetostratigraphic Epoch 14, some 13 m.y. ago, as shown by the time-scale adopted for this volume (see chapter 1). The marl oozes and red clays deposited during the dissolution-maximum contain up to 84% IR and of 70% BF. One thin lamina of carbonate-free sediment in NN 6/7 has been identified during the detailed studies of Hole 521 samples for nannofossil stratigraphy (von Salis, 1984), indicating that the CCD rose above the paleodepth of 3500 m sometime between 12 and 10 m.y. A second period of reduced dissolution began when NN 10 sediments were about to be deposited at the beginning of Epoch 7 (Chron C-4). The sedimentation rate was about 1.8 m/m.y. for the mesolytic and pleistolytic Miocene sediments, but became 6.2 m/ m.y. for the NN 10 sediments (Table 3).

The second dissolution event during the latest Miocene and early Pleistocene produced sediments that contain almost as much IR (up to 80%), but much less BF (17% or less) than those of the Middle Miocene, even though the site had subsided considerably more since that time; the paleodepth at the end of the Miocene here should have been almost 4000 m. The mesolytic and pleistolytic sediments were deposited at an average rate of about 2.7 m/m.y. (Table 3). The IR accumulation rates for

Table 3 *Sedimentation rates at Site 521.*

Datum/interval	Depth (m)	Age (m.y.)	Sedimentation rate (m/m.y.)	
			Bulk	IR
Sea bottom	0	0		
Plio-Quaternary			10.6	1.3
Cochiti, top	40.25	3.86		
Mio-Pliocene			2.7	1.0
Epoch 7, top	47.5	6.54		
'NN 10'			6.2	1.1
Epoch 7, bottom	55.8	7.88		
Middle to Late Miocene			1.8	1.1
Epoch 15, top		13.2		
NN 5			7.0	1.4
Chron C-5C(N), top	84.2	15.8		

the three Miocene and Pliocene intervals were almost the same as the rates at Site 519 (approximately 1 m/m.y.).

The sedimentation rate since the Middle Pliocene at Site 521 is 10.6 m/ m.y. for the bulk sediment and 1.3 m/m.y. for IR; both rates are somewhat less than the equivalent rates at Site 519. Light eolytic oozes and darker oligolytic marly oozes were deposited during interglacial and glacial epochs of the Plio-Quaternary (Weissert *et al.*, 1984).

Site 15 on the western flank of the Mid-Atlantic Ridge was drilled into a slightly older crust, yet the sediments of the equivalent facies there contain less IR on the whole, and were deposited at slightly higher rates (Hsü & Andrews, 1970). The first mesolytic marl at Site 15, signifying a rapid increase of dissolution, occurs just above the *Orbulina* Datum, and the timing seems thus to have been slightly earlier there than at this site.

Three significant increases of the *N. umboniferus* percentage have been noted. The first increase (up to 12%) may simply signify that the site sank below the top of AABW some 14 m.y. ago, when the paleodepth was 3400 m. A second significant increase to more than 20% took place during latest Miocene some 5 m.y. ago, and a third to more than 30% during late Pliocene at about 2.5 m.y. ago. Each of the three increases coincided more or less in timing to an oxygen-shift or climatic cooling (see Douglas & Savin, 1975), but not necessarily to episodes of elevated CCD. The data at this site served to reinforce our conclusion after drilling 519: namely, the corrosive waters of AABW may result in locally increased dissolution, but the overall CCD of the world's oceans seems to have been controlled by some other factor.

Site 522

We obtained a good suite of cores down to the Upper Eocene at this site. The magnetostratigraphy is clear down to the Cochiti Event of the Gilbert Epoch. The lower Pliocene and Miocene section cannot be accurately dated. The Miocene/Oligocene boundary has, however, been identified and is correlative to Chron C-6 (R-1) at 23.6 m.y age. All Oligocene biostratigraphic zones have been recognized and correlated to magnetostratigraphic chrons. The Oligocene/Eocene boundary is placed within Chron C-13(R) with an age of 37.1 m.y. The Site 522 sequence is entirely pelagic. The Plio-Quaternary and Oligocene consist of chalk oozes, whereas the late Quaternary (Brunhes) and the condensed Plio-Miocene section are made up largely of marls and red clays.

The crust at Site 522 is older than that at previously discussed sites;

Fig. 5. Paleoceanographic data, Hole 522. Polarity epochs as in Fig. 3; also shown are magnetostratigraphic chrons with their numbering corresponding to that of seafloor anomalies. Two solid arrows (B) show occurrences of two *Braarudosphaera* chalk layers. C.S. indicates horizon of carbon-shift at 6.2 m.y.

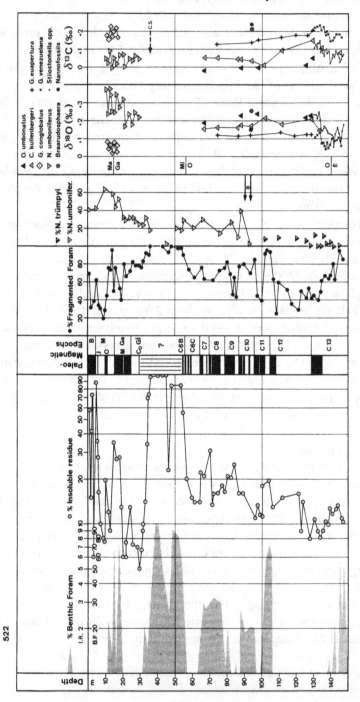

the IR content shows that the pleistolytic and hololytic sediments were deposited during a longer time span from Early Miocene to middle Pliocene. The benthic foraminifers are abundant in those marls and clays, reaching a maximum of 100% (Fig. 5). The period of extensive dissolution extended from about 23 to about 4 m.y. BP, but a short 'pause' or episode of CCD depression was indicated by the relatively low IR and BF contents of the sediments between 15 and 16 m.y. age (NN 5). The maximum dissolution was reached immediately after this 'pause', during Middle to Late Miocene, or from Chron C-5B(R) to Chron C-4' (R-2), time 15 to 18 m.y. ago; the sediments deposited during this interval are mainly hololytic and barren of nannofossils. The Miocene sediments older and younger than those deposited during the maximum dissolution are mainly pleistolytic, containing up to 85 and 98% of IR and 92 to 99% BF, respectively. The bulk rate of the Miocene sedimentation ranges from 0.7 to 1.3 m/m.y., not much different from the rate of IR accumulation of 0.7–0.8 m/m.y. (Table 4).

The late Quaternary marls and clays also showed signs of severe dissolution when the site had subsided to lysocline depth, but those sediments, for some reason, do not contain abundant benthic microfossils.

Peaks in abundance of BF and FFT suggest dissolution events during (a) Chron 13(R), of 37–38 m.y., (b) Chrons C-11(R) to Chron 12(N), between 33–34 m.y., (c) during Chron C-10 (30–32 m.y.) when *Braarudosphaera* blooms occurred, (d) Chrons C-7 to C-8 (26–29 m.y.), and (e) late Pliocene, coinciding in timing approximately to the 3.2 and 2.4 m.y. cooling events. Although the maximum BF exceeds 70%, the

Table 4. *Miocene sedimentation rate at Site 522*

Datum/interval	Depth (m)	Age (m.y.)	Sedimentation rate (m/m.y.)	
			Bulk	IR
C3'N-4	36	4.5		
Younger Miocene and earliest Pliocene			1.3	0.77
C4'R	40.6	8		
Upper and Middle Miocene			0.7	0.7
C5BR	45.5	15		
Older Miocene			1.3	0.75
C6BR-1	55.5	23		

insoluble residue content of the sediments deposited during those events did not rise above 40%. Compared to the Miocene crisis, those five were apparently subsidiary events during the Oligocene and Pliocene, when the level of CCD remained relatively high.

Site 522 had a depth close to lysocline during the Plio-Quaternary so that a minor shift in the lysocline depth resulted in cyclic sedimentation of light and dark oozes. Those dissolution events of short duration are related to glacial–interglacial climatic changes.

We have calculated the rates of accumulation for the various magnetostratigraphic chrons in terms of g/cm² per thousand years, and those rates can be translated into m/m.y. by a factor of 10/1.8. As shown by Fig. 6, the decrease of the bulk rates from 7 to 2 m/m.y. during the Oligocene is very impressive. There seems to be a perturbation from the downward trend during Chron C-8, but the BF and FFT data do not indicate a degree of dissolution in those sediments much different from those in other upper Oligocene sediments. The rates of IR accumulation during the Oligocene varied, but the average is not much different from that of the Miocene.

The steady increase of the calcite dissolution in the Oligocene sediments at ths site is certainly related to the subsidence. However, the effect of the changes in CCD must have been superimposed on top of the depth changes. The CCD rose to above 4000 m depth at about 15 m.y. and probably reached its highest levels during the late Middle Miocene

Fig. 6. Sedimentation rates, Hole 522.

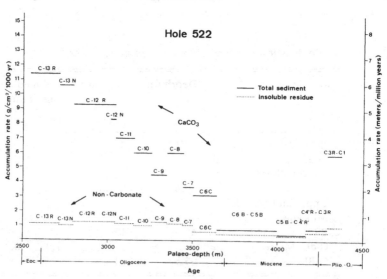

and/or early Late Miocene. The CCD then dropped below 4200 m 8 m.y. ago, but the lysocline remained above this depth for a few million years more. The rapid depression of the CCD during the early Pliocene is reflected by the facies change from red clay to nannofossil ooze, as the lysocline must have dropped below 4400 m depth some 4.5 m.y. ago.

The long-range variation of CCD was probably not related to the AABW activities. The first increase of N. *umboniferus* during Chron C-10 (30–31 m.y.) was the time when the site sank into the zone of ancestral AABW, although the paleodepth then was only about 3100 m. Perhaps an increase in the vigour of the AABW resulted in a rise of its upper boundary from about 3400 or 3300 m to about 3100 m during the middle Oligocene.

The variation of CCD, as shown by the data at this site, cannot be simply related to worldwide temperature changes. Detailed studies of Pliocene samples indicated more dissolution during glacial epochs of more active AABW (Weissert *et al.*, 1984). However, the rise of CCD from late Oligocene to Early Miocene coincided in timing with a warming trend. Our observations at this site thus confirm again our tentative conclusions that we must look for the supply side of the equation to understand the shifts of CCD during the Tertiary. The Oligocene and Plio-Quaternary depression of the CCD are probably more directly related to ocean fertility during those epochs than to ocean temperature or worldwide sealevel changes.

Site 523

The 190-m thick sequence at this site is entirely pelagic and consists of Plio-Quaternary oozes, Miocene red clays, and Eocene–Oligocene oozes. This three-fold division is a manifestation of the effect of the Miocene CCD-crisis. The hydraulic piston-cores of the Neogene sediments have been disturbed because of unfavourable weather conditions at the time of coring; those sediments have been dated biostratigraphically, but not magnetostratigraphically. Precision stratigraphy has been possible in dating the lower Oligocene and Eocene cores. The Oligocene/Eocene boundary falls within Chron C-13(R). The hole was bottomed in Middle Eocene.

The Neogene record of dissolution is obscured by the slow sedimentation rate. Compared to other Leg 73 sites, the seafloor here had subsided to greater depths during the Miocene so that the sediments of the whole epoch belong to the pleistolytic or hololytic red clay facies. One specimen barren of nannofossils was recognized from NN 11/12, indicating the rise of CCD above the paleodepth of the site, which was

Fig. 7. Paleoceanographic data, Hole 523. Only magnetostratigraphic chrons are shown. B$_1$ and B$_2$ indicate two *Braadudosphaera* chalk layers.

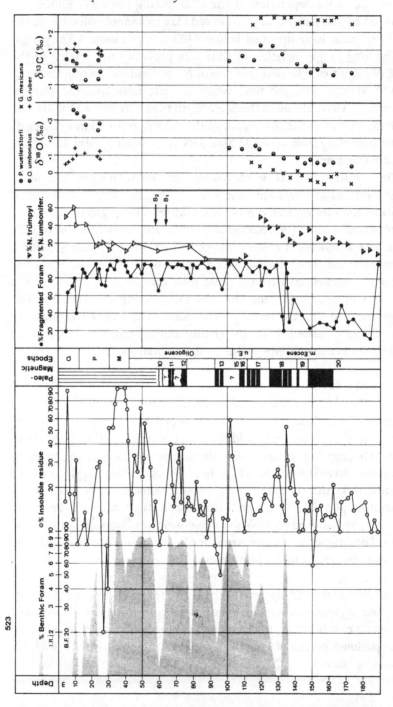

then 4200 m. Nannofossil zones NN 2–4 and NN 7–10 were not identified. The marker-fossils for those zones are apparently destroyed by dissolution, and those were times of very intense dissolution.

Insoluble-residue concentration indicates that dissolution events took place during the early Oligocene shortly before the *Braarudosphaera* blooms (Chrons C-11 and C-12). A more significant event approximately coincided in timing with the Eocene–Oligocene boundary (Chron C-13R). This boundary event produced marls with 60% IR, 90% BF, and almost 100% FFT. The Late Eocene equivalent facies at Site 19 on the other side of the Mid-Atlantic Ridge has been named Gazelle Ooze, and has an average IR of about 32% and a maximum close to 50% (Hsü & Andrews, 1970, p. 449). These Paleogene dissolution events have been recorded also by the Site 522 sequence; they are second-order events compared to the major CCD crisis during the Miocene.

Another sharp dissolution event during the late Middle Eocene has been indicated by the sediment at 135 m subbottom, a horizon that has been dated as the Chron C-18-2N, near the top of NP 16, and within the *T. rohri* foraminiferal zone. The IR increased from about 10 to 50%, the BF from less than 10 to more than 90%, and the FFT from about 50 to more than 90%. As Fig. 7 shows this event coincides in timing with the beginnings of a shift to more positive $\delta^{18}O$ and $\delta^{13}C$ values. At the same time the percentage of *N. trümpyi* (the Paleogene ancestor of *N. umboniferus*) increased significantly in the benthic assemblage. The paleodepth at that time was about 3050 m, and this event probably marked a brief but significant excursion of the lysocline to a depth of 3000 m or less. This late Middle Eocene dissolution event has also been found at Site 19 on the western side of the Mid-Atlantic Ridge marking the base of the Gazelle Ooze. The mesolytic sediment there lies within NP 16 and the *T. rohri* Zone and has an IR of about 50% (see Maxwell *et al.*, pp. 305 and 502).

The calcite dissolution has greatly affected the sedimentation rate. The Plio-Quaternary sediments were deposited at an average rate of 6 m/m.y. (1.2 m/m.y. IR). The Miocene is less than 12 m thick and has been deposited at a rate of about 0.65 m/m.y. (0.5 m/m.y. IR). The Oligocene has been deposited at a rate of 4.5 m/m.y. (1.0 m/m.y. IR). The Eocene rate was about 8 m/m.y. (1.2 m/m.y. IR). The rates of IR accumulation at this site are comparable to those at other Leg 73 sites.

A significant change in oceanic environments took place during the late Middle Eocene some 45 m.y. ago (Chron C-18-2N). Both the benthic and planktonic foraminifers show oxygen shifts of about 1‰, indicating rapid cooling of ocean temperatures and/or the first significant accumulation of Antarctic ice. The temperatures remained little changed during the Late Eocene and began to decrease again toward the beginning of the Oligocene (as shown by the evidence from Site 522). The increase of *N. umboniferus* took place when the site sank to 3500 m paleodepth during the early Oligocene. The first abundant occurrence of this species here was some 3–4 m.y. earlier than that at Site 522 and was apparently related to the subsidence history.

We found no clear correlation between the dissolution events and paleotemperatures. While the beginning of the late Middle Eocene oxygen shift was marked by a dissolution event, the CCD-change seemed to have reversed its direction soon after the rise that most of the upper Middle and Upper Eocene sediments are oligolytic ($<20\%$ IR). The early Oligocene cooling was also marked by a very brief dissolution event, followed by a dramatic depression of CCD. Oligocene dissolution events took place before and after the *Braadrudosphaera* blooms. Apparently the seafloor here was positioned near the Oligocene lysocline so that slight changes in CCD levels produced the rapid facies changes.

Two *Braarudosphaera* chalk layers have been identified at this site, both during Chron C-10 with ages of about 30.5 and 31.5, respectively; they are correlative to the upper two *Braarudosphaera* chalk layers at Site 522. The blooms took place at times when the CCD was deep (B_1 and B_2 for Fig. 7).

During the Miocene, the seafloor here always remained below the lysocline, and at times below the CCD, so that only pleistolytic and hololytic sediments were accumulated. The lysocline was rapidly depressed during the Early Pliocene, so that the sediments were changed from pleisto- to meso- to oligolytic within a span of a million years or so. Site 523, like Site 522, lay close to lysocline during the Plio-Quaternary. Periodic rises and falls of the lysocline led to cyclic sedimentation of light chalk oozes and dark marl oozes.

Site 524

The site located at the foot of Walvis Ridge, on a bench-like feature just above the abyssal plain of the Cape Basin. Bathymetric surveys prior to drilling delineated a submarine canyon, heading NW–SE, winding its way from a saddle at a height of about 3000 m subsea down to the abyssal plain where the water depth is greater than 5000 m. A large submarine fan at 4800–5000 m below sealevel is present at the mouth of the canyon, and Site 524 is located on this fan.

The youngest sediments are early Eocene pelagic oozes belonging to nannofossil Zones NP 11 to 13 or Chrons C-23 and C-24(N). The Paleocene consists of pelagic oozes, chalks, cherty limestones, and numerous turbidites. The upper Cretaceous consists of marly claystones, volcanoclastic sandstones, volcanic breccias, and basalt flows of Maastrichtian age.

Back-tracking suggests that the paleodepth at this site was less than 3500 m before the Eocene. Rapid burial under turbidites further enhanced the preservation of $CaCO_3$. A significant dissolution event took place at the beginning of Tertiary. Calcareous nannofossils and planktonic foraminifers are largely dissolved in the uppermost Cretaceous sediments. There is a remarkable decrease of $CaCO_3$ content (from about 40 to 2%) in the $\frac{1}{2}$-m interval below the Cretaceous/Tertiary contact. One could postulate a gradual rise of lysocline during the late Maastrichtian, resulting at first in the dissolution of planktonic foraminifers and finally of all the calcium carbonate at the Cretaceous/Tertiary boundary. However, the global record has indicated that the terminal Cretaceous event took place without forewarning (see Hsü *et al.*, 1982). We prefer, therefore, to assume seafloor-dissolution of shallow subbottom sediments during the terminal Cretaceous event.

The presence of a much undersaturated bottom water led to the dissolution of the topmost sediment on the ocean bottom and consequently induced a steep gradient of the concentration of dissolved carbonates in the interstitial water. Such a gradient in dissolved silica is known today, and the gradient is responsible for the transfer of dissolved biogenic silica from sediments a few meters below the bottom to the bottom waters of the oceans. The analogy suggests that the dissolution of planktonic foraminifers and the poor preservation of the nannofossils in the uppermost Maastrichtian sediments were the results of grossly undersaturated bottom water during and shortly after the terminal Cretaceous event.

The sequence became mainly pelagic toward the end of the Paleocene. No sediments younger than Early Eocene are present; dissolution may have removed much calcite, but the absence of insoluble residues indicates active erosion by bottom-currents coming down from a canyon from the Walvis Ridge. In Hole 20 drilled into a somewhat younger crust on Anomaly 30 we found condensed Paleocene and Eocene, and a fairly complete Oligocene sequence; the site there only sank below the CCD during the Early Miocene.

The late Cretaceous sedimentation rate was 34 m/m.y. The Paleocene sedimentation rate was similar at 28 m/m.y. These rates are three or four times more than the normal Tertiary pelagic sedimentation rates in the

South Atlantic, and have resulted mainly from the accumulation of turbidites. The Lower Eocene has a sedimentation rate of less than 10 m/ m.y.

Discussion of results

The calcite-compensation depth is the level at which the rate of supply and rate of dissolution of calcareous tests is approximately equal (Bramlette, 1961). If all the calcium carbonate descended from an overlying water column is dissolved, only the insoluble residue is deposited. Therefore, the calcite-compensation surface should mark the upper limit of the red-clay sedimentation on the ocean bottom at that time. However, a sample of pelagic sediment is an accumulative record of oceanographic conditions over a period of thousands, or many thousands of years. Therefore, the facies boundary on the present ocean bottom may deviate from the momentary CCD by a depth difference of 200 m or so (Berger, 1974).

The lysocline is the upper limit of the zone where dissolution rate increases markedly over a short depth interval. The lysocline in the Atlantic lies hundreds of meters above the calcite-compensation surface (Berger, 1974). The foraminiferal lysocline is defined as the upper limit of the boundary zone between well-preserved and poorly preserved foraminiferal assemblages on the seafloor. This lysocline has the properties of a compensation depth in that it can exist independently of the particular shape of a dissolution profile as long as the rate increases with depth. In the central Atlantic, where the lysocline was originally defined, there is evidence for an acceleration in dissolution rate associated with the foraminiferal lysocline (Berger, 1974, 1976).

The study of the dissolution facies at each site permitted us to estimate the levels of past CCDs and lysoclines. Increased dissolution is manifested by increases of (1) insoluble residue (IR); (2) fragmentation of foraminiferal tests (FFT); and (3) benthic foraminifers (BF).

A high percentage of terrigenous insoluble residue results from the dissolution of calcareous skeletal tests. The fragmentation of formaniferal tests finds its expression in grain-size analyses. The degrees of dissolution of a pelagic sediment can, therefore, be defined by two numerical parameters: carbonate percentage (or conversely insoluble-residue percentage) and percents of size fractions greater than 62 μm metres (Hsü & Andrews, 1970). Using these parameters to classify the South Atlantic sediments devoid of siliceous plankton, we found that the eolytic and oligolytic pelagic sediments contain more than 1% sand fraction and less than 30% terrigenous detritus. The mesolytic contains 30–70% terrigenous detritus and the pleistolytic more than 70%, whereas the hololytic is an abyssal clay devoid of calcium carbonate; all these more dissolved pelagic sediments contain less than 1% sand fraction (Violanti *et al.*, 1979). A study by Finger (1984) on Leg 73 cores confirmed the pattern.

Variations of CCD and of Foraminiferal lysoclines

The planktonic foraminiferal faunas are fairly well preserved in the sediments of the eolytic and oligolytic facies, but are much fragmented and poorly preserved in mesolytic and more dissolved sediments. We may thus conclude that the upper limit of the paleodepth of mesolytic facies may define the foraminiferal lysocline; eolytic and oligolytic sediments were deposited at depths above the lysocline.

The hololytic sediments are deposited below the CCD and their paleodepths give an estimate of the minimum CCD. A sample of pelagic clay or pleistolytic sediment from a 1 cm interval represents, however, a span of time of about 10^4 years, during which the CCD may have varied somewhat. A pleistolytic sample, therefore, may include sediments deposited both above and below the CCD, and its paleodepth should represent a good average of the CCD for the time-span represented by the sample.

Using the dissolution data summarized in the site sections, and the back-tracking method to determine the paleodepth of the samples studied, the temporal variations and the depth distribution of the dissolution facies are shown by Fig. 8. Assuming that the lysocline and the CCD are defined by mesolytic and pleistolytic depths respectively, the depth variations of these surfaces with time can be estimated. Also shown by the figure are the paleodepths of the samples yielding the first benthic microfauna containing significant amounts (10–20%) of *Nuttallides umboniferus*.

The temporal variations of the lysocline and of the CCD in the South Atlantic show the same general trend as those observed previously by van Andel *et al.* (1977) and by Melguen (1978). Two broad CCD 'peaks' have been identified corresponding to the very shallow depths of CCD during the Middle Eocene and the Middle Miocene. However, a close comparison, as illustrated by Fig. 9, reveals significant differences. Oddly enough, the Neogene part of our CCD curve closely simulates that of Melguen, with a CCD-'peak' at about 3 km depth some 12 my ago, whereas the Paleogene part of our curve simulates that of van Andel *et al*, with a broad 'valley' (CCD at about 4.3 km) for the Oligocene.

We believe that van Andel *et al.* underestimated the CCD-rise during the Middle Miocene. Their interpretation of the Neogene CCD–variations was based primarily upon Leg 3 data (Sites 13 to 20), and from analyses of Miocene sequences that were not continuously cored. We now have a close sampling from a well-dated and complete sequence, and the high quality of our data justifies a revision. For example, the presence of Middle Miocene hololytic sediments at Site 521 indicates a CCD not shallower than 3.5 km 12 m.y. ago. The presence of mesolytic sediments on newly formed ridge-crest at Site 519 indicates a foraminiferal lysocline as shallow as 2.6 km some 10 m.y. ago. Considering that the CCD is commonly several hundred meters deeper than the lysocline (Berger, 1976), we conclude that the Middle Miocene (11–13 m.y.) CCD should have been about 3 km deep at 30° S latitude and our estimate confirms the conclusion by Melguen (1978).

The upper Miocene Messinian sediments are chalk oozes at Site 16 on the west flank of the Mid-Atlantic Ridge, whereas the sediments of the same age on its east flank deposited on bottom 200 or 300 m deeper are oligolytic or mesolytic marl oozes. This facies difference first suggested to us that the lysocline may have been deeper on the west flank, and this lack of symmetry might explain the discrepancy between our postulate

Fig. 8. Temporal variations of the CCD and lysocline in the South Atlantic. Empty symbols are eolytic and oligolytic sediments, which should lie above the lysocline. Half-solid symbols are mesolytic sediments, which should lie at or near the lysocline depth. Solid symbols are pleistolytic sediments which should lie at or near the CCD. Crosses are hololytic sediments which should lie below CCD. Numbers designate Leg 73 sites.

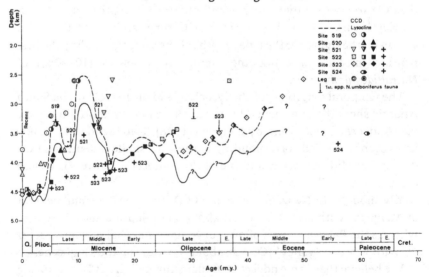

and that by van Andel *et al.* (1977). However, after we plotted the Leg 3 data on our diagram, we noted that the temporal–spatial plots of the Site 16 oligolytic samples lie above our lysocline curve. The facies difference in the Messinian could thus signify that the sediments of this age at Site 16 were deposited just above, and those at Site 519 just below, the lysocline. A mesolytic sediment of Langhian age from Site 15 was deposited at a paleodepth of 3.4 km on the west flank of the ridge, and the data point lies on our lysocline curve for the southeastern Atlantic. This fact suggests that there was little difference in the lysocline depth between the two sides of the ridge.

Fig. 9. A comparison of three different postulates of the temporal variation of CCD in South Atlantic. The top of figure shows temporal variations of $\delta^{18}O$. The van Andel (1977) curve was published in van Andel *et al.* (1977).

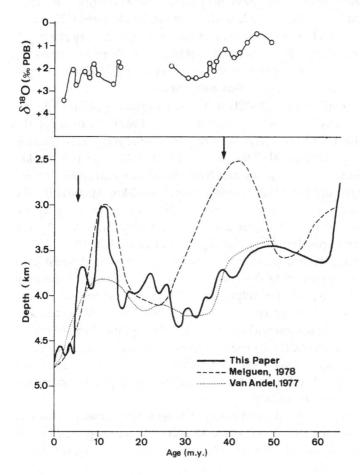

The discrepancy between our estimate and Melguen's estimate of the Paleogene CCD may be a manifestation of a geographical difference. Her conclusions were based on data from Sites 360 and 361 in Cape Basin, and our curve described the CCD-change in the Mid-Atlantic area along the present 30°S latitude. There could have been sufficient paleocean-ographical factors to result in a significant difference of CCD between the two regions. However, we believe that Melguen overestimated the degree of dissolution during the Paleogene. The Cape Basin sites were drilled on a continental margin, where the high clay-content of the Paleogene nannofossil marls and clays may have resulted from a large component of detrital input from land, rather than from dissolution alone.

Whereas we are very confident of the reliability of our CCD-curve for the last 50 my, we are less certain of the CCD-variations during the first 15 my of the Cenozoic. We have no pleistolytic sediments older than Oligocene at Sites 522 and 523, so that the early Paleogene CCD-curve has been constructed on the basis of an assumption that it lay at a depth of about 400 m beneath the foraminiferal lysocline. The close similarity between our estimate and that by van Andel et al. (1977) (on the basis of Leg 73 data) seems to justify this assumption.

The apparently very high CCD at the very beginning of the Cenozoic has been discussed in another paper (Hsü et al., 1982). We question the postulate that the CCD was raised to the photic zone at this time because the boundary clays are absent at a number of DSDP sites (see Supko, Perch-Nielsen et al., 1977). The data from Site 524 suggest an abrupt and temporary rise of CCD at that time to a level above 3 km. Meanwhile, the hololytic boundary clay at various sites on land (Denmark, Italy, Spain, Tunisia) may have resulted from dissolution by very corrosive CO_2-rich waters in an expanded zone of oxygen minimum (Hsü, 1984).

Our detailed record permitted us to recognize several subsidiary 'peaks' and 'valleys' of CCD-variations; we also could prove the very rapid rate of changes. For example, the CCD rose more than 1 km during the first few million years after the start of Middle Miocene, and dropped more than 1 km within the first million years of the Pliocene. The lowering of the CCD after the Middle Eocene 'peak' was, however, a stepwise change, simulating somewhat the oxygen-isotope (or paleotemperature) shift (Fig. 9). There was no sharp plunge across the Eocene–Oligocene boundary.

Aside from the general trend of CCD-changes, very rapid oscillations of CCD during the late Pliocene and Quaternary are suggested by the alternations of dark, marly and light, chalky oozes. The dark sediments are of glacial origin, characterized by signs of more advanced degrees of dissolution (higher insoluble residue, higher fragmentation of foraminiferal tests, higher benthic percentages). The dark–light couplet has a cyclicity of about 10^5 years, and has been related to oscillations of CCD during the late Neogene glaciation by Weissert *et al.* (1984).

The benthic microfaunas could be classed into two major types. After the first appearance of abundant (10–20%) *N. umboniferus*, the benthic fauna is moderately diverse. Aside from this species, which may constitute a maximum of 60% of the benthic assemblage, the fauna includes *Cassidulina subglobosa, Oridorsalis umbonatus, Epistominella exigua, Planulina wuellerstorfi*, and *Pullenia* spp. The assemblage dominated by *N. umboniferus* is similar to that found today on ocean bottom under Antarctic Bottom Water (AABW). Older faunas at Sites 519, 521 and 522, contain little or no *N. umboniferus* and are dominated by *E. exigua* and *P. wuellerstorfi*. Those Miocene faunas of the South Atlantic Deep Water (MSADW) are similar to a modern fauna typical of lower North Atlantic Deep Water (NADW), but we are not certain if MSADW could be considered the Miocene equivalent of NADW. Assuming that the *N. umboniferus* fauna of the Tertiary also lived on ocean bottom under AABW, the first appearance of this fauna at each site should mark the time and paleodepth when the sea bottom at this site subsided to a depth below the top of a deep water mass equivalent to the present AABW.

Fig. 8 shows that the *N. umboniferus* fauna first appeared at every site when it subsided to a depth of about 3.25 to 3.75 km. The species first occurred at Site 523 when it had subsided to 3.5 km depth, but this deeper (or later) first occurrence may have resulted from evolutionary development, because the older benthic faunas at this site were dominated by ancestral forms of the same genus, *N. trümpyi*. If we accept the face-value of the data from Sites 523 and 522, however, we conclude that the top of the AABW rose some 250 m during the first 5 my of the Oligocene because of an intensification of the bottom-current activity. The Oligocene lysocline lay below the top of the AABW, similar to the situation in the South Atlantic today, where the dominance of *N. umboniferus* is noticeable below 3500 m, at a level considerably above the lysocline depth of more than 4 or 4.5 km. On the other hand, the first appearance of the *N. umboniferus* fauna during the Miocene occurred when a site subsided to a depth near the top of the lysocline. It seems that the Miocene lysocline may have followed the boundary between the AABW and NADW as it does in the Central Atlantic today (Berger, 1974).

Cause of CCD variations

The calcite-compensation level is determined by supply and by loss of calcium carbonate through dissolution. The supply is related to the fertility of the calcareous plankton of the open oceans, whereas the loss is influenced to a large extent by the chemistry of the bottom water mass with which a newly deposited sediment is in contact. In our analyses of CCD-changes we see two types of variations: first-order variations involving changes from one geological epoch to another (10^7 year cycles), and second-order variations involving fractions of a million years (10^4 or 10^5 year cycles).

The second-order variations are exemplified by the light–dark couplets in the Plio-Quaternary sediments of the South Atlantic. Weissert et al. (1984) found that foraminifers in the more marly dark sediments have undergone more advanced degrees of dissolution and cited isotope evidence in their conclusion that the dark sediments were deposited during glacial stages. They also found that the AABW, characterized by predominance of an N. umboniferus fauna, was intensified during glacial times, and that this water mass was then enriched in biogenic CO_2 to cause increased dissolution. The increase of dissolution must have prevailed over a possible increase of production in the South Atlantic during the glacial stages so that there was a rise in lysocline and CCD during the deposition of more marly dark sediments.

Our graphical summary of the CCD-change during the Tertiary (Fig. 9), as well as several reconstructions by previous workers (e.g. Hay, 1970; Berger & Roth, 1975; van Andel et al., 1977; Melguen, 1978) show first-order changes. Melguen (1978) attempted to invoke the same mechanism for dissolution events of all kinds, she also related the first-order rises of the CCD to intensified AABW. We found, however, only small differences in the AABW activities during the Cenozoic, if the paleodepth of the first appearance of the N. umboniferus fauna is any indication. In fact, it seems that the AABW first started in late Eocene or Oligocene time and became somewhat intensified during the Plio-Quaternary, when the CCD was depressed; the correlation was opposite to the one postulated by Melguen.

An alternative suggestion emphasizes ocean fertility. Van Andel et al. (1977), for example, related the drop of CCD during the Late Eocene/ Early Oligocene and during Early Pliocene to a sharp increase of plankton productivity. This is a more likely explanation of the first-order changes in CCD during the Cenozoic.

The productivity of calcareous plankton in open oceans depends upon the supply of nutrients, particularly phosphorus and nitrogen, and this supply is controlled by (1) fluxes from the continents and (2) consumptions by competing sinks for nutrients.

The first-order 'highs' for the CCD occurred during the Eocene and Miocene. Berger *et al.* (1981) pointed out that the high stands of CCD are correlated to times of transgression, of warmer temperatures, and of enriched $\delta^{13}C$ in dissolved carbonates of the oceans. Arthur & Jenkyns (1981) further noted a correlation of CCD to the accumulation rate of economic phosphate deposit: the shallow CCD during the Eocene and Miocene corresponded to the epochs of 'phosphatic giants', whereas the deep Oligocene and Plio-Quaternary CCD prevailed at the time when relatively few phosphate deposits were formed.

It is difficult to judge the temporal changes of the fluxes from continents. Whereas more nutrients might have been set free by weathering under warmer conditions, the enlarged land areas during lower stands of the sea-level may have made more source terranes available for nutrient-delivery. A first estimate indicates, however, that the consumption by competing sinks for nutrients was definitely larger during the Eocene and Miocene. Much of the phosphorous was fixed as phosphate deposits on submerged shelves during those epochs of high sea-level stands. Furthermore, the Eocene and Miocene were epochs of very high productivity of siliceous plankton, which were completing with calcareous plankton for nutrients. The Eocene was an epoch when the fertility of radiolarians in the equatorial regions reached its zenith, whereas the Miocene was an epoch when continental margins with upwelling waters became sites of diatomite sedimentation. We suggest, therefore, low productivity of calcareous plankton in the open oceans as the cause of high CCD during the Eocene and Miocene Epochs.

The correlation of CCD to the isotope records or to ocean temperatures is not as straightforward. Fig. 9 shows the record of oxygen-isotope changes in global oceans. The $\delta^{18}O$ values decreased stepwise during the Paleogene and reached a minimum in early Oligocene. The trend was reversed until Middle Miocene, when another positive trend began. Briefly reversed again at the beginning of the Pliocene, the $\delta^{18}O$ values continued the overall positive trend with glacial and interglacial fluctuations till the present. The record is one of stepwise cooling and of successive expansions of polar glaciers during the Cenozoic.

The CCD stood high during the Eocene (warmer than Oligocene) and Miocene (warmer than Plio-Quaternary). Those epochs were also times of transgression. We could interpret the oxygen-isotope and the CCD records in terms of global temperatures, ice volume, phosphate formation, and nutrient supply. The late Eocene was a time of rapid cooling, and the Antarctic Ice Cap may have started to grow in early Oligocene. The change of bottom circulation consequent upon the climatic change apparently resulted in high productivity of calcareous plankton and depression of CCD in the beginning of the Oligocene Epoch. Conversely, the warming trend in late Oligocene or Early Miocene may have had the opposite effect. Yet a closer look at the record reveals a more complex pattern (cf. Fig. 9 and 10); the CCD continued to rise during the Middle Miocene when the climate again turned colder and the polar ice volume increased. It seems that the intensified Antarctic glaciation, in this case, caused a rise, not a depression, of the CCD. We have to postulate competition for limited nutrients because of increased productivity in coastal waters. Vincent & Berger (1984) noted a significant shift in carbon-isotope composition of the foraminifers toward the end of Early Miocene, during Epoch 16 (Chron C-5C) some

Fig. 10. A comparison of temporal variations of oxygen-isotope trends in three oceans. Leg 73 results are compared to those of Southern Oceans by Shackleton & Kennett (1975), and those of Central Pacific by Savin et al. (1975).

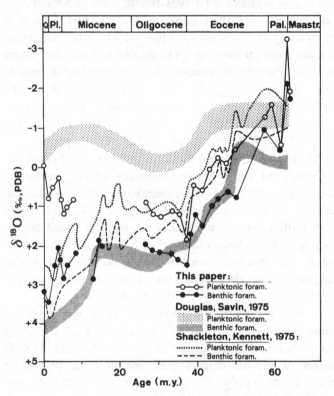

16.5 my ago. They related this shift to the rapid deposition of organic carbon in coastal upwelling regions of high productivity. The carbon-isotope signal is a manifestation of the preferential burial of this carbon which is enriched in ^{12}C. Another consequence of the carbon extraction was to deplete the dissolved carbon dioxide in the oceans and to reduce the partial pressure of CO_2 in the atmosphere. This in turn should have weakened the CO_2 'greenhouse-effect' and caused further cooling and still more expansion of the Antarctic Ice Cap. A corollary of the Vincent–Berger hypothesis is that nutrients were also rapidly extracted and buried with the organic carbon. The depletion of the nutrients caused a reduction of fertility of calcareous plankton in open oceans and resulted in a continued rise of CCD even though the climatic trend reversed itself from Early Miocene warming to Middle Miocene cooling.

We can use the same line of reasoning to explain the sharp drop of CCD during the Pliocene, when sea-level rose and Antarctic glaciers retreated. A warming trend at the beginning of Pliocene should have decreased the latitudinal temperature gradient, weakened the oceanic thermocline structure, and reduced coastal upwelling. The reduction of fertility in nearshore areas released more nutrients for plankton production in open oceans, and caused the lowering of CCD.

The CCD varied little during the Plio-Quaternary, when cooling continued and polar ice caps expanded. The effect of increased productivity during the glacial stages may have been compensated, at least in part, by the more corrosive bottom waters of those ages.

To conclude, the climate has no direct effect on CCD, but the paleoceanographical consequences of climatic changes did cause CCD changes. While the climate continued to deteriorate during the Cenozoic, the calcite-compensation depth rose and fell periodically.

The correlation of the CCD and the carbon-isotope records are complicated by many factors (cf. Figs 9 and 11). Organic productivity in the oceans produced a fractionation of carbon isotopes in ocean waters: carbon-13 is enriched in surface water (Broecker, 1974). Our records indeed show such a 1 to 2‰ difference between the $\delta^{13}C$ values of the planktonic and benthic foraminifers. The overall trend of this fraction-ation has, however, not changed much during the Neogene despite the big differences in CCD between the Late Miocene and the Plio-Quaternary oceans. Factors other than plankton-production rates must have influenced the apparent fractionation. We note, for example, a very large difference in the $\delta^{13}C$ values (3‰) between the planktonic and benthic foraminifers of the Eocene. Using the Recent case as a guide, one might conclude that the carbon-isotope fractionation indicated

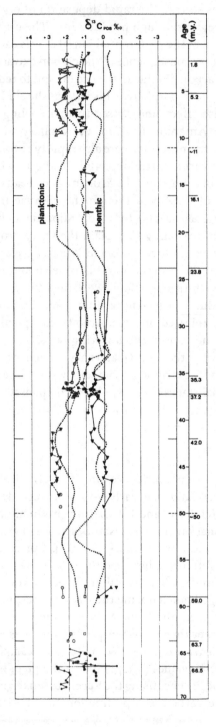

Fig. 11. A comparison of temporal variations of the carbon-isotope trends. Selected values of the South Atlantic are shown by solid (benthic) and empty (planktonic) symbols. The curves for the Southern Oceans are shown by dotted lines, based upon three-point sliding averages of the data points in Shackleton & Kennett (1975)

greater biologic fertility in surface waters. This might indeed be correct as the common and widespread occurrences of Eocene radiolarian oozes bear witness to this fertility. On the other hand, fractionation of carbon isotopes is also related to turnover time of the oceans; sluggish mixing should accentuate the difference between the compositions of the surface and bottom waters. We could thus alternatively consider the Eocene carbon-isotope record a sign of the relatively long mixing time of the ocean waters; the sudden decrease in the carbon-isotope fractionation at the beginning of the Oligocene should thus indicate a change to more vigorous oceanic circulations consequent upon the expansion of the Antarctic Ice Cap.

Conclusions

Our study of the calcite-dissolution record of the Cenozoic sediments of the southeastern Atlantic indicates that the calcite-compensation depth rose and fell between 3000 m and 5000 m depth during the Cenozoic. The CCD reached higher levels during the Eocene and Miocene and lower levels during the Oligocene and Plio-Quaternary. We find, however, no simple correlation of the CCD record to the oxygen- and carbon-isotope records. The preservation of calcareous fossils on ocean bottom is determined by many factors, including climate, ocean temperatures, ocean circulations, nutrient supplies, regional variations in productivity (ocean/ocean, ocean/shelf fractionations), etc. Long-term variations from epoch to epoch are related to productivity of calcareous plankton in open oceans, but short-term fluctuations apparently resulted from differential dissolution, in response to the changing chemistry of bottom waters during glacial and interglacial stages.

Acknowledgements

The writers are indebted to many of their colleagues, particularly Judy McKenzie, Helmut Weissert, Hedi Oberhänsli, Kerry Kelts, and Murlene Clark for their cooperation. We have relied very much upon the stratigraphic framework and the paleoceanographic data published in the Leg 73 cruise report, and we are grateful to our shipboard colleagues for making this information available to us prior to its publication. This paper is contribution No. 245 of the Laboratory of Experimental Geology, ETH, Zürich.

REFERENCES

Arthur, M.A. & Jenkyns, H.C. 1981. Phosphorites and paleoceanography, Oceanol. Acta, Special Publication, pp. 83–96.

Berger, W.H. 1974. Dissolution of deep-sea carbonates: an introduction, Cushman Found. Foram. Res., Spec. Publ., 13, 7–10.

Berger, W.H. 1976. Biogenous deep-sea sediments: production, preservation and interpretation. In: J.P. Riley & R. Chester (eds.), Chemical Oceanography vol. 5, 2nd edition, Academic Press, London, pp. 265–388.

Berger, W.H. & Ruth, P.H. 1975. Oceanic micropaleontology: progress and prospects. Rev. Geophys. Space Phys., 13, 561–636.

Berger, W.H. & Winterer, E.L. 1974. Plate stratigraphy and the fluctuating carbonate line. In: K.J. Hsü & H.C. Jenkyns (eds), Pelagic sediments on land and under the sea, Int. Ass. Sedimentol., Spec. Publ., 1, pp. 11–48.

Berger, W.H., Vincent, E. & Thierstein, H.R. 1981. The deep-sea record: major steps in Cenozoic ocean evolution. Soc. Econ. Pal. Min., Spec. Publ., 32, pp. 489–504.

Bolli, H.M., Ryan, W.B.F. et al. 1978. Angola continental margin sites 364 and 365. Initial Repts. of DSDP, 40, 357–455.

Bramlette, M.N. 1961. Pelagic sediments. In: M. Sears (ed.), Oceanography. Am. Assoc. Adv. Sci. Pub., vol. 67, pp. 345–366.

Broecker, W.S. 1974. Chemical Oceanography. Harcourt & Brace Jovanovich, New York, 214 pp.

Douglas, R.G. & Savin, S.M. 1975. Oxygen and carbon isotope analyses of Tertiary and Cretaceous microfossils from Shatsky Rise and other sites in the North Pacific Ocean. In: R.L. Larson, R. Moberly, et al., Initial Repts. of DSDP, 32, 509–520.

Dreyfus, M. & Ryan, W.B.F. 1972. Geochronology of late Neogene boundaries. In: Symposium on late Neogene epoch boundaries, International Geological Congress, Montreal, unpublished report cited by Ryan and others, 1974.

Finger, W. 1984. Calcite dissolution facies. In: K.J. Hsü, J.L. La Brecque, et al., Initial Repts. of DSDP, 73, 765–776.

Foster, J.H. & Opdyke, N.D. 1970. Upper Miocene to Recent magnetic stratigraphy in deep-sea sediments. J. Geophys. Res., 75, 4464–4473.

Gombos, A.M. 1984. Late Neogene diatoms and diatom oozes in the central South Atlantic. In: K.J. Hsü, J.L. La Brecque, et al., Initial Repts. of DSDP, 73, 487–494.

Hay, W.W. 1970. Calcium-carbonate compensation. In: R.G. Bader, et al., Initial Repts. of DSDP, 4, 672–673.

Herbin, J.P. & Deroo, G. 1984. Organic geochemistry of Micoene–Pliocene laminated diatomites on the eastern flank of the Mid-Atlantic Ridge (Site 520, Leg 73). In: K.J. Hsü, J.L. La Brecque, et al., Initial Repts. of DSDP, 73, 537–538.

Hsü, K.J. 1984. A scenario for terminal Cretaceous Event. In: K.J. Hsü, J.L. La Brecque, et al., Initial Repts. of DSDP, 73, 755–764.

Hsü, K.J. & Andrews, J.E. 1970. Lithology. In: A.E. Maxwell, et al., Initial Repts. of DSDP, 3, 445–453.

Hsü, K.J. He, Q., McKenzie, J.A. et al. 1982. Mass mortality and its environmental and evolutionary consequences. Science, 216, 249–256.

Hsü, K.J., La Brecque, J.L. et al. 1984. Initial Repts. of DSDP, 73, 789 pp.

La Brecque, J.L., Kent, D.V. & Cande, S.C. 1977. Revised magnetic polarity time scale for late Cretaceous and Cenozoic time. Geology, 5, 330–335.

Maxwell, A.E. et al. 1970. Initial Repts. of DSDP, 3, 806 pp.

Melguen, M. 1978. Facies evolution, carbonate dissolution cycles in sediments of the eastern South Atlantic (DSDP 40) since the Early Cretaceous. In: H.M. Bolli, W.B.F. Ryan, et al., Initial Repts. of DSDP, 40, 981–1024.

Savin, S.M., Douglas, R.G. & Stehli, F.G. 1975. Tertiary marine paleotemperatures. Geol. Soc. Am. Bull., 86, 1499–1510.

Schrader, H.-J. 1978. Diatoms in Leg 41 sites. In: Y. Lancelot, E. Seibold, *et al.*, *Initial Repts. of DSDP*, **41**, 791–812.

Sclater, J.G., Anderson, R.N. & Bell, M.L. 1971. The elevation of ridges and the evolution of the Central Eastern Pacific. *J. Geophys. Res.*, **76**, 7883–7915.

Shackleton, N.J. & Kennett, J.P. 1975. Paleotemperature history of the Cenozoic and initiation of the Antarctic glaciation: Oxygen and carbon analyses in DSDP Sites 277, 279, 281. In: J.P. Kennett, R.E. Houtz, *et al.*, *Initial Repts. of DSDP*, **29**, 743–755.

Shackleton, N.J. & Opdyke, N.D. 1977. Oxygen isotope and paleomagnetic evidence for early Northern Hemisphere glaciation. *Nature*, **270**, 216–219.

Supko, P.R., Perch-Nielsen, K.A. *et al.* 1977. *Initial Repts. of DSDP*, **39**, 1139 pp.

Theyer, F. & Hammond, S.R. 1974. Cenozoic magnetic time scale in deep-sea cores: completion of the Neogene. *Geology*, **2**, 487–492.

van Andel, T.H., Thiede, J., Sclater, J.G. & Hay, W.W. 1977. Depositional history of the South Atlantic ocean during the last 125 million years. *J. Geol.*, **85**, 651–698.

Vincent, E. & Berger, W.H. 1984. CO_2 and Antarctic ice-buildup in the Miocene: time, Monterey Hypothesis. In: E. Sundquist (ed.), Chapman Conference on Natural Variations of CO_2, Tarpon Springs, Florida, Proc., US Geol. Survey, Reston, Virginia.

Violanti, D., Premoli-Silva, I., Cita, M.B., Kersey, D. & Hsü, K.J. 1979. Quantitative characterization of carbonate dissolution facies on the Atlantic Tertiary sediments – an attempt. *Riv. Ital. Paleont.*, **85**, 517–548.

von Salis, A.K. 1984. Miocene calcareous nannofossil biostratigraphy of DSDP Hole 521A, SE Atlantic. In: K.J. Hsü, J.L. La Brecque, *et al.*, *Initial Repts. of DSDP*, **73**, 425–428.

Weissert, H., McKenzie, J.A., Wright, R., Clark, M., Oberhänsli, H. and Casey, M. 1984. Paleoclimatic record of the Pliocene at DSDP sites 519, 521, 522, and 523 (Central South Atlantic). In: K.J. Hsü, J.L. La Brecque, *et al.*, *Initial Repts. of DSDP*, **73** 701–716.

11

Cenozoic carbon-isotope record in South Atlantic sediments

K. J. HSÜ, J. A. McKENZIE, and H. J. WEISSERT

Introduction

The carbon-isotope composition of dissolved carbonate in marine water is primarily controlled by the photosynthesis–respiration cycle. Carbonate precipitates of marine organisms are close to isotopic equilibrium with the dissolved carbonate, and monitor the carbon-isotope signature of the ambient water mass. Changes in global carbon budget, in biogenic productivity, or in biomass on land therefore are, among other factors, reflected in the carbon-isotope record of marine calcareous fossils. An excellent record of carbon-isotope variations during the Cenozoic has been obtained from South Atlantic sediments cored during DSDP Legs 73 and 74. The data are published in articles by Hsü *et al.* (1984), Poore & Matthews (1984), Shackleton & Hall (1984), Shackleton *et al.* (1984), McKenzie & Oberhänsli (this volume, chapter 8), and Oberhänsli & Toumarkine (this volume, chapter 9). In this summary we present composite diagrams of the carbon-isotope records of Leg 74 (Shackleton *et al.*, 1984) and of Leg 73 (Hsü *et al.*, 1984) illustrated graphically by Figs. 1 and 2.

Cenozoic carbon-isotope record

The benthic and planktic carbon-isotope curves show parallel trends during wide parts of the Cenozoic. Notable variations during the Eocene and early Oligocene resulted in significant steepening or flattening of the carbon-isotope gradient in these epochs. The data from the benthic foraminifers provide an indication of the trend of total dissolved carbonate because the values of the planktic skeletons are much influenced by biological productivity.

Previous studies of samples from the southern oceans and from the central Pacific have established two broad peaks of carbon isotopes since the beginning of the Eocene (Savin *et al.*, 1975, Shackleton & Kennett, 1975). An earlier and a still more prominent peak in middle Paleocene was suggested by a few analyses of Pacific cores published by Douglas & Savin (1975). Shackleton & Hall (1984) in their detailed analysis of Leg 74 sediments confirmed a $\delta^{13}C$ maximum in Paleocene sediments. Oberhänsli & Toumarkine (this volume, chapter 9) found the same Paleocene peak through their study of Leg 73 cores. They could assign an age of foraminiferal Zone P 4 and nannofossil Zone NP 9 to the positive peak. The possible cause of the Paleocene anomaly has been discussed by Shackleton & Hall (1984) and by Oberhänsli & Toumarkine (this volume).

The carbon-isotope peak of Eocene age shows two maxima during early and late Eocene, at about 53 and 42 m.y., respectively. Shackleton & Hall's (1984) record is similar, but their information on the middle

Fig. 1. Carbon-isotope record of benthic foraminifers (upper curve) and planktic foraminifers from DSDP Sites 525–529. Modified after Shackleton *et al.* (1984).

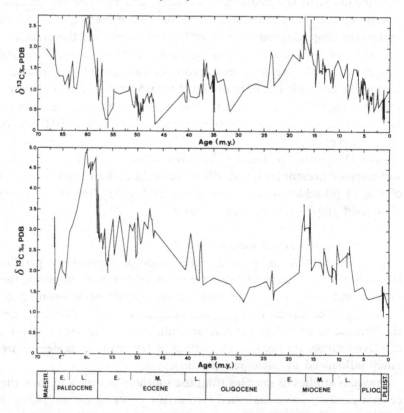

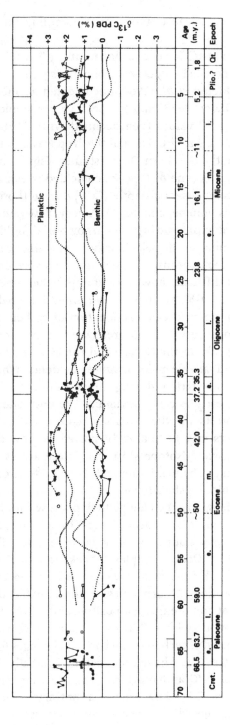

Fig. 2. Carbon-isotope record from DSDP Sites 519–524. Solid symbols denote benthic foraminifers, empty symbols planktic foraminifers. The curves for the southern oceans are shown by heavy dotted lines, which are based upon three-point sliding averages of the data in Shackleton & Kennett (1975). Figure adapted from Hsü *et al.* (1984).

Eocene is less perfect, because of inadequate sampling during the DSDP Leg 74 cruise.

Intense dissolution of Miocene carbonate at Leg 73 sites did not allow us to establish a detailed Miocene carbon-13 isotope record. On the other hand, Shackleton et al. (1984) documented an increase of $\delta^{13}C$ values from late Oligocene (33 m.y.) to peak values in middle Miocene (17–15 m.y.). The $\delta^{13}C$ values then declined with short-term fluctuations from about $+2.5$ to $+0.5$‰ and the drop was slightly more than that registered by the Southern Ocean studies (Figs. 1 and 2).

The Miocene and Eocene $\delta^{13}C$ peaks coincided in timing with high stands of calcite-compensation depth, (CCD) during those epochs (see Hsü & Wright, this volume, chapter 6). A very slow rate of carbonate deposition, coupled with rapid organic carbon burial could well explain the Miocene maximum, as the pelagic sediments of this age were largely marls or clays in open oceans, and diatomaceous ooze on continental margins. It seems difficult, however, to apply the same rationale to explain the Eocene $\delta^{13}C$ maximum. We should probably consider changes in the continent/ocean fractionation of carbon.

Land plants constitute a major reservoir of preferentially light carbon; the $\delta^{13}C$ peaks in the Eocene and the Miocene might thus be explained by the available paleobotanical evidence which indicates extensive global forestation during those times. The $\delta^{13}C$ values dropped after the early Eocene maximum. We postulate a significant destruction of forests on the Antarctic continent to explain this anomaly, as glaciers may have already started to grow then.

A remarkable carbon-isotope shift took place across the Eocene–Oligocene boundary. The maximum shift is $+1.5$‰ (from about 0 to 1.5‰) in the benthic and slightly more than $+1$‰ (1.2 to 2.3‰) in the planktic foraminifers, and the change took place within half a million years (Oberhänsli et al., 1984). Carbon shifts of similar magnitude across the Eocene–Oligocene boundary have been reported from investigations of Southern Ocean samples (Shackleton & Kennett, 1975; Keigwin, 1980; Shackleton et al., 1984). We have no evidence of a preferential dissolution of carbonates, or preferential deposition of organic carbon that could explain a positive shift. In fact, the isotope shift was synchronous with a CCD depression which resulted in an accelerated deposition of carbonate carbon. The changes took place in a very short time interval and have been recorded globally. We interpret this enrichment of $\delta^{13}C$ as having resulted from an increase of the biomass on land. The Oligocene trend was contrary to the trend of the carbon-isotope changes during the Quaternary, when a advancing glacier led to destruction of forests and thus an enrichment of ^{12}C, not ^{13}C in the dissolved carbonates of ocean waters (Shackleton, 1977). However, if we interpret the Eocene/Oligocene oxygen-isotope signal as a manifestation of increased ice-volume, corresponding to a drop of sea-level by about 100 m or more, the consequent exposures of the marginal shelves could lead to an enlargement of land areas for plant growth. Such a scenario could thus explain both the carbon and oxygen shifts across the Eocene–Oligocene boundary. It should be pointed out that the Quaternary glaciers of the Northern Hemisphere covered large tracts of land in the northern temperate climatic zone where forests grew during the interglacial stages. The Quaternary glaciation also brought about a change from forest to steppe vegetation, thus decreasing the total biomass on land. If the terminal Eocene glaciers grew mainly in the Antarctic, where there had been little forest vegetation, and if the early Oligocene coastal plains were largely sites of forest growth, one could imagine that the Oligocene and the Quaternary decreases of sea-level may have had exactly the opposite effects on the δ^{13}C values of dissolved carbonates in the oceans.

Poore & Matthews (1984) noted that a negative shift in the δ^{13}C value of benthic foraminfers at Site 522 took place at about 30 m.y. ago. They interpreted this change as a local phenomenon, occurring at a time when the seafloor there subsided from an intermediate water mass into a more corrosive bottom water mass enriched in ^{12}C. Yet, Shackleton's curve also showed a minimum in late Oligocene (about 33 m.y.). We can offer no adequate explanation for the carbon isotope minimum. A quantitative evaluation of the forested areas during that time could yield the revealing clues.

The amelioration of climate toward the Miocene is a well known fact. Coral reefs were extended to high latitudes, and rain forests grew in East Africa and western Asia, where steppe or desert landscape now predominate. The increase of δ^{13}C values from Oligocene to middle Miocene could be interpreted in terms of an increase of terrestrial biomass. A severe global cooling started in the middle Miocene (15 m.y.) and has continued, with short-term fluctuations, till today. The paleontological record indicated a deforestation of East Africa, and a change toward desert in western and southern Asia. The change was probably linked to the severance of connection between the Indian and Atlantic Oceans during the middle Miocene (Hsü et al., 1977). Instead of a maritime climate, the warm and dry climate on the west side of the continent became the characteristic of the Mediterranean. Fossil primates adapted to rain forest habitats are found in the deserts of

Pakistan and Iran today. We believe, therefore, the decrease of 1.5 to 2.0 per mill of ^{13}C value since the middle Miocene is a manifestation of the overall trend of biomass destruction on land. The short-term fluctuations are, on the other hand, related to alternations of glacial and interglacial climates and have been adequately explained by Shackleton et al. (1984).

Carbon-isotope gradient

Our results revealed a striking parallelism between the benthic and planktic carbon-isotope trends for the Oligocene and the Neogene times but a divergence for the Eocene, as Shackleton & Kennett (1975) noted previously. Furthermore, there seems to be a convergence of carbon-isotope values (of the bottom and surface faunas) during the terminal Cretaceous and the terminal Eocene crises.

The Plio-Quaternary parallelism between the carbon-isotope trends of the planktic and benthic foraminifera took place during the time when the oxygen-isotope trends of the two diverged, whereas the Eocene divergence of the carbon-isotope trends is synchronous to the parallelism of the oxygen-isotope trends. At the time of crisis, however, both the carbon and oxygen values seem to converge.

The difference in the $\delta^{13}C$ values of the planktic and benthic foraminifers is attributed to fractionation of carbon isotopes by planktic organisms in surface waters.

High fertility leads to a fractionation so that dissolved carbonates in ocean surface waters become impoverished in ^{12}C or enriched in ^{13}C (Broecker, 1982). This difference between the surface and bottom waters is reflected by the difference of isotopic composition of the foraminiferal shells: the benthic and planktic species in the same sediment may exhibit a $\delta^{13}C$ difference, ranging from about 1 to 2 per mil. The Oligocene and Neogene samples from our sites show such a difference, attesting to a normal steady-state fractionation by planktic productivity. After the terminal Cretaceous event, however, there was a mass mortality (Hsü et al., 1982); the plankton-fertility of the oceans may have been greatly suppressed for a period comparable in duration to the overturn time of the ocean waters (10^3 years). This results in the anomaly that the earliest Tertiary planktic foraminifers have $\delta^{13}C$ values equal to or less than those of benthic foraminifers (Hsü & McKenzie, 1984).

An unusually large difference of 2.5 to 3 per mil in $\delta^{13}C$ values exists between the planktic and benthic foraminifers in Middle Eocene sediments. The high CCD argues against an unusually high production rate of calcareous plankton, but the abundant Middle Eocene cherts suggest a fertile ocean for siliceous plankton. Aside from the influences of productivity, the rate of oceanic mixing must also be considered. A sluggish ocean with an overturn rate twice as long as the present may be more stratified and may allow an isotope fractionation twice as advanced as it is today. We may thus explain the large differences in vertical gradient of carbon isotopes in the Eocene oceans in terms of sluggish mixing. During the Oligocene and Neogene epochs, vigorous circulations of Antarctic Bottom Water and North Atlantic Deep Water contributed to accelerate the turnover time of ocean waters.

Conclusions

Two drilling cruises to the South Atlantic have provided us with a detailed record of the Cenozoic variations of carbonate isotopes in ocean sediments. This is a preliminary attempt to relate the record to well known global paleoclimatic trends. We have not been able to make quantitative evaluations of changes of land and oceanic biomasses and their effect on carbon isotopes. Such a computer-modelling project could only be accomplished after sufficient paleogeographic data are collected.

Acknowledgements

This paper is contribution No. 247 of the Laboratory of Experimental Geology, ETH, Zürich.

REFERENCES

Broecker, W.S. 1982. Glacial to interglacial changes in ocean chemistry. *Progr. Oceanogr.*, 11, 151–197.

Douglas, R.G. & Savin, S.M. 1975. Oxygen and carbon isotope analysis of Tertiary and Cretaceous microfossils from Skatsky Rise and other sites in the North Pacific Ocean. *Init. Repts. DSDP*, 32, 509–520.

Hsü, K.J. & McKenzie, J.A. 1985. A 'strangelove' ocean in the earliest Tertiary. In: Sundquist, E. & Broecker, W.S. (eds.), The carbon cycle and atmosphere CO$_2$: Natural variations Archean to Present. *Geophys. Monograph*, 32, 487–492.

Hsü, K.J., Montadert, L., Bernoulli, D., Cita, M.B., Erickson, A., Garrison, R.B., Kidd, R.B., Mélières, F., Müller, C. & Wright, R. 1977. History of Mediterranean Salinity Crisis. *Nature*, 267, 399–403.

Hsü, K.J. *et al.* 1982. Mass mortality and its environmental and evolutionary consequences. *Science*, 216, 249–256.

Hsü, K.J., McKenzie, J.A., Oberhänsli, H., Weissert, H. & Wright, R.C. 1984. South Atlantic Cenozoic paleoceanography. *Init. Repts. DSDP*, 73, 771–785.

Keigwin, L.D. 1980. Oxygen and carbon isotope analyses from Eocene/Oligocene boundary at DSDP site 277. *Nature*, 287, 722–728.

Oberhänsli, H., McKenzie, J.A., Toumarkine, M. & Weissert, H. 1984. A paleoclimatic and paleoceanographic record of the Paleogene in the Central South

Atlantic. (Leg 73, Sites 522, 523, 524). *Init. Repts. DSDP*, **73**, 737–748.

Poore, R.Z. & Matthews, R.K. 1984. Late Eocene–Oligocene oxygen and carbon isotope record from the South Atlantic DSDP Site 522. *Init. Repts. DSDP*, **73**, 725–736.

Savin, S.M., Douglas, R.G. & Stehli, F.G. 1975. Tertiary marine paleotemperatures. *Bull. Geol. Soc. Am.*, **86**, 1499–1510.

Shackleton, N.J. 1977. Carbon-13 in Uvigerina: Tropical rainforest history and the Equatorial Pacific carbonate dissolution cycles. In: Anderson, N.R. & Malakoff, A. (eds.) The fate of fossil fuel CO_2 in the ocean, *Mar. Sci.*, **6**, 401–427.

Shackleton, N.J. & Hall, M.A. 1984. Carbon isotope data from Leg 74 sediments. *Init. Repts. DSDP*, **74**, 613–620.

Shackleton, N.J. & Kennett, J.P. 1975. Paleotemperature history of the Cenozoic and the initiation of Antarctic glaciation: Oxygen and carbon isotope analyses in DSDP Sites 277, 279 and 281. *Init. Repts. DSDP*, **29**, 743–755.

Shackleton, N.J., Hall, M.A. & Boersma, A. 1984. Oxygen and carbon isotope data from Leg 74 foraminifers. *Init. Repts. DSDP*, **74**, 599–612.

12

Hiatuses in Mesozoic and Cenozoic sediments of the Zaire (Congo) continental shelf, slope, and deep-sea fan

J. H. F. JANSEN

Abstract

Sedimentation-rate curves for the Zaire deep-sea fan show that 70% of its sediment mass had already been supplied during the Late Cretaceous, although the Cenozoic sedimentation was also relatively strong. Hiatuses in shelf and slope sediments are correlated with African denudation surfaces and South Atlantic deep-sea hiatuses. They coincide with the worldwide low sea-levels assumed by Vail *et al.* (1977), but are at least partly of tectonic origin.

Introduction

The Zaire or Congo river drainage area is one of the largest sediment sources of the eastern South Atlantic Ocean. Its sediment supply has built a deep-sea fan extending more than 1000 km westward from the continental shelf into the Angola Basin. Two expeditions, in 1978 and 1980, by the Netherlands Institute for Sea Research, were undertaken to study the sediment transport processes and sedimentation patterns of the Zaire deep-sea fan. Many high resolution seismic recordings were collected. A number of multichannel seismic reflection profiles were put at our disposal by Shell Exploration, in The Hague. This paper shows some preliminary results of the seismic investigations (Fig. 1).

Mesozoic and Cenozoic stratigraphy

The maximum age of the Zaire deep-sea fan is determined by the initial opening of the South Atlantic Ocean, where the oldest oceanic crust, identified near South Africa, is of Early Cretaceous age or 130 million years old (Larson & Ladd, 1973; Rabinowitz, 1976; Rabinowitz & La Brecque, 1976). Opening shifted to the north and at the latitude of the Zaire fan the spreading of the ocean floor started a few million years later (Ojeda, 1982). In the young ocean north of the Walvis Ridge and Rio Grande Rise it was followed by important salt accumulation during the Middle Aptian, which later resulted in extensive zones of diapiric structures along the continental margins of Angola, Zaire, Congo, and Gabon in the east, and Brazil in the west. Recent reviews of the development of the South Atlantic Ocean are given by Rabinowitz & La Brecque (1979) and Reyment (1980).

In the seismic profiles from two *Tyro* cruises in 1978 and 1980, diapirs form the basement for a sequence of oceanic sediments in which several major widespread horizons occur. Several Shell seismic profiles have been published by Beck & Lehner (1974: Figs. 6 and 8) and Lehner & de Ruiter (1977: Fig. 16). Two other profiles enabled us to determine the compressional wave velocities for various units of the fan area (Fig. 2). These velocities can be correlated with velocities from a number of Deep Sea Drilling Project (DSDP) holes in the South Atlantic Ocean, and with

Fig. 1. Locations of seismic record (A–A') across the Gabon continental shelf (Fig. 3) and two multichannel profiles of the Zaire deep-sea fan (B–B' and C–C').

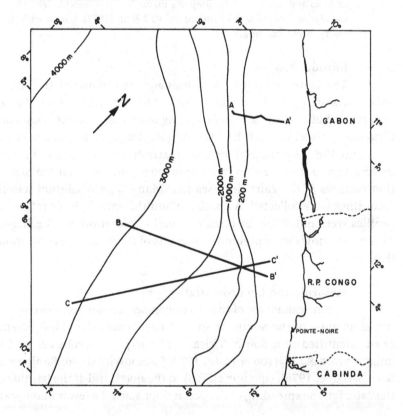

transmission velocities measured by Leyden *et al.* (1972) along the margin from Angola to Gabon. In the southern Angola Basin they include Site 530 (preliminary results obtained from the JOIDES Office, see also Hay, Sibuet *et al.*, 1981) and Site 364 (Bolli, Ryan *et al.*, 1978); at the Saõ Paulo Plateau and the Rio Grande Rise they include the Sites 356 and 357 (Perch-Nielsen, Supko *et al.*, 1975a,b).

The velocities below the salt (5.2–5.7 km/s) are distinctive of ocean Layer 2, and indicate deep-sea pillow basalts (Nairn & Stehli, 1973; Ewing & Houtz, 1979; O'Connell, 1979; Talwani & Langseth, 1981). In addition, the continuation of the profiles towards the Congo-Gabon shelf provides a perfect fit with the Cretaceous and Tertiary layers there (Jansen *et al.*, 1984: Fig. 3), thus corroborating the stratigraphic interpretation. Another check is provided by the lithology of DSDP holes. For the entire post-salt sequence, the sedimentary facies of the sites 364 and 530 are both correlated with more shallow facies at the Congo–Gabon shelf and mainland, as summarized by Jansen *et al.* (1984).

Densities and sedimentation rates

The good lithological correlation and the similarity of compressional wave velocities indicate that we can reasonably apply

Fig. 2. Compressional wave velocities of the seismic units of the deep sea, continental slope, and continental shelf off R.P. Congo and Zaire, as calculated from multichannel profiles B–B' and C–C'. For locations see Fig. 1.

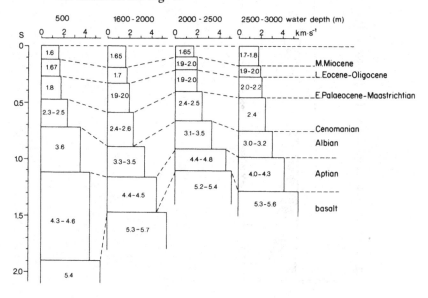

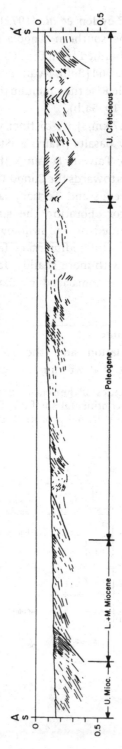

Fig. 3. Seismic reflection record (A–A') across the continental shelf off Gabon (Jansen et al., 1984). For location see Fig. 1.

the densities measured on the DSDP samples to the associated Zaire fan seismic units (Fig. 4). Sedimentation rates were calculated by using these densities, travel times, and seismic wave velocities. In accordance with the interpretations of the DSDP Shipboard Scientific Parties and geologic time scales by van Hinte (1976) and Lowrie & Alvarez (1981) we selected the following ages: top of the Aptian evaporities – 108 m.y., Cenomanian hiatus – 95 m.y., Cretaceous–Tertiary boundary – 66 m.y., Early Oligocene hiatus – 32 m.y., and Miocene discontinuity – 13 m.y. (Fig 4). The development of sedimentation rates in the course of time shows that the greatest interval of deposition (40% of the total) dates to the Albian, the first period after the opening of the ocean. The greater part, or 70%, of the sediments in the Zaire fan area were already supplied before the Tertiary boundary, 66 million years ago. This is not surprising because ocean opening will clearly lead to a reorganization of the fluvial drainage patterns on the continent and result in vigorous new incisions and strong erosion.

Comparison of formation thicknesses with those in nearby profiles from outside the reach of the fan (Uchupi & Emery, 1972, 1974; Emery *et al.*, 1975) leads to the conclusion that during both the Cretaceous and Cenozoic, sedimentation in the Zaire fan was relatively rapid. Seismic velocities were measured on profiles positioned just alongside a lobe of massive recent turbidite sedimentation (Jansen, *et al.* 1984). Here Late Tertiary sedimentation rates $(2–4\,\mathrm{g.cm^{-2}.(10^3)y^{-1}}$ compare well with Quaternary values.

Fig. 4. Calculated Mesozoic and Cenozoic sedimentation rates and estimated sediment densities ($\mathrm{g.cm^{-3}}$) of the Zaire fan and slope (500–3200 m water depth).

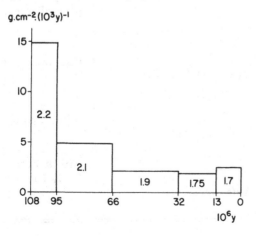

Discussion

Ocean-wide intervals with common hiatuses are known to occur in the Cenomanian, at the Cretaceous–Tertiary boundary, in the Late Eocene–Oligocene, and in the Late Miocene epochs (Moore et al., 1978). The latter three intervals are visible as regression levels at the Congo–Gabon shelf, and possibly, as continent-wide denudation surfaces in Africa (Jansen et al., 1984). Vail et al. (1977) attributed these intervals to low global sea-levels. At the shelf, however, the hiatuses separate seismic units of different tectonic character in west equatorial Africa (Fig. 3). The Upper Cretaceous unit is built up by relatively intensively folded strata while the Paleogene unit is only gently folded. There is also a shift of 20° in the direction of the fold axes. The Lower and Middle Miocene unit is monoclinal and dips oceanwards (c.2°) like the Upper Miocene one (c.4°). These tectonic differences are the result of Late Cretaceous tensional activity, and a subsequent Early Paleocene compression which attained its climax during the Middle Eocene and which gave way again to a tensional regime during the Miocene (Jansen et al., 1984). The geomorphological development of Africa shows several continent-wide denudation surfaces being formed since the Late

Table 1. *Proposed correlation of marine sedimentary hiatuses and geomorphological events in southern and equatorial Africa (after King, 1962).*

Marine hiatus	Denudation surface	Pediplanation cycles (age of initial uplift)
		↑ Congo cycle (Late Pliocene)
		↑ (Late Miocene)
Middle Miocene	Coastal plain surface	
		↑ (Late Oligocene)
Early Oligocene	African surface	
		↑ African cycle (Early Paleocene)
Early Paleocene/ Late Maastrichtian	Late Cretaceous surface	

Cretaceous. On the basis of the chronology by King (1962) these surfaces could tentatively be correlated with the regression surfaces at the shelf, an initial uplift starting a period of renewed marine sedimentation (Table 1). The tectonic changes recorded in the shelf sediments as well as the possible connection with the African denudation surfaces indicate that the hiatuses at the equatorial Atlantic shelves and margins are at least partly caused by large-scale tectonic processes in Africa.

The same is probably true for the hiatuses in the South Atlantic pelagic sediments that can be related with the shelf record.

Acknowledgements

Thanks are due to Bert Aggenbach for the drawings and Joke Hart for typing the manuscript. Dr P. Lehner (Shell Exploration, The Hague) is particularly acknowledged for access to the multichannel seismic records. This work was financially supported by the Netherlands Council of Oceanic Research, who chartered the RV *Tyro* for the expeditions.

REFERENCES

Beck, R.H. & Lehner, P. 1974. Oceans, new frontier in exploration. *Am. Ass. Petrol. Geol. Bull.*, **58** (3), 376–395.
Bolli, H.M., Ryan, W.B.F. *et al.*, 1978. Angola continental margin sites 364 and 365. *Initial Repts. DSDP*, **40**, 357–455.
Emery, K.O., Uchupi, E., Phillips, J., Bowin, C. & Mascle, J. 1975. Continental margin off western Africa: Angola to Sierra Leone. *Am. Ass. Petrol. Geol. Bull.*, **59** (12), 2209–2265.
Ewing, J. & Houtz, R. 1979. Acoustic stratigraphy and structure of the oceanic crust. In: M. Talwani, C.G. Harrison & D.E. Hayes. *Deep Drilling Results in the Atlantic Ocean: Ocean Crust*. American Geophysical Union, Washington DC, 14 pp.
Hay, W.W, Sibuet J.-C. *et al.*, 1981. Leg 75, Walvis Ridge. *JOIDES Journal*, **7** (1), 17–29.
Hinte, J.E. van, 1976. A Cretaceous time scale. *Am. Ass. Petrol. Geol. Bull.*, **60** (4), 269–287.
Jansen, J.H.F., Giresse, P. & Moguedet, G. 1984. Structural and sedimentary geology of the Congo and southern Gabon continental shelf; a seismic and acoustic reflection survey. *Neth. J. Sea Res.*, **17** (2–4), 364–384.
King, L.C. 1962. *The Morphology of the Earth*. Oliver & Boyd, Edinburgh, 726 pp.
Larson, R.L. & Ladd, J.W. 1973. Evidence for the opening of the South Atlantic in the Early Cretaceous. *Nature*, **246** (5430), 209–212.
Lehner, P. & de Ruiter, P.A.C. 1977. Structural history of Atlantic margin of Africa. *Am. Ass. Petrol. Geol. Bull.*, **61** (7), 961–981.
Leyden, R., Bryan, G. & Ewing, M. 1972. Geophysical reconnaissance on African shelf: 2. Margin sediments from Gulf of Guinea to Walvis Ridge. *Am. Ass. Petrol. Geol. Bull.*, **56** (4), 682–693.
Lowrie, W. & Alvarez, W. 1981. One hundred million years of geomagnetic polarity history. *Geology*, **9** (9), 392–397.
Moore, T.C., van Andel, T.H., Sancetta, C. & Pisias, N. 1978. Cenozoic hiatuses in pelagic sediments. *Micropaleontol.*, **24** (2), 113–138.

Nairn, A.E.M. & Stehli, F.G. 1973. A model for the South Atlantic. In: A.E.M. Nairn & F.G. Stehli. *The Ocean Basins and Margins. 1. The South Atlantic.* Plenum, New York, pp. 1–24.

O'Connell, S. 1979. Ophiolites: ocean crust and land. *Oceanus,* **22** (3), 23–32.

Ojeda, H.A.O. 1982. Structural framework, stratigraphy and evolution of Brazil marginal basins. *Am. Ass. Petrol. Geol. Bull.,* **66** (6), 732–749.

Perch-Nielsen, K., Supko, P.R. *et al.*, 1975a. Site 356: São Paulo Plateau. *Initial Repts. DSDP,* **39**, 141–230.

Perch-Nielsen, K., Supko, P.R. *et al.*, 1975b. Site 357: Rio Grande Rise. *Initial Repts. DSDP,* **39**, 231–327.

Rabinowitz, P.D., 1976. Geophysical study of the continental margin of southern Africa. *Geol. Soc. Am. Bull.,* **87** (11), 1643–1653.

Rabinowitz, P.D., & La Brecque, J. 1979. The Mesozoic South Atlantic Ocean and evolution of its continental margins. *J. Geophys. Res.,* **84** (B11), 5973–6002.

Reyment, R.A., 1980. Paleo-oceanology and paleobiogeography of the Cretaceous South-Atlantic Ocean. *Oceanol. Acta,* **3** (1), 127–133.

Talwani, M. & Langseth M. 1981. Ocean crustal dynamics. *Science,* **213** (4503), 22–31.

Uchupi, E. & Emery K.O. 1972. Seismic reflection, magnetic, and gravity profiles of the eastern Atlantic continental margin and adjacent deep-sea floor. I: Cape Francis (South Africa) to Congo canyon (Republic of Zaire). *Woods Hole Oceanogr. Inst. Techn. Rep.,* **72–95**, 1–9.

Uchupi, E. & Emery, K.O. 1974. Seismic reflection, magnetic, and gravity profiles of the eastern Atlantic continental margin and adjacent deep-sea floor. II: Congo canyon (Republic of Zaire) to Lisbon (Portugal). *Woods Hole Oceanogr. Inst. Techn. Rep.,* **74–19**, 1–14.

Vail, P.R., Mitchum R.M. & Thompson, S., III, 1977. Seismic stratigraphy and global changes of sea level, Part 4: Global cycles of relative changes of sea level. In: C.E. Payton. Seismic stratigraphy applications to hydrocarbon exploration. *Am. Ass. Petrol. Geol. Mem.,* **26**, 83–97.

13

Synthesis of late Cretaceous, Tertiary, and Quaternary stable isotope records of the South Atlantic based on Leg 72 DSDP core material

D. F. WILLIAMS, R. C. THUNELL, D. A. HODELL
and C. VERGNAUD-GRAZZINI

Abstract

Stable oxygen and carbon isotope records for three threshold time periods during the late Cretaceous/Tertiary provide a comprehensive paleoceanographic history. Sedimentary sequences from the Rio Grande Rise region (DSDP Leg 72) and the Falkland Plateau (DSDP Leg 71) are used to determine how the southwestern Atlantic responded to global climatic changes during the Pliocene through the Quaternary and at the Cretaceous/Tertiary and Eocene/Oligocene boundaries. Evidence is found for two modes of glaciation during the Pliocene, suggesting that glaciations were more severe and extensive in the early Quaternary than in the latest Pliocene. A significant increase in the surface water temperature gradient between the Rio Grande Rise and Falkland Plateau is found from the late Eocene to early Oligocene time. It is also suggested that the production of a cold intermediate water mass was initiated in the vicinity of the Falkland Plateau by the early Oligocene. The $\delta^{13}C$ and calcium carbonate records across the Cretaceous/Tertiary boundary in Site 516F (Leg 72) suggest that a significant reduction occurred in primary productivity, possibly accompanied by rapid shoaling of the carbonate compensation depth. The large isotopic anomalies associated with the Cretaceous/Tertiary boundary may have had significant impact extending beyond the boundary by as much as 1 to 2 million years. The Leg 72 and 71 isotopic records, when combined with those from the other Atlantic DSDP Legs, should lead to a comprehensive understanding of the Atlantic's response to the climatic and productivity fluctuations which characterize the Tertiary.

Introduction

The South Atlantic has been the target of an unprecedented drilling effort by the Deep Sea Drilling Project (Legs 71–75) dedicated in large part to paleoenvironmental and paleoceanographic reconstructions. Both at present and in the past, the South Atlantic has played an important role in global thermohaline processes. Today the South Atlantic is the principal conduit by which North Atlantic Deep Water (NADW) is exchanged with the Southern Ocean (and thereby the Indian and Pacific Oceans) and by which Antarctic Bottom Water (AABW) passes into the North Atlantic. The Leg 72 drilling effort (which concentrated on the Rio Grande Rise and the Brazil Basin) provided core materials for stable isotopic, biostratigraphic, micropaleontologic and lithologic studies. These studies are aimed at better understanding the circulation history of the western South Atlantic, as well as providing records of globally important events that accompanied the glaciation of the Northern Hemisphere beginning in the late Pliocene and the development of the psychrosphere near the Eocene/Oligocene boundary, and the dramatic oceanographic events which occurred at the Cretaceous/Tertiary boundary.

The purpose of this contribution is to synthesize the Leg 72 South Atlantic stable isotope studies in terms of both global and regional paleoceanographic and paleoclimatic events. Three threshold time periods provide a comprehensive record of the subtropical South Atlantic during its developmental history. Core materials from Rio Grande Sites 516A, 517, and 518 were used to investigate the onset and character of Northern Hemisphere glaciation during the Pliocene and Quaternary periods. Cores from Site 516 were also used to examine the Eocene/Oligocene and Cretaceous/Tertiary boundaries. The isotopic results discussed from the Quaternary section are from Vergnaud-Grazzini et al. (1983), the Pliocene isotopic results are from Hodell et al. (1983) and Leonard et al. (1983), and the Cretaceous/Tertiary boundary data are from Williams et al. (1983).

Leg 72 of the Deep Sea Drilling Project cored three sites along a depth transect down the western flank of the Rio Grande Rise (Fig. 1 and Table 1). Site 516A is presently at 1313 m water depth near the present-day transition between Antarctic Intermediate Water (AAIW) and upper Circumpolar Deep Water (CPDW) (Reid et al., 1977). Site 517 was drilled at 2963 m water depth in the core of NADW. Site 518 is situated at 3944 m in the benthic thermocline between NADW and AABW. The Pliocene and Quaternary sections of Sites 515A, 517, and 518 provide a combined record from approximately 4.6 Ma to present with which to evaluate climatic changes related to ice volume fluctuations on Northern Hemisphere continents. Initial build-up of ice on Northern Hemisphere continents (Shackleton & Opdyke, 1977) and conspicuous global coolings have been suggested to occur during the period from 3.2 to 2.5 million years before present (Backman, 1979; Prell, 1982a;

Shackleton *et al.*, 1982, 1984; Leonard *et al.*, 1983; Thunell & Williams, 1983; Hodell *et al.*, 1984). The rapid climatic deterioration which began at approximately 3.2 Ma previously has been interpreted as representing the onset of Northern Hemisphere glaciation (Shackleton & Opdyke, 1977). Recent evidence now suggests that extensive ice-rafting into the North Atlantic did not begin until approximately 2.5 to 2.4 Ma

Table 1. *Locations and water depths of Leg 72 DSDP sites, southwestern Atlantic.*

Site	Latitude (°S)	Longitude (°W)	Water depth (m)	Water mass
516A, F	30°16.59′	35°17.10′	1313	CPDW or AAIW
517	30°56.81′	38°02.47′	2963	NADW
518	29°58.42′	38°08.12′	3944	AABW

Fig. 1. Location map of DSDP Sites 515, 516, 517, and 518 cored from the Rio Grande Rise during Leg 72 in the Southwestern Atlantic. Also shown are DSDP Sites 356 and 357 from Leg 39, and the mean flow of Antarctic Bottom Water through the Vema Channel is shown by the dark arrows.

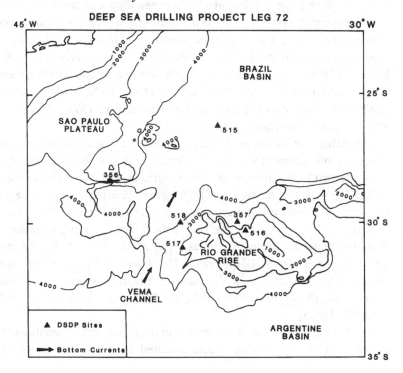

(Backman, 1979; Backman *et al.*, 1983) and isotopic evidence suggests that the build-up of Northern Hemisphere ice sheets also occurred at this later time (Prell, 1982b; Shackleton *et al.*, 1982, 1984; Thunell & Williams, 1983; Hodell *et al.*, 1984).

Another critical step in the evolution of our modern oceanographic regime occurred near the Eocene/Oligocene boundary, with the transformation of the oceans from a 'thermospheric' state to a 'psychrospheric' state (Brunn, 1957; Benson, 1975). This change in the thermal structure of the global ocean had a direct effect on biogeographic and evolutionary patterns (Haq *et al.*, 1977, Kennett, 1978; Corliss, 1979, 1981; Sancetta, 1979; Thunell, 1981a, Muza *et al.*, 1983) and sediment accumulation (Davies *et al.*, 1977; van Andel *et al.*, 1977; Davies & Worsley, 1981; Thunell & Corliss, 1983). Based on an oxygen isotopic study of DSDP Site 277 from the Subantarctic, Kennett & Shackleton (1976) estimated that bottom water temperatures dropped by 4–5 °C within a period of less than 100 000 years in the early part of the Oligocene. They attributed this cooling to the onset of cold bottom water formation in high southern latitudes, which in turn resulted in the development of a two-layered ocean. Keigwin (1980) later recalculated the bottom water temperature drop at this site to be about 3 °C and placed the cooling at the Eocene/Oligocene boundary. This enrichment at the Eocene/Oligocene boundary in benthic foraminiferal $\delta^{18}O$ records appears to be a global event considering that the isotopic change has been identified in sites from the Pacific (Savin *et al.*, 1975; Savin, 1977; Keigwin, 1980; Murphy & Kennett, 1983), North Atlantic (Vergnaud-Grazzini *et al.*, 1978; Letolle *et al.*, 1979; Miller & Curry, 1982) and South Atlantic (Boersma & Shackleton, 1977; Muza *et al.*, 1983; Poore & Matthews, 1984).

A synthesis of available Eocene/Oligocene oxygen isotopic records based on planktonic foraminifera demonstrates a distinct latitudinal trend, with the magnitude of the ^{18}O enrichment near the boundary increasing with increasing latitude (Keigwin, 1980). This trend suggests that strong latitudinal temperature gradients were being established at this time and these gradients are reflected in a reorganization of plankton biogeographic provinces (Haq *et al.*, 1977, 1979; Kennett, 1978; Sancetta, 1979; Haq, 1981). The Eocene/Oligocene boundary cooling also resulted in a decrease in the diversity of planktonic microfossils (Berggren, 1969; Cifelli, 1969; Lipps, 1970; Frerichs, 1971; Thunell, 1981a).

The Eocene/Oligocene envigoration of deep-sea circulation due to the production of cold bottom water resulted in substantial submarine erosion and widespread hiatuses in pelagic sequences (Rona, 1973; Moore *et al.*, 1978). The pattern of carbonate sedimentation in the oceans also changed dramatically. Throughout much of the middle and late Eocene, the calcite compensation depth (CCD) was relatively shallow in all of the ocean basins causing most of the carbonate to accumulate on the shelves rather than in the deep sea. Near the Eocene/ Oligocene boundary, the CCD dropped precipitously (Heath, 1969; Berger, 1973; van Andel, 1975; van Andel *et al.*, 1975, 1977), and carbonate accumulation in the deep sea increased (Thunell & Corliss, 1983). Others have suggested that substantial quantities of ice may account for part of the isotopic shift near the boundary (Poore & Matthews, 1984).

Site 516F also provided material for a detailed study of the Cretaceous/Tertiary (K/T) boundary of the South Atlantic. Previous isotopic studies in the South Atlantic at Site 356 (Thierstein & Berger, 1978) and Site 524 (Hsü *et al.*, 1982) had shown evidence for a widespread reduction in primary productivity associated with a large carbon isotope excursion at the K/T boundary. Detailed biostratigraphic studies of Site 516F (Pujol, 1983; Weiss, 1983) give no evidence of any hiatus between the late Maestrichtian and early Paleocene sections recovered in core 89 of Site 516F. The paleomagnetic work of Hamilton (1983) also indicates that Chrons 28, 29, and 30 are present in Site 516F. While the K/T boundary section of Site 356 did not contain a prominent increase in iridium, Sites 516F (Leg 72) and 524 (Leg 73) did contain well-defined iridium peaks reaching a maximum of 1–3 parts per billion (ppb) (Hsü *et al.*, 1982; Michael *et al.*, 1983). The purpose of our isotopic study of the Site 516F K/T boundary section is to compare the magnitude of the oxygen and carbon stable isotopic anomalies recorded in Site 516F with those recorded in other K/T sections (Thierstein & Berger, 1978; Arthur *et al.*, 1979; Hsü *et al.*, 1982; Perch-Nielsen *et al.*, 1982).

Materials and methods

Specific details about the standardized procedures used in the stable isotope laboratories at South Carolina and Rhode Island can be found in Williams *et al.* (1977); Hodell *et al.* (1983) and Vergnaud-Grazzini *et al.* (1983). Briefly, however, the planktonic species, *Globigerinoides quadrilobatus*, from the 300–350 µm size fraction was selected for the Pliocene isotopic studies of Sites 516A, 517, and 518. This species is a near-surface dweller which deposits its test with a constant offset from carbon and oxygen isotopic equilibrium (Williams et al., 1977). Planktonic isotopic analyses for the Quaternary of Site 517 were performed on Globigerinoides ruber from the > 150 μm size fraction. Benthic isotopic analyses were performed on Planulina wuellerstorfi and/ or Cibicidoides kullenbergi from the Pliocene of Sites 516A, 517, and 518. Whenever possible, monospecific samples of P. wuellerstorfi were analyzed from the > 150 μm size fraction, but the paucity of this species in some samples occasionally required using a mixture of P. wuellerstorfi and C. kullenbergi. These two species have been shown to be isotopically similar and may be used interchangeably within experimental error (Duplessy et al., 1980; Woodruff et al., 1981; Keigwin, 1982a; Leonard et al., 1983). Unfortunately, benthic isotopic analyses from the Quaternary of Site 517 were made on a variety of benthic foraminiferal species because the scarcity of benthic specimens in that site precluded the analysis of monospecific samples. We have therefore retained only those analyses of P. wuellerstorfi and mixed benthics from the original data set of Vergnaud-Grazzini et al. (1983). All isotopic results are expressed as the per mil (‰) difference from PDB. None of the data are corrected for departure from either oxygen or carbon isotopic disequilibria. Based upon replicate analyses of B-1 powdered standard, analytical precision (1σ of the mean) at URI was ±0.10‰ for $\delta^{18}O$ and ±0.09‰ for $\delta^{13}C$. Analytical precision at USC was ±0.09 for $\delta^{18}O$ and ±0.07 for $\delta^{13}C$ based upon routine analysis of NBS-20 and several other standard carbonate powders. Vergnaud-Grazzini et al. (1983) report an analytical precision of ±0.07‰ for $\delta^{18}O$ and 0.06‰ for $\delta^{13}C$ based upon repeated analysis of a standard carbonate powder.

The late Eocene–early Oligocene interval has been studied isotopically in South Atlantic DSDP Sites 511 (Leg 71) and 516F (Leg 72). At site 511 on the Falkland Plateau, a 175 m thick sequence of nannofossil–diatomaceous ooze extending from planktonic foraminiferal zones P 15 to P 20 was examined. The Eocene/Oligocene boundary at this site occurs between cores 16 and 17 at a subbottom depth of approximately 147.5 m (Krasheninnikov & Basov, 1983). There is, however, a hiatus near the boundary, with the latest Eocene foraminiferal zone P 17 missing. Despite this, the recovery of a Paleogene carbonate-bearing sequence in this area is significant due to the general lack of similar sequences from other high-latitude regions. At Site 516F on the Rio Grande Rise, a 40 m late Eocene–early Oligocene sequence (approximately 510–550 m subbottom depth) was examined consisting predominantly of foraminiferal–nannofossil chalks. Based on nannoplankton studies, the Eocene/Oligocene boundary at this site has been placed at a subbottom depth of approximately 527–529 m with no apparent hiatus. The major drawback for both Sites 511 and 516F is the relatively poor core recovery (approximately 30%) for the late Eocene–early Oligocene time interval.

Both planktonic and benthic foraminifera were separately analyzed from each site. For Site 511, the planktonic analyses were performed on mixed assemblages of *Globigerinia angiporoides* and *G. linaperta*, while the benthic analyses were made on mixed assemblages. Monospecific samples of the planktonic foraminifera *G. ampliapertura* were used throughout the late Eocene–early Oligocene interval at Site 516F. For the benthic foraminiferal analyses at Site 516F, a mixed assemblage of species belonging to the genus *Cibicidoides* was used. The resultant isotopic records can be used not only to evaluate the paleoclimatic and paleoceanographic history of the South Atlantic for this portion of the Paleogene, but also for comparison with similar records from the Pacific Ocean (Keigwin, 1980; Murphy & Kennett, 1983), Southern Ocean (Shackleton & Kennett, 1975; Keigwin, 1980), and North and South Atlantic (Boersma & Shackleton, 1977; Vergnaud-Grazzini *et al.*, 1978; Letolle *et al.*, 1979; Miller & Curry, 1982; Poore & Matthews, 1984).

Cores (88, 89, 90) from Site 516F containing early Paleocene and late Maestrichtian sediments were sampled at roughly 20 cm intervals for isotopic study of the Cretaceous–Tertiary boundary. Oxygen and carbon isotopic analyses were performed on bulk sediment that was dried and ground to a powder (Williams *et al.*, 1983). Replicate analyses of the powders yielded a precision (1σ about the mean) of better than ± 0.1 for both $\delta^{18}O$ and $\delta^{13}C$ values.

Pliocene–Quaternary interval

Pliocene–Quaternary oxygen isotopic records of Site 517
The combined isotopic data from the Pliocene and Quaternary sections of Site 517 provide 50 m long record extending from about 3.2 Ma to present (Fig. 2). There are, however, numerous problems associated with the synthesis of these data. The planktonic isotopic analyses for the Pliocene section were made on *Globigerinoides quadrilobatus* from the 300–350 μm size fraction, whereas the Quaternary planktonic data were based upon analyses of *G. ruber* from the > 150 μm size fraction. Although most species of planktonic foraminifera are believed to calcify their tests in oxygen isotopic equilibrium (Curry & Matthews, 1981), the broader size fraction analyzed for *G. ruber* undoubtedly increases the variability of the

Fig. 2. The oxygen isotopic records for Site 517 based on surface
dwelling planktonic foraminifera (left-hand column)
Globigerinoides ruber (> 150 μm) and *Globigerinoides quadrilobatus*
(300–350 μm size fraction) and the benthic record (right-hand
column) based on mixed benthics or analyses of *Planulina
wuellerstorfi* and/or *Cibicidoides kullenbergi*. The odd numbers
refer to the inferred oxygen isotope stages in the Quaternary
(Vergnaud-Grazzini *et al.*, 1983).

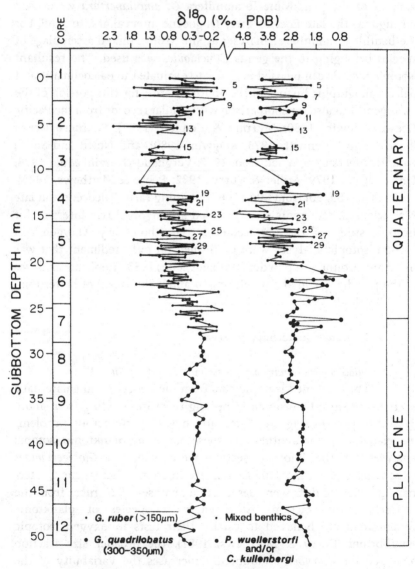

Quaternary data. The benthic oxygen isotopic data is complicated by the fact that the Pliocene section consists of monospecific analyses of *P. wuellerstorfi* and/or *C. kullenbergi* (> 150 μm) while the Quaternary section contains data from analyses of both monospecific samples of *P. wuellerstorfi* and samples composed of mixed benthic species. The effect of the aforementioned species and size fraction problems is to produce a noisy planktonic and benthic oxygen isotopic record for the Quaternary (Fig. 2). Despite the high noise level of the isotopic record from Site 517, the data appear to be reliable since: (1) the planktonic and benthic isotopic curves are parallel throughout most of the record indicating a consistent response to glacial/interglacial changes in the $\delta^{18}O$ composition of seawater and (2) the Quaternary isotopic record of Site 517 is correlatable to the standard oxygen isotope stratigraphy of Emiliani (1955), Shackleton & Opdyke (1973, 1976), and van Donk (1976).

Twenty-nine oxygen isotope stages have been recognized in the benthic $\delta^{18}O$ record of Site 517 (Vergnaud-Grazzini *et al.*, 1983). Although the planktonic and benthic curves are generally parallel throughout most of the record, isotopic stage assignments were based upon the benthic record which is considered primarily to reflect ice volume fluctuations in the Quaternary. The planktonic record at this site may be highly susceptible to surface water temperature changes induced by fluctuations in the velocities of the Brazil and Falkland Currents (Vergnaud-Grazzini *et al.*, 1983).

The benthic isotopic record of Site 517 can be divided into three parts:

(1) Late Quaternary (0.9 Ma to present) – From isotopic stage 23 (15 m subbottom depth) to the top of the core. The Jaramillo Subchron (0.9 Ma), which brackets isotopic stage 24, has been identified in Site 517 as occurring between 15.3 and 15.8 m subbottom depth (Hamilton & Suzyumov, 1983). This magnetic event agrees well with the placement of isotopic stage 23 at 15 m.

(2) Early Quaternary (1.8 to 0.9 Ma) – From the Pliocene/Pleistocene boundary at about 25 m to isotopic stage 23 near the Jaramillo Subchron. The placement of the Pliocene/Pleistocene boundary in Site 517 is based upon the First Appearance Datum (FAD) of *Globorotalia truncatulinoides*.

(3) Late Pliocene (3.2 to 1.8 Ma) – From the bottom of the core (51 m) to the Pliocene/Pleistocene boundary at 25 m.

The top 23 stages in Site 517 have been correlated to the standard oxygen isotopic sequence of Pacific Core V28–239 (Vergnaud-Grazzini

Table 2. *Mean and standard deviation of late Quaternary, early Quaternary, and late Pliocene* $\delta^{18}O$ *values of planktonic and benthic foraminifera from DSDP Site 517*

A. Planktonic

	late Quaternary (0.9–0 Ma)	early Quaternary (1.8–0.9 Ma)	late Pliocene (3.2–1.8 Ma)
$\delta^{18}O$ $\bar{x}$	0.74‰[a]	0.64‰[a]	0.10‰[b]
σ	0.55	0.43	0.26

[a] G. ruber (>150 μm size fraction).
[b] G. quadrilobatus (300–350 μm size fraction).

B. Benthic

	late Quaternary (0.9–0 Ma)	early Quaternary (1.8–0.9 Ma)	late Pliocene (3.2–1.8 Ma)	early late Pliocene (3.2–2.5 Ma)	late late Pliocene (2.5–1.8 Ma)
$\delta^{18}O$ $\bar{x}$	3.67‰[a]	2.81‰[a]	2.56‰[c]	2.44‰[c]	2.70‰[c]
σ	0.81	0.98	0.43	0.30	0.51
$\bar{x}$		3.28[b]			
σ		0.71			

[a] Mixed benthics and P. wuellerstorfi/C. kullenbergi.
[b] Mixed benthics only.
[c] P. wuellerstorfi/C. kullenbergi only.

et al., 1983). Although the major stages can be identified, there are missing intervals related to coring disturbance and lack of core recovery. The Holocene (stage 1) is apparently absent and stages 14, 16, and 17 are partially missing due to incomplete core recovery.

Prell (1982a) has statistically demonstrated two modes of oxygen isotopic variability during the Quaternary. The early Quaternary (1.8 to 0.9 Ma) is marked by relatively depleted $\delta^{18}O$ values with low variability whereas the late Quaternary (0.9 Ma to present) exhibits enriched $\delta^{18}O$ values and high variability. Table 2 shows mean $\delta^{18}O$ values and standard deviations of planktonic and benthic foraminifera for the late Quaternary, early Quaternary, and late Pliocene of Site 517. The characteristic increase in mean $\delta^{18}O$ values is observed in both the planktonic and benthic records. No conclusions can be drawn about changes in $\delta^{18}O$ variability between early and late Quaternary time because of multi-specific benthic analyses and size fraction variability in the planktonic foraminifera. The maximum value of benthic $\delta^{18}O$ increases from 4‰ in the early Quaternary section to about 5‰ in the late Quaternary (Vergnaud-Grazzini *et al.*, 1983). This increase in isotopic amplitude at the stage 22/23 transition has been recognized in many sites (502, 397, V28–238, V28–239), and suggests that glaciations were more extreme during the past 900 000 years (Shackleton & Opdyke, 1976; Prell, 1982a; Thunell & Williams, 1983). Williams *et al.* (1981) have suggested that the transition in glacial mode at 0.9 Ma can be explained by the co-occurrence of Arctic Ocean and continental ice sheets during the late Quaternary.

Only four detailed oxygen isotopic records are available for the early Quaternary: Site 502 (Prell, 1982a), V16–205 (van Donk, 1976), V28–239 (Shackleton & Opdyke, 1976), and Site 132 (Thunell & Williams, 1983). Site 517 contains a complete early Quaternary section which has been isotopically analyzed at 10 cm intervals. Distinct isotopic events are evident in the early Quaternary, and isotopic stages have been tentatively numbered to stage 29. Confirmation of these and earlier isotopic events will have to await analyses of other high-resolution core material from the early Quaternary. The early Quaternary section of Site 517 exhibits oxygen isotopic events of smaller amplitude and greater frequency compared to the character of the late Quaternary isotopic record. The apparent benthic $\delta^{18}O$ enrichment between stages 33 and 31 is an artifact produced by a change from the analysis of *P. wuellerstorfi* below 21 m to mixed benthic species above 21 m.

Both the planktonic and benthic records show an abrupt increase in both mean $\delta^{18}O$ value and standard deviation between late Pliocene and

Quaternary time (Table 2). This change undoubtedly represents a change in the mode of glaciation which included more severe and extensive glaciations during the Quaternary period. Unfortunately, part of the change in variability over the Pliocene/Pleistocene boundary is an artifact of combining two different data sets. In the planktonic data, a change from analysis of *G. quadrilobatus* (300–350 μm) in the late Pliocene to the analysis of *G. ruber* (> 150 μm) in the Quaternary increases the variability of the data. In the benthic data, monospecific samples of *P. wuellerstorfi/C. kullenbergi* were analyzed between 51 and 21 m. These benthic $\delta^{18}O$ values also show increased variability beginning near the Pliocene/Pleistocene boundary. Aliasing of the data, caused by a reduced sampling density in the late Pliocene, may also account for part of the decreased variability of the late Pliocene isotopic data. The Quaternary was sampled about every 10 cm, whereas the late Pliocene section was sampled every 50 cm. There is also an indication of two modes of glaciation within the late Pliocene. Late late Pliocene benthic $\delta^{18}O$ values are enriched by an average of 0.26‰ over early late Pliocene values (Table 2). Glaciation during the late Pliocene is discussed in more detail in the next section.

Pliocene oxygen isotopic records – relationship to Northern
Hemisphere Glaciation
    The combined isotopic curves from the Pliocene of Sites 516A, 517, and 518 provide a continuous record from ∼4.6 to 1.7 Ma. Fig. 3 represents a composite of the planktonic and benthic oxygen isotopic curves from all three sites. Both planktonic and benthic records are relatively unchanging during the interval from 4.6 to 3.2 Ma. The two major isotopic events in both the planktonic and benthic oxygen isotope records of Sites 516A, 517, and 518 occur at ∼3.2 Ma and ∼2.6 Ma.
    During the 3.2 Ma event, planktonic $\delta^{18}O$ values in Site 516A show a very slight depletion trend from ∼3.4 to 3.2 Ma, while Sites 517 and 518 become more negative by about 0.65‰ between 3.2 and 2.8 Ma. In contrast, all Leg 72 benthic records show an ^{18}O enrichment near 3.2 Ma, and the magnitude of the enrichment appears to increase with water depth from 0.5‰ in the shallower Site 516A (1313 m) to 1.0‰ in the deepest Site 518 (3944 m).
    The isotopic event at ∼2.6 Ma is evidenced by a parallel enrichment in both planktonic and benthic $\delta^{18}O$ values in Site 517. The planktonic oxygen isotopic event in Site 517 is a double-peaked enrichment centered about 2.6 and 2.4 Ma, but the remainder of the Site 517 record

is damped, exhibiting variations of less than 0.2‰. The benthic $\delta^{18}O$ record at Site 517 shows a mean $\delta^{18}O$ increase of 0.26‰ between the early late Pliocene and late late Pliocene (Table 2). Between 2.8 and 2.2 Ma, the planktonic and benthic oxygen isotopic records of Site 518 show high variability but no permanent enrichment in mean $\delta^{18}O$ values.

The enrichment in benthic $\delta^{18}O$ near 3.2 and 2.5 Ma have been observed in numerous deep-sea sedimentary sequences (Shackleton & Opdyke, 1977; Keigwin, 1979, 1982a,b; Weissert & Oberhänsli, this volume, chapter 7). Shackleton & Opdyke (1977) initially interpreted the [18]O enrichment at 3.2 Ma as representing the onset of Northern Hemisphere glaciation, while the 2.5 Ma isotopic event was inferred to represent an increase in the intensity of glaciation. Recent work, however, has now established that large ice sheets did not form in the Northern Hemisphere until ~ 2.5 Ma (Backman, 1979; Prell, 1982b; Shackleton *et al.*, 1983, 1984; Thunell & Williams, 1983).

If the 3.2 Ma paleoceanographic event included a permanent increase in global ice volume, then a parallel [18]O enrichment would be expected in both planktonic and benthic foraminifera (Prell, 1982b) An enrich-

Fig. 3. A comparison of the oxygen isotopic records for Sites 516A, 517, and 518 based on the surface dwelling planktonic foraminifer, *Globigerinoides quadrilobatus*, and analyses of benthic foraminifera comprised of *Planulina wuellerstorfi* or *Cibicidoides kullenbergi*.

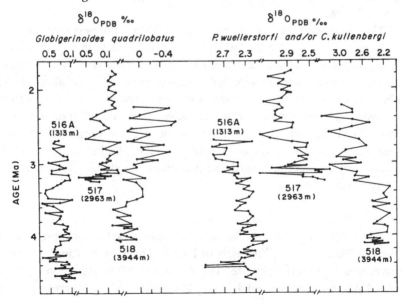

ment in planktonic $\delta^{18}O$ is absent in all Leg 72 sites. In fact, a depletion in planktonic $\delta^{18}O$ at 3.2 Ma indicates a decrease in global ice volume and/or a warming of surface waters in the vicinity of the Rio Grande Rise. However, the benthic $\delta^{18}O$ values in all Leg 72 sites show an enrichment near 3.2 Ma, and the magnitude of the enrichment appears to increase with greater water depth. These data support the suggestion that the 3.2 Ma isotopic event did not include a large, permanent increase in the Earth's global ice volume, but rather resulted from a circulation event which included a decrease in bottom water temperatures. From analyses of benthic foraminiferal assemblages from Site 518 and benthic isotopic gradients among Leg 72 sites, Hodell et al. (1984) suggest that the 3.2 Ma benthic isotopic event resulted from a pulse-like increase in the formation of AABW.

The oxygen isotopic event at ~ 2.6 Ma does show a parallel ^{18}O enrichment of $\sim 0.5\%_0$ in both planktonic and benthic foraminifera of Site 517 indicating a temperature decrease and/or an increase in global ice volume. The 2.5 Ma isotopic event is believed to have resulted from the initial, rapid build-up of Northern Hemisphere ice sheets. This isotopic event is coincident with the inception of ice-rafted detritus in DSDP Sites 111, 116, and 522A (Backman, 1979; Shackleton et al., 1982). In Site 522A, Shackleton et al (1982) found that high-amplitude orbital frequencies (100 K power) in the oxygen isotope and calcium carbonate records only begin at about 2.4 Ma. Thus, it appears that permanent growth of ice sheets in the Northern Hemisphere did not occur until about 2.5 Ma.

The planktonic $\delta^{18}O$ enrichment at ~ 2.6 Ma is not permanent in Site 517, but rather an ^{18}O depletion is observed at 2.3 Ma followed by a damped record between 2.3 and 1.8 Ma. A surface water temperature increase, resulting from increased transport of warm, tropical waters by the Brazil Current, may have offset the expected enriched $\delta^{18}O$ values between 2.3 and 1.8 Ma. Vergnaud-Grazzini et al. (1983) have shown that increases in the percentages of right-coiling *Globorotalia truncatulinoides* are coincident with depleted planktonic $\delta^{18}O$ values indicating strong southward excursions of the Brazil Current in the early Quaternary.

Pliocene carbon isotopic records

The composite planktonic and benthic carbon isotopic records for the Pliocene are illustrated in Fig. 4. The base of the planktonic carbon isotopic record in Site 516A exhibits a 1.0‰ depletion from 4.6 to 3.9 Ma. Between 3.9 and 3.4 Ma, planktonic $\delta^{13}C$ values are relatively unchanging in both Sites 516A and 518. A depletion trend is observed in the records of Sites 516A and 518 from about 3.3 to 3.0 Ma. This depletion is absent in the base of the Site 517 record. In Sites 517 and 518, a gradual depletion of 0.5 ‰ is observed from 2.95 to 2.45 Ma, while the remainder of the records are marked by fluctuations of about 0.4 ‰.

The benthic carbon isotopic record of Site 516A is relatively unchanging throughout the Pliocene fluctuating about a mean of 0.8 ‰. Carbon isotopic values in Site 517, between 3.2 and 1.8 Ma, also show no distinct trends and fluctuate about a mean of 0.9 ‰. Unlike the records at the shallower sites, benthic carbon isotopic values in the deepest Site 18 exhibit distinct events. An apparent enrichment occurs at 3.9 Ma, though this enrichment may be artificial since it occurs across a coring gap. A sharp depletion trend of about 1.0 ‰ begins at ~3.4 Ma and culminates at 3.0 Ma. This is followed by further enrichment at 2.9 Ma, although part of the enrichment occurs over an interval of no core recovery.

Fig. 5 shows that the depletion trend in the benthic $\delta^{13}C$ record of Site 518 between 3.3 and 3.0 Ma is coincident with the oxygen isotopic enrichment at ~3.2 Ma. In fact, a linear correlation exists ($r = -0.65$)

Fig. 4. A comparison of the carbon isotopic records for sites 516A, 517, and 518 based on the surface dwelling planktonic foraminifer, *Globigerinoides quadrilobatus* and analyses of benthic foraminifera comprised of *Planulina wuellerstorfi* or *Cibicidoides kullenbergi*.

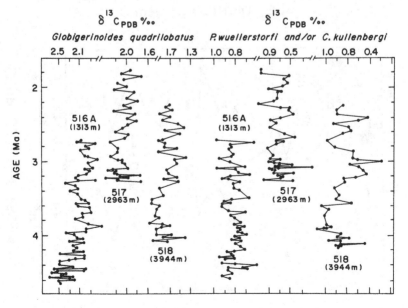

between enriched $\delta^{18}O$ values and depleted $\delta^{13}C$ values throughout the entire record of Site 518. This synchronous enrichment in benthic $\delta^{18}O$ and depletion in benthic $\delta^{13}C$ during the 3.2 Ma event supports the suggestion that the production of AABW increased over the 3.2 Ma event, since AABW is both cold (heavy $\delta^{18}O$) and high in total dissolved CO_2 (light $\delta^{13}C$) (Hodell et al., 1984).

Late Eocene–early Oligocene interval

Oxygen and carbon isotopic records
The results of the oxygen and carbon isotopic analyses of foraminifera from Site 511 are presented in Fig. 6. The major trends in the planktonic $\delta^{18}O$ and $\delta^{13}C$ records appear to be mirrored in the benthic records, although the magnitudes of the changes differ. From the lower portion of Site 511 (below core 16) the $\delta^{18}O$ values for planktonic and benthic foraminifera from the same sample are fairly similar, and range from -0.2 to 1.2‰. Above core 16, benthic $\delta^{18}O$ values are systematically heavier than the equivalent planktonic $\delta^{18}O$ values by an average of 1.1‰.

Both $\delta^{18}O$ records contain an enrichment of approximately 1.0‰ just

Fig. 5. A comparison between the late Pliocene carbon and oxygen isotopic records for DSDP Site 518 from within the Antarctic Bottom Water mass.

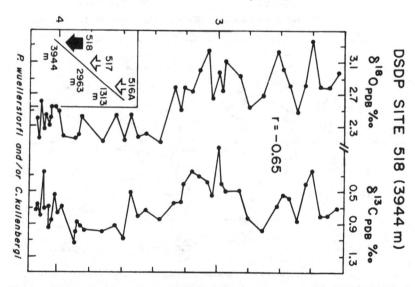

below the Eocene/Oligocene boundary. Specifically, this enrichment occurs over an 8 m interval (18-1, 24–26 cm to 17-1, 87–89 cm) in the planktonic $\delta^{18}O$ record, and within a 1.7 m interval (17-3, 10–12 cm to 17-1, 87–89 cm) in the benthic $\delta^{18}O$ record. Within core 12, there is an abrupt enrichment of 1.1‰ in planktonic $\delta^{18}O$, with values above this level tending to be heavier than values below it. The benthic record reveals only a small ^{18}O enrichment (0.3‰) within core 12, but a longer-term enrichment of over 2.0‰ is observed between cores 14 and 12. In the upper portion of the benthic $\delta^{18}O$ record, two significant depletions exist, one in core 6 (1.0‰) and the other in core 3 (0.7‰). In the planktonic record, a minor depletion (0.3‰) occurs in core 6, but a coincident 1.0‰ depletion occurs in core 3.

Two striking features are present in the $\delta^{13}C$ record for Site 511 (Fig. 6). First, both the planktonic and benthic $\delta^{13}C$ records seem to covary

Fig. 6. The late Eocene to early Oligocene oxygen and carbon isotopic records for benthic and planktonic foraminifera from Leg 71 DSDP Site 511 (Muza *et al.*, 1983).

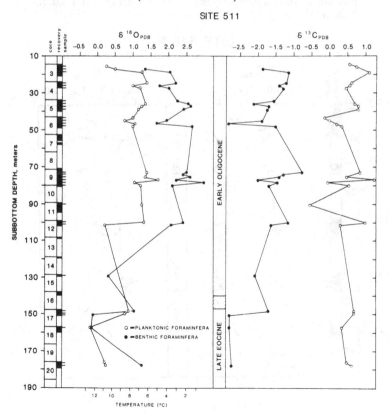

with the $\delta^{18}O$ records. The largest fluctuations in the $\delta^{13}C$ records are associated with large enrichments or depletions in ^{18}O. Second, a large difference exists throughout the section in planktonic and benthic $\delta^{13}C$ values from the same sample, with the benthic values always being lighter. Benthic $\delta^{13}C$ values range from -2.9 to -0.8‰, while the planktonic values vary between 0.5 and 1.3‰. The greatest deviation between planktonic and benthic $\delta^{13}C$ values is 3.2‰ (18-1, 24–26 cm), while the smallest difference is 1.4‰ (9-5, 96–98 cm).

The planktonic and benthic $\delta^{18}O$ and $\delta^{13}C$ records for Site 516F are very different from those for the higher latitude Site 511, both in absolute values and overall character (Fig. 7). The planktonic and benthic $\delta^{18}O$ records for Site 516F are marked by very little variability through time. With the exception of one sample, planktonic $\delta^{18}O$ values vary between -1.4 and -0.6‰. The benthic $\delta^{18}O$ values have an even smaller range, varying between -0.2 and 0.3‰. Except for core 41,

Fig. 7. The oxygen and carbon isotopic records for benthic and planktonic foraminifera across the Eocene/Oligocene Boundary from Site 516F.

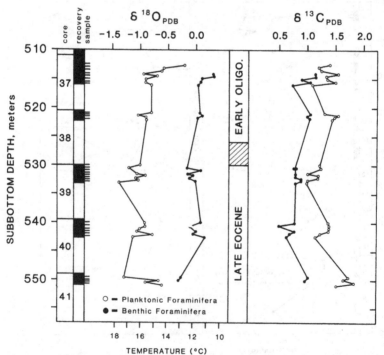

SITE 516 F

fairly consistent offset exists between the two records, with the benthic $\delta^{18}O$ values typically being about 1.0‰ heavier. Overall, there do not appear to be any significant short-term changes or long-term trends in either the planktonic or benthic $\delta^{18}O$ records for Site 516F (Fig. 7).

The benthic and planktonic $\delta^{13}C$ records for Site 516F demonstrate a high degree of covariance (Fig. 7). These $\delta^{13}C$ records are similar in character to the $\delta^{18}O$ records for this site in that they do not demonstrate a great deal of variability through time. The planktonic values range from 1.0 to 1.86‰, while benthic values fluctuate between 0.5 and 1.14‰. The planktonic $\delta^{18}O$ values are consistently heavier than the benthic value for the same sample, with this difference being as much as 0.9‰ (41-1, 130–133 cm) and as little as 0.2‰ (39-2, 100–103 cm).

Paleotemperature estimates

Assuming little if any isotopic disequilibrium effects, the planktonic and benthic oxygen isotope results have been used to calculate surface and bottom water paleotemperatures, utilizing the paleotemperature equation of Shackleton (1974):

$$T^\circ C = 16.9 - 4.38\,(\delta_c - \delta_w) + 0.10\,(\delta_c - \delta_w)^2$$

where $\delta_c = \delta^{18}O$ calcite relative to PDB and $\delta_w = \delta^{18}O$ water relative to PDB. This equation is an expression of the isotopic equilibrium between water and calcite (O'Neil *et al.*, 1969) wherein a value of -1.2‰ for δ_w has been used as an average isotopic composition for the oceans prior to the build-up of a substantial Antarctic ice sheet in the middle Miocene (Shackleton & Kennett, 1975; Savin, 1977; Woodruff *et al.*, 1981). Poore & Matthews (1984) argued, however, that the Shackleton equation and the assumption of an ice-free δ_w of -1.2‰ yields paleotemperature estimates that are too cold for both surface and bottom waters in the central South Atlantic (Leg 73). Although an ice volume correction may later prove to be necessary, the paleotemperature estimates at Sites 511 and 516F appear to be reasonable when compared with modern surface and bottom water temperatures at these locations.

For Site 511, bottom water temperature estimates for this portion of the Paleogene range from approximately 12 to 1 °C (Fig. 6). Surface water paleotemperatures based on the planktonic $\delta^{18}O$ results fluctuate between approximately 12 and 5 °C (Fig. 6). For the lower portion of the Site 511 record (below core 16), the surface and bottom water paleotemperatures estimates are fairly similar. Both records indicate a temperature drop in the latest Eocene (core 17), with the decrease being about 4 °C in both surface and bottom waters. Another significant cooling occurs in the lower part of the early Oligocene. In particular, bottom water temperatures decreased by nearly 8 °C between cores 15 and 12, resulting in an average temperature difference between surface and bottom waters of about 4 °C that persists throughout the early Oligocene (Fig. 6).

The isotopic paleotemperatures for both surface and mid-depth waters at Site 516F are warmer and less variable than those estimated for Site 511. Late Eocene–early Oligocene mid-water paleotemperatures for Site 516F range from about 13 to 10 °C (Fig. 7). The middle portion of this record is extremely stable, with bottom temperatures fluctuating by no more than 1 °C. The planktonic $\delta^{18}O$ results indicate a 3 °C warming of surface waters at the base of the record for Site 516F (core 41). Above this point, a cooling trend exists, with surface water temperatures dropping from a maximum of about 17 °C in the late Eocene to about 13 °C in the early Oligocene (Fig. 7).

Discussion

A discussion of the Eocene–Oligocene oxygen isotopic or paleotemperature results for Sites 511 and 516F must bear in mind that both of these sites were drilled on topographic highs and, as such, are impinged by intermediate to shallow depth water masses and not by the abyssal or bottom water masses. This fact is particularly critical for Site 516F, which was drilled at a present-day water depth of only 1313 m Since the Rio Grande Rise appears to have subsided in a fashion similar to that for normal oceanic crust (Detrick *et al.*, 1977; Thiede, 1977), the late Eocene–early Oligocene paleodepth for this site would have been substantially shallower than its present depth.

The similarity between the benthic and planktonic $\delta^{18}O$ values in the latest Eocene of Site 511 indicates either little or no thermal gradient between intermediate and surface waters at this time on the Falkland Plateau (Fig. 6) or recrystallization. A cooling at the Eocene/Oligocene boundary seems to have had a similar effect on both surface and intermediate water masses. The 4 °C cooling recorded in the benthic foraminifera is also seen in the planktonic $\delta^{18}O$ record suggesting that there was still no appreciable difference in surface and intermediate water temperatures at this time.

A second, more dramatic cooling event occurs within the early Oligocene and appears to be responsible for the development of a thermocline in this region of the Falkland Plateau (Fig. 6). At this time, intermediate water temperatures dropped to about 2 °C, and were significantly colder than surface water temperatures (~7 °C). The pronounced cooling at this time of intermediate-level water masses over the Falkland Plateau most likely reflects the onset of cold deep water formation in the high southern latitudes (Kennett & Shackleton, 1976; Keigwin, 1980). A well-established thermocline persists throughout the remainder of the early Oligocene, with surface and intermediate water temperatures fluctuating in phase with each other. These conclusions, however, must be tempered with the realization that the planktonic and benthic values may have been diagenetically altered and that the mixed benthics analyzed may not yield $\delta^{18}O$ values in equilibrium with seawater.

A very different scenario is recorded at Site 516F on the Rio Grande Rise which is located in the subtropics 20° to the north of Site 511. The separation of benthic and planktonic $\delta^{18}O$ values at Site 516F indicates that a thermocline was in existence at this location throughout the late Eocene–early Oligocene (Fig. 7). The thermal gradient between surface and intermediate water masses at Site 516F is fairly uniform at 4–5 °C throughout the section, and is thus comparable in magnitude to the thermal gradient that developed in the early Oligocene on the Falkland Plateau.

A comparison of estimated surface water temperatures for Sites 511 and 516 indicates that during the late Eocene an average surface temperature gradient of about 4–5 °C existed between these two sites, in good agreement with the previous work of Shackleton & Boersma (1981). This latitudinal surface temperature gradient increased significantly in early Oligocene time averaging about 8 °C (7 °C at Site 511 and 15 °C at Site 516F), primarily due to cooling at the higher latitude site. A similar relationship was observed by Keigwin (1980) and Keigwin & Corliss (1982, 1983) in a comparison of high and low-latitude sites from the Atlantic, Pacific, and Indian Oceans.

During most of the late Eocene, 'bottom' or intermediate water temperatures at Sites 511 and 516F were quite similar, fluctuating between 11 and 13 °C (Figs. 6 and 7). While these relatively warm temperatures were maintained at Site 516F during the early Oligocene, a two-stage cooling of bottom waters at Site 511 (one decrease near the Eocene/Oligocene boundary and another in the early Oligocene) resulted in an overall drop in temperature of 9–10 °C in the vicinity of the Falkland Plateau. The resulting temperature difference between these two sites during the early Oligocene would seem to suggest that the two topographic highs were being bathed by distinctly different water masses. In particular, 'bottom' temperatures of 11–12 °C during the early Oligocene at Site 516F would seem to indicate a very shallow paleodepth. Subsurface temperatures of 11–12 °C are presently found at a depth of about 500 m in the vicinity of the Rio Grande Rise, associated with a water mass that is very much different from Antarctic Intermediate Water (AAIW) or Circumpolar Deep Water (CPDW). In contrast, the 2–3 °C 'bottom' water found on the Falkland Plateau during the early Oligocene is similar, at least in thermal character, to present-day CPDW. These isotopic temperature estimates clearly indicate that, by early Oligocene time, cold intermediate waters were being generated in the high southern latitudes, in association with the production of cold bottom waters (Shackleton & Kennett, 1975; Keigwin, 1980).

DSDP Site 357, also drilled on the flank of the Rio Grande Rise, has a somewhat different late Eocene–early Oligocene $\delta^{18}O$ record (Boersma & Shackleton, 1977) from that described for Site 516F. The late Eocene benthic $\delta^{18}O$ records for both sites indicate that 'bottom' temperatures were about 12 °C. While the benthic record for Site 516F suggests that these temperatures persisted through the early Oligocene (Fig. 7), the equivalent record for Site 357 indicates the 'bottom' temperatures dropped to 6 °C by early Oligocene time. This difference in early Oligocene 'bottom' temperatures is not due to geographic differences (both sites were drilled at approximately 30 °S), but rather to a bathymetric difference. Sites 357 and 516 were drilled at depths of 2109 m and 1313 m, respectively. Although both sites obviously had shallower late Eocene–early Oligocene paleodepths, the 800 m differ- ence between the two sites was sufficient for them to be impinged by different water masses. The trends in the 'bottom' temperature record for Site 357 are quite similar to that found at Site 511 on the Falkland Plateau, and suggest that this site was being bathed by cold intermedi- ate waters that began to form in latest Eocene time at high southern latitudes.

Late Cretaceous–early Tertiary interval

Oxygen and carbon isotopic and carbonate records
    The $\delta^{18}O$, $\delta^{13}C$, and $CaCO_3$ records for cores 88–90 from Site 516F are plotted relative to time and subbottom depth in Fig. 8. The $\delta^{18}O$ values for the bulk $CaCO_3$ range from −1.5 to −3.2‰ and exhibit several long-term and numerous short-term changes. In the late Maestrichtian from core 90–6 to the lower part of 90–1, $\delta^{18}O$ values decrease steadily by approximately 1‰ with at least five rapid 0.5‰ decreases occurring along this trend. From 90–1 to a level near the Cretaceous/Tertiary (K/T) boundary in 89–5, $\delta^{18}O$ values increase by

Fig. 8. A comparison between the stable isotopic (δ^{18}O, δ^{13}C) and total carbonate records across the Cretaceous/Tertiary boundary in Site 516F (from Williams *et al.*, 1983).

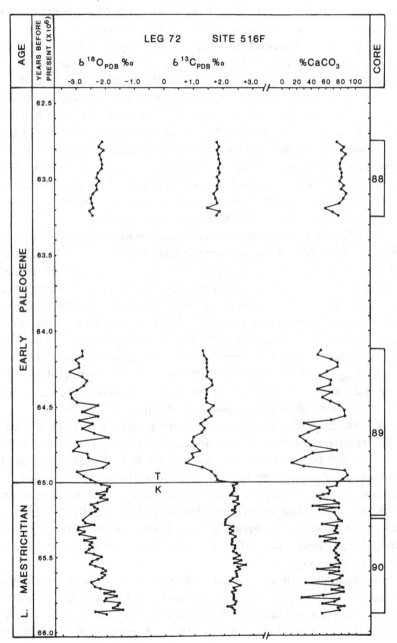

~1‰ with four rapid 0.5‰ increases. In the early Paleocene, respresented by planktonic foraminiferal zone P 1, the long-term $\delta^{18}O$ trend is again toward lighter values with five to six rapid fluctuations of magnitudes ranging from 0.5 to 1.5‰ (Fig. 8). In planktonic foraminiferal zone P 2 of core 88, $\delta^{18}O$ values become steadily more positive again and exhibit very little short-term change.

The character of the carbon isotopic record is quite different from the trends exhibited in the $\delta^{18}O$ record (Fig. 8). The total range in $\delta^{13}C$ values 2.3‰ from +0.5 to +2.8‰) is slightly greater than that of the $\delta^{18}O$ record (1.6‰), with the largest change clearly associated with the K/T boundary. In the late Maestrichtian section from cores 90–6 to 89–5, the $\delta^{13}C$ values are remarkably constant, varying less than 0.5‰ about a mean value of +2.4‰. A dramatic 1.8‰ depletion in ^{13}C occurs across the K/T boundary from core 89–5 into 89–4. The $\delta^{13}C$ values then increase steadily in the early Paleocene, but the values in foraminiferal zone P 2 of core 88 remain depleted in ^{13}C by at least 0.5‰ relative to the mean $\delta^{13}C$ value for the late Maestrichtian (Fig. 8).

The character of the total calcium carbonate record is distinctly different from that of the two isotope records (Fig. 8). During the late Maestrichtian, the sediment of Site 516F had a mean $CaCO_3$ content of 70% and underwent at least two rapid $CaCO_3$ decreases exceeding 40%. However, no long-term trends are discernible. In core 89–6, $CaCO_3$ content steadily increases across the K/T boundary to a maximum of 82% in the earliest Paleocene of 89-5. Above this point, two prominent $CaCO_3$ minima occur representing decreases of 80 and 70%, respectively (Fig. 8). After recovering at the core 89-3/2 boundary, the remainder of the early Paleocene record is characterized by carbonate fluctuations of 10 to 20% about a mean value of 70 to 80%.

A detailed plot of the three geochemical indices through the late Maestrichtian and early Paleocene sections of core 89 illustrates the relationships between the records across the K/T boundary (Fig. 9). A lithological examination of core 89 shows that it is primarily a calcareous nannofossil chalk with only six marly intervals and minimal signs of bioturbation. Biostratigraphic identification of the K/T boundary in the upper part of section 5 is well below the first of three closely spaced marly intervals in which $CaCO_3$ contents decrease to 10, 23, and 30%, respectively.

Interestingly, a lead-lag relationship appears to exist among the records (Fig. 9). Approximately 40 cm below the K/T boundary, the $\delta^{18}O$ record begins a prominent decrease of >1.2‰ across the boundary. The start of this ^{18}O depletion precedes the beginning of the 1.8‰ ^{13}C

Fig. 9. Detail of the stable isotope ($\delta^{18}O$, $\delta^{13}C$) and total carbonate records across the Cretaceous/Tertiary boundary in core 89 of Site 516F (from Williams *et al.*, 1983).

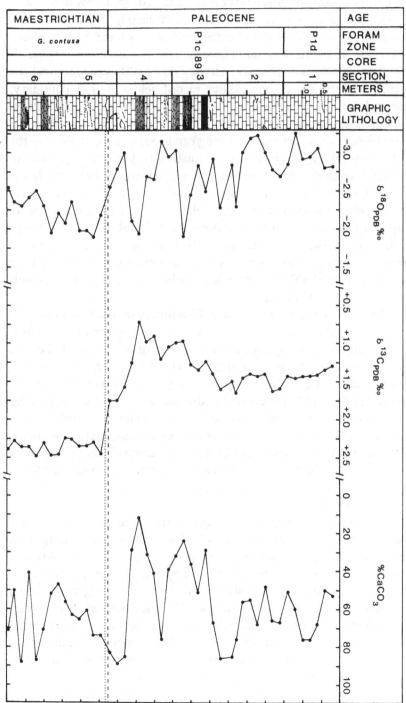

depletion across the K/T boundary by 20 cm. Both of these isotopic depletions occur within the broad $CaCO_3$ maximum ($> 70\%$) spanning the K/T boundary and precede the major $CaCO_3$ drop of 80% in the early Paleocene by a significant interval of 60 to 80 cm (Fig. 9). The carbonate minimum of 10% in 89–4 coincides precisely with the most depleted $\delta^{13}C$ value, and both these minima occur at a time when the $\delta^{18}O$ values have become 1‰ more positive. Although the $CaCO_3$ record shows little correlation with the carbon isotope record below the K/T boundary, they exhibit a high degree of covariance (low $CaCO_3$, less positive $\delta^{13}C$ values) throughout the Paleocene section of core 89. Intervals where the correspondence is particularly strong can be seen in Fig. 9. Prior to the K/T boundary, $\delta^{13}C$ values are systematically positive regardless of carbonate content, while above the boundary a strong linear correlation ($r = 0.84$; $r^2 = 0.71$) exists between $\delta^{18}O$ and $CaCO_3$ values with regard to stratigraphic position above or below the K/T boundary. More important is the fact that more negative $\delta^{18}O$ values above the K/T boundary are associated with high carbonate values and not vice versa. However, it is possible that early or late cementation of chalk may have an effect on $\delta^{18}O$ values during high carbonate intervals (M. Arthur, personal communication).

After a 60 cm interval across the K/T boundary in cores 89–5 and 89–4 where the $\delta^{18}O$ and $\delta^{13}C$ records are both becoming negative, the $\delta^{18}O$ record undergoes a significant ^{18}O enrichment of $\sim 1‰$ as the $\delta^{13}C$ record continues to decrease (Fig. 9). The $\delta^{18}O$ record continues to exhibit rapid 0.5 to 1.3‰ positive and negative events while the $\delta^{13}C$ record begins a gradual increase but does not achieve values as positive as those characteristic of pre-boundary samples (Fig. 9). Additionally, little direct correlation is evident between individual fluctuations in the $\delta^{18}O$ and $CaCO_3$ records, although a general trend exists toward more negative $\delta^{18}O$ and higher $CaCO_3$ values in sections 1–3 of core 89.

Discussion

The stable isotope and carbonate records from the late Maestrichtian–Paleocene section of Site 516F clearly imply that significant changes in productivity occurred in the South Atlantic across the Cretaceous–Tertiary boundary. An age model has been constructed using the estimated ages of the biostratigraphic zonal boundaries to approximate the rates and duration of these changes (Fig. 8). Sedimentation rates for the early Paleocene section were estimated principally using the planktonic foraminiferal zonation of Pujol (1983), the calcareous nannofossil stratigraphy of C. Cepek (personal communication), and the magnetostratigraphy of Hamilton (1983). Absolute ages of 58, 60, and 62 Ma were taken for the tops of zones P 3, P 2 and P 1, respectively (Hardenbol & Berggren, 1978). The K/T boundary was placed at core 89, section 5, 33.5 interval, based on planktonic foraminiferal and calcareous nannofossil biostratigraphy (Pujol, 1983; C. Cepek, personal communication), and the age of the K/T boundary was assumed to be 65 Ma. For the late Cretaceous section of Site 516F, the late Maestrichtian/early Maestrichtian boundary between cores 92 and 93 was assumed to be 67.5 Ma as used by other authors (Arthur *et al.*, 1979). This age model is, however, preliminary and subject to revision as more detailed biostratigraphic and paleomagnetic data become available and accumulation-rate estimates can be made using the bulk density determinations.

According to our age model, the upper 900 000 years of the late Maestrichtian is represented by our data, with a sample spacing of approximately 20 000 years (Fig. 8). Only the $\delta^{18}O$ record shows any trends through this interval with an apparent warming (depleted values) near 65.3 Ma and two relative cold periods at 65.8 Ma and just prior to the K/T boundary. The $\delta^{18}O$ and $\delta^{13}C$ changes associated with the K/T boundary actually precede the boundary by $\sim 60\,000$ years and $\sim 20\,000$ years, respectively. The shift toward lighter $\delta^{13}C$ values continues for $\sim 100\,000$ years after the K/T boundary before beginning a slow, steady recovery toward pre-boundary values over the next 800 000 years (Fig. 8). In contrast, the $\delta^{18}O$ record undergoes numerous increases and decreases every 100 000 to 200 000 years. Total carbonate values steadily increase for a 175 000 year period across the K/T boundary, reaching a maximum 50 000 to 75 000 years after the K/T boundary. The major $CaCO_3$ minimum does not occur until at least 100 000 years after the K/T boundary (Fig. 8). Over the next 700 000 year period, the $CaCO_3$ record approaches pre-boundary values several times, with another broad $CaCO_3$ maximum lasting 50 000 to 75 000 years at approximately 64.55 Ma (Fig. 8). A lack of core recovery in core 88 produced a break in the records from 64.1 to 63.2 Ma. During the interval from 63.2 to 62.7 Ma, all three records are unchanging. Perhaps it is significant, however, that 2 million years after the K/T boundary, the $\delta^{13}C$ record remains depleted by at least 0.5‰ compared to pre-boundary $\delta^{13}C$ values (Fig. 8).

Numerous workers have used the isotopic records from either DSDP or land-based sections to postulate on the cause(s) of the major biotic changes heralding the transition from the Mesozoic to the Cenozoic (Douglas & Savin, 1971; Anderson & Schneidermann, 1973; Coplen &

Schlanger, 1973; Douglas & Savin, 1973; Letolle *et al.*, 1978; Thierstein & Berger, 1978; Arthur *et al.*, 1979; Letolle, 1979; Scholle & Arthur, 1980; Hsü *et al.*, 1982, among others). Based on anomalously high amounts of iridium in these and other sedimentary sections containing the K/T boundary, some type of extraterrestrial event, involving an asteroid or cometary impact, has gained widespread appeal as the cause of the terminal Cretaceous extinctions (Alvarez *et al.*, 1980; Emiliani, 1980; Ganapathy, 1980; Hsü, 1980; Smit & Hertogen, 1980; Hsü *et al.*, 1982). A prominent iridium anomaly of 0.95 ±0.18 ppb is associated with the K/T boundary of Site 516F (Michael *et al.*, 1983) (Fig. 10). As can be seen in an expanded section across the K/T boundary (Fig. 10), no isotopic data are presently available at the detail shown for the whole-rock iridium abundance in Site 516F. At the scale of Fig. 10, the $\delta^{13}C$ and $\delta^{18}O$ offset is not as prominent as illustrated in Fig. 9 and further isotopic work is planned to resolve whether large or small-scale isotopic changes are associated with both the major and minor (>0.4 ppb) iridium anomalies. Fig. 10 shows that the biostratigraphic, trace element, and paleomagnetic evidence strongly suggests a fairly complete Cretaceous/Tertiary sequence.

For at least 800 000 years prior to the K/T boundary, the South Atlantic was characterized by a relatively deep, stable calcium carbonate compensation depth (CCD) and a diverse calcareous nannoplankton flora with $\delta^{13}C$ values typical of marine biogenic carbonate (Fig. 8). Preservation of the nannofossil assemblages is good while preservation of the planktonic foraminifera is only moderately good (Pujol, 1983). A strict interpretation of the late Maestrichtian $\delta^{18}O$ data in terms of isotopic paleotemperatures (Epstein *et al.*, 1953) suggests that the South Atlantic was warm with temperature extremes ranging between 25 and 18 °C (assuming the diagenetic effects were minimal and the isotopic composition of seawater in the South Atlantic, δ_w, was -1.0‰).

The cooling just prior to the K/T boundary and the subsequent temperature increase across the boundary amounted to a rapid 5–6 °C increase (~ -1.2‰) lasting slightly more than 100 000 years (Figs. 8 and 10). Such large oxygen isotopic changes, however, may be due to the influence of cementation in the chalk sequence. Although the total $CaCO_3$ content through this interval remains high, few foraminifera are present in the >150 μm fraction. Beginning with the iridium anomaly at the end of the Cretaceous and continuing for a little over 100 000 years, the sharp $\delta^{13}C$ decrease, the covariance between the $\delta^{13}C$ and $CaCO_3$ record above the K/T boundary, and the elimination of the

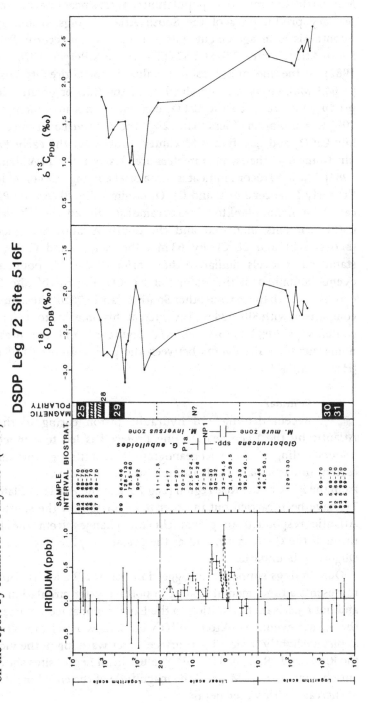

Fig. 10. A comparison of the iridium anomaly recorded at the Cretaceous/Tertiary boundary in Site 516F (Michael *et al.*, 1983) with the oxygen and carbon isotopic record at a similar scale. The ranges of biostratigraphically important foraminifera and nannofossils are based on reports by Pujol (1983) and Weiss (1983) and the magnetic polarity is based on the report by Hamilton & Suzyumov (1983).

Maestrichtian nannofossil population suggest a near total collapse in the primary productivity of the South Atlantic (Figs. 8 and 10). This hypothesis is in agreement with carbon isotope records from other South Atlantic sites, 356 and 524 (Thierstein & Berger, 1978; Hsü *et al.*, 1982). At the time of lightest $\delta^{13}C$ values in the entire Site 516F record ($\sim$64.9 Ma), either a rapid shoaling of the CCD or greatly decreased productivity reduced the $CaCO_3$ content from a maximum of nearly 90% to a minimum of less than 12%. The positive covariance between low $CaCO_3$ and negative $\delta^{13}C$ values throughout the early Paleocene illustrates how the two parameters are closely related (Williams *et al.*, 1983). The $\delta^{18}O$ record indicates that a brief cooling was coincident with the early Paleocene $\delta^{13}C$ and $CaCO_3$ minima (Fig. 8). As the Paleocene calcareous nannoplankton flora became established, the $\delta^{13}C$ values and $CaCO_3$ contents increased and the South Atlantic remained warm (between 21 and 28°C). By 63 Ma, the isotope and $CaCO_3$ records stabilized at levels similar to those prior to the K/T boundary. The events contained in the isotope and $CaCO_3$ records of Site 516F are consistent with those from other South Atlantic DSDP sites, but a direct comparison with Site 524 is prevented at this time due to the low $CaCO_3$ contents (<20%) across the K/T boundary and the presence of numerous turbidite events between the pelagic intervals of that site (Hsü *et al.*, 1982).

Summary

Recent drilling and hydraulic piston coring in the South Atlantic by the Deep Sea Drilling Project has led to a much better understanding of the paleoceanographic evolution of this important region. In particular, stable isotopic studies of sedimentary sequences from the Rio Grande Rise region (Leg 72) and the Falkland Plateau (Leg 71) have been carried out in order to evaluate how the southwestern Atlantic responded to global climatic changes from the Pliocene through the Quaternary, and at the Cretaceous/Tertiary and Eocene/Oligocene boundaries.

Our findings support the suggestion that the 3.2 Ma isotopic event represents a fundamental change in bottom water circulation patterns and not a permanent addition to the Earth's global ice volume. Across the 3.2 Ma event, planktonic $\delta^{18}O$ values show no change or become depleted slightly indicating a surface water warming in the vicinity of the Rio Grande Rise. Benthic $\delta^{18}O$ values in all Leg 72 sites show an ^{18}O enrichment near 3.2 Ma, and the magnitude of the enrichment appears to increase with water depth.

Two modes of glaciation are inferred from the Pliocene benthic $\delta^{18}O$ record of Site 517. Late late Pliocene (2.5–1.8 Ma) $\delta^{18}O$ values are enriched by 0.26‰ over early late Pliocene (3.2–2.5 Ma) values. An abrupt increase in both planktonic and benthic mean $\delta^{18}O$ values and isotopic variability occurred over the Pliocene/Pleistocene boundary, suggesting that glaciations were more severe and extensive in the early Quaternary than in the latest Pliocene. Part of the increased $\delta^{18}O$ variability in the Quaternary is an artifact of combining two different data sets. The Quaternary planktonic and benthic $\delta^{18}O$ record of Site 517 also supports the dichotomous character of Quaternary glaciation. The early Quaternary (1.8 to 0.9 Ma) is marked by relatively depleted values with low variability whereas the late Quaternary (0.9 Ma to present) exhibits enriched $\delta^{18}O$ values and high variability (Williams *et al.*,1981; Prell, 1982a).

During the late Eocene, bottom water conditions appear to have been very similar between the Rio Grande Rise region and the Falkland Plateau, with a surface water temperature gradient of about 4 °C existing between these areas. In early Oligocene time, this surface water temperature gradient increased significantly, due primarily to a high-latitude cooling (Site 511). Bottom water temperatures at Site 511 also dropped dramatically by the early Oligocene (from approximately 1 to 2 °C), suggesting the production of a cold intermediate water mass at this time. Similar early Oligocene bottom temperatures are recorded on the Rio Grande Rise at Site 357 but not at the mid-water depths of Site 516, which remained much warmer. This finding places a constraint on the upper depth limit of this cold, northward flowing, intermediate water mass during the early Oligocene.

An isotopic study of the late Maestrichtian–early Paleocene section of Site 516F reveals that large isotopic anomalies are associated with the Cretaceous/Tertiary boundary. Across the K/T boundary, total carbonate content reaches a maximum of 80% before rapidly decreasing in covariance with the carbon isotope record. This strong covariance between $\delta^{13}C$ and $CaCO_3$ content suggests either a significant reduction in primary productivity or a rapid shoaling of the carbonate compensation depth. Importantly, the $\delta^{13}C$ record remains depleted in ^{13}C by at least 0.5‰ for a period of 2 million years after the K/T boundary as compared to the late Maestrichtian.

Acknowledgements
Samples were provided through the Deep Sea Drilling Project with funding from the US National Science Foundation. We appreciate

the efforts of Ken Hsü, Helmi Weissert and their colleagues at the ETH in Zürich for organizing the First International Conference on Paleoceanography, at which most of the isotopic results in this synthesis were presented by the authors. Donna Black and David Mucciarone at South Carolina are thanked for their dedicated technical assistance in producing this manuscript. This research was partially funded by NSF Grant OCE80–25208 to Robert Thunell and Douglas F. Williams, NSF Grant OCE82–14937 to James P. Kennett, and CNRS–ATP G.G.O. to C. Vergnaud-Grazzini.

REFERENCES

Alvarez, L. W., Alvarez, W., Asaro, F. & Michel, H. V. 1980. Extraterrestrial cause for the Cretaceous–Tertiary extinction. *Science*, **208**, 1095–1098.

Anderson, T. F. & Schneidermann, N. 1973. Stable isotope relationships in pelagic limestones from the central Caribbean: Leg 15, Deep Sea Drilling Project. *Initial Reports of the Deep Sea Drilling Project*, **15**, 795–803.

Arthur, M. A., Scholle, P. A. & Hasson, P. 1979. Stable isotopes of oxygen and carbon in carbonates from Sites 398 and 116 of the Deep Sea Drilling Project. *Initial Reports of the Deep Sea Drilling Project*, **47**, 447–491.

Backman, J. 1979. Pliocene biostratigraphy of DSDP Sites 111 and 116 from the North Atlantic Ocean and the age of Northern Hemisphere Glaciation. In: *Stockholm Contributions in Geology*, **32** (3), 115–137.

Backman, J., Shackleton, N. J. & Pestiaux, P. 1983. Data on Pliocene paleoclimates from the North Atlantic Ocean. *First International Conference on Paleoceanography, Zürich, 1983*, (abstracts), p. 10.

Benson, R. A. 1975. The origin of the psychrosphere as recorded in changes of deep-sea ostracode assemblages. *Lethaia*, **8**, 69–83.

Berger, W. H. 1973. Cenozoic sedimentation in the eastern tropical Pacific. *Geol. Soc. Am. Bull.*, **84**, 1941–1954.

Berggren, W. A. 1969. Rates of evolution in some Cenozoic planktonic foraminifera. *Micropaleontol.*, **15**, 351–365.

Berggren, W. A. 1977. Late Neogene planktonic foraminiferal biostratigraphy of the Rio Grande Rise (South Atlantic). *Mar. Micropaleontol.*, **2**, 265–313.

Berggren, W. A., Aubry, M.-P. & Hamilton, N. 1983. Neogene magneto-biostratigraphy of DSDP Site 516 (Rio Grande Rise, South Atlantic). *Initial Reports of the Deep Sea Drilling Project*, **72**.

Boersma, A. & Shackleton, N. 1977. Tertiary oxygen and carbon isotope stratigraphy, Site 357 (mid-latitude South Atlantic). *Initial Reports of the Deep Sea Drilling Project*, **39**, 911–924.

Bremer, M. L. & Lohmann, G. P. 1981. Evidence for primary control of the distribution of certain Atlantic Ocean benthonic foraminifera by degree of carbonate saturation. *Deep Sea Res.*, **12**, 987–988.

Brunn, A. F. 1957. Deep sea and abyssal depths. *Geol. Soc. Am. Mem.*, **67**, 641–672.

Brunner, C. A. 1978. Late Neogene and Quaternary Paleoceanography and Biostratigraphy of the Gulf of Mexico. Unpublished doctoral thesis, University of Rhode Island, 341 pp.

Cifelli, R. 1969. Radiation of Cenozoic planktonic foraminifera. *Syst. Zool.*, **18**, 154–168.

Coplen, T. B. & Schlanger, S. O. 1973. Oxygen and carbon isotope studies of carbonate sediments from Site 167, Magellan Rise, Leg 17. *Initial Reports of the Deep Sea Drilling Project*, **17**, 505–509.

Corliss, B. H. 1978a. Studies of deep-sea benthonic foraminifera in the Southern Ocean. Unpublished doctoral thesis, University of Rhode Island.

Corliss, B.H. 1978b. Studies of deep-sea benthonic foraminifera in the southeast Indian Ocean. *Antarctic J.*, **13**, 116–117.

Corliss, B.H. 1979. Recent deep-sea benthonic foraminiferal distributions in the southeast Indian Ocean: Inferred bottom-water routes and ecological implications. *Mar. Geol.*, **31**, 115–138.

Corliss, B.H. 1981. Deep-sea benthonic foraminiferal faunal turnover near the Eocene/Oligocene boundary. *Mar. Micropaleontol.*, **6**, 367–384.

Corliss, B.H. 1983. Distribution of Holocene deep-sea benthonic foraminifera in the southeast Indian Ocean. *Deep Sea Res.*, **30**, 47–61.

Curry, W.B. & Matthews, R.K. 1981. Equilibrium ^{18}O fractionation in small size fraction planktic foraminifera: evidence from Recent Indian Ocean sediments. *Mar. Micropaleontol.*, **6**, 327–337.

Davies, T.A. & Worsley, T.R. 1981. Paleoenvironmental implications of oceanic carbonate sedimentation rates. *Soc. Econ. Pal. Min. Spec. Publ.*, **32**, 169–180.

Davies, T.A., Hay, W.W., Southam, J.R. & Worsley, T.R. 1977. Estimates of Cenozoic oceanic sedimentation rates. *Science*, **197**, 53–55.

Detrick, R.S., Sclater, J.G. & Thiede, J. 1977. The subsidence of aseismic ridges. *Earth Planet. Sci. Lett.*, **34**, 185–196.

Douglas, R.G. & Savin, S.M. 1971. Isotopic analyses of planktonic foraminifera from the Cenozoic of the Northwest Pacific, Leg 6. *Initial Reports of the Deep Sea Drilling Project*, **6**, 1123–1127.

Douglas, R.G. & Savin, S.M., 1973. Oxygen and carbon isotope analysis of Cretaceous and Tertiary foraminifera from the central North Pacific. *Initial Reports of the Deep Sea Drilling Project*, **17**, 5591–5605.

Duplessy, J.C., Moyes, J. & Pujol, C. 1980. Deep water formation in the North Atlantic Ocean during the last ice age. *Nature*, **286**, 479–482.

Emiliani, C. 1955. Pleistocene temperatures. *J. Geol.*, **63**, 538–573.

Emiliani, C. 1980. Death and renovation at the end of the Mesozoic. *EOS, EPSL paper*, **61**, 505–605.

Epstein, S., Buchsbaum, R., Lowenstam, H. & Urey, H.C. 1953. Revised carbonate-water isotopic temperature scale. *Geol. Soc. Am. Bull.*, **64**, 1315–1325.

Frerichs, W.E. 1971. Evolution of planktonic foraminifera and paleotemperatures. *J. Paleontol.*, **45**, 963–968.

Ganapathy, R. 1980. A major meteorite impact on the Earth 65 million years ago: evidence from the Cretaceous–Tertiary Boundary clay. *Science*, **209**, 921–923.

Gofas, S. 1978. Une approche du paléoenvironment océanique: les foraminifères benthiques calcaires, traceurs de la circulation abyssale. Doctoral thesis, Université de Bretagne Occidentale, Brest, France.

Hamilton, N. 1983. Cretaceous/Tertiary boundary studies at Deep Sea Drilling Project Site 516, Rio Grande Rise, South Atlantic: a synthesis. *Initial Reports of the Deep Sea Drilling Project*, **72**, 949–952.

Hamilton, N. & Suzyumov, A.E. 1983. Late Cretaceous magnetostratigraphy of Site 516, Rio Grande Rise, southwestern Atlantic Ocean, Deep Sea Drilling Project, Leg 72. *Initial Reports of the Deep Sea Drilling Project*, **72**, 723–730.

Haq, B.U. 1981. Paleogene paleoceanography: early Cenozoic oceans revisited. *Oceanol. Acta*, Special Publication, 71–82.

Haq, B.U., Premoli-Silva, I. & Lohmann, G.P. 1977. Calcareous plankton paleobiogeographic evidence for major climatic fluctuations in the early Cenozoic Atlantic Ocean. *J. Geophys. Res.*, **82**, 3861–3876.

Haq, B.U. Okada, H. & Lohmann, G.P. 1979. Paleobiogeography of the Paleocene/Oligocene calcareous nannoplankton from the North Atlantic Ocean. *Initial Reports of the Deep Sea Drilling Project*, **43**, 617–629.

Hardenbol, J. & Berggren, W.A. 1978. A new Paleogene numerical time scale. In: Contributions to the Geologic Time Scale, ed. G.V. Cohee, M.F. Glaessner & H.D. Hedberg, *Am. Ass. Petrol. Geol. Bull.* pp. 213–234.

Heath, G.R. 1969. Carbonate sedimentation in the abyssal equatorial Pacific during the past 50 million years. *Geol. Soc. Am. Bull.*, **80**, 689–694.

Hodell, D.A., Kennett, J.P. & Leonard, K. 1983. Climatically-induced changes in vertical water mass structure of the Vema Channel during the Pliocene: Evidence from DSDP Sites 516A, 517 and 518. *Initial Reports of the Deep Sea Drilling Project*, **72**, 907–919.

Hodell, D.A., Williams, D.F. & Kennett, J.P. 1984. Reorganization of deep vertical water mass structure in the Vema Channel at 3.2 Ma: Faunal and isotopic evidence from DSDP Let 72 (South Atlantic). *Geol. Soc. Am. Bull.*, 1984.

Hsü, K.J. 1980. Terrestrial catastrophe caused by cometary impact at the end of the Cretaceous. *Nature*, **285**, 201–203.

Hsü, K.J. 1981. Origin of geochemical anomalies at Cretaceous–Tertiary boundary. Asteroid or cometary impact? *Oceanol. Acta*, **4**, 129–133.

Hsü, K.J., He, Q., McKenzie, J.A., Weissert, H., Perch-Nielson, K., Oberhänsli, H., Kelts, K., La Brecque, J., Tauze, L., Krähenbühl, U., Percival, S.F., Wright, R., Karpoff, A.M., Petersen, N., Tucker, P., Pooer, R.Z., Gombos, A.M., Pisciotto, K., Carman, M.F. & Schrieber, E. 1982. Mass mortality and its environmental and evolutionary consequences. *Science*, **216**, 249–256.

Keigwin, L.D., Jr 1979. Late Cenozoic stable isotopic stratigraphy and paleoceanography of DSDP sites from the east equatorial and central north Pacific Ocean. *Earth Planet. Sci. Lett.*, **45**, 361–382.

Keigwin, L.D., Jr 1980. Paleoceanographic change in the Pacific at the Eocene–Oligocene boundary. *Nature*, **287**, 722–725.

Keigwin, L.D., Jr 1982a. Isotopic paleoceanography of the Caribbean and East Pacific: Role of Panama uplift in Late Neogene time. *Science*, **217**, 350–353.

Keigwin, L.D., Jr 1982b. Stable isotope stratigraphy and paleoceanography of Sites 502 and 503. *Initial Reports of the Deep Sea Drilling Project*, **68**, 445–453.

Keigwin, L.D., Jr & Corliss, B.H. 1982. Eocene–Oligocene isotopic paleoceanography. *Geol. Soc. Am., Abstracts with Programs*, **14**, 527.

Keigwin, L.D., Jr & Corliss, B.H. 1983. Preliminary synthesis of Eocene–Oligocene stable isotope data from Atlantic, Indian and Pacific Ocean sites. *Am. Ass. Petrol. Geol. Annual Convention*, (abstracts), p. 105.

Keigwin, L.D. & Thunell, R.C. 1979. Middle Pliocene climatic change from the western Mediterranean from faunal and oxygen isotopic trends. *Nature*, **282**, 294–296.

Kennett, J.P. 1978. The development of planktonic biogeography in the Southern Ocean during the Cenozoic. *Mar. Micropaleontol.*, **3**, 301–345.

Kennett, J.P. & Shackleton, N.J. 1976. Oxygen isotopic evidence for the development of the psychrosphere 38 m.y. ago. *Nature*, **260**, 513–515.

Kennett, J.P., Shackleton, N.J., Margolis, S.V., Kroopnick, S. & Goodney, S. 1979. Late Cenozoic oxygen and carbon isotopic history and volcanic ash stratigraphy: DSDP Site 284, South Pacific. *Am. J. Sci.*, **279**, 53–69.

Krasheninnikov, B.A. & Basov, I.V. 1983. Cenozoic planktonic foraminifers of the Falkland Plateau and Argentine Basin, Deep Sea Drilling Project Leg 71. *Initial Reports of the Deep Sea Drilling Project*, **71**, 821–858.

Kroopnick, P. 1980. The distribution of $\delta^{13}C$ in the Atlantic Ocean. *Earth Planet. Sci. Lett.*, **49**, 469–484.

Ledbetter, M.T. 1981. Paleoceanographic significance of bottom-current fluctuations in the Southern Ocean. *Nature*, **294**, 554–556.

Ledbetter, M.T. & Huang, T.C. 1980. Reduction of manganese micronodule distribution of the South Pacific during the last three million years. *Mar. Geol.*, **36**, M21–28.

Ledbetter, M.T., Williams, D.F. & Ellwood, B.B. 1978. Late Pliocene climate and southwest Atlantic abyssal circulation. *Nature*, **272**, 237–239.

Leonard, K.A., Williams, D.F. & Thunell, R.C. 1983. Pliocene paleoclimatic and paleoceanographic history of the South Atlantic Ocean: Stable isotopic records from Leg 72 DSDP Sites 516A and 517. *Initial Reports of the Deep Sea Project*, **72**, 895–906.

Letolle, R. 1979. Carbon and oxygen isotopes of limestones, Deep Sea Drilling Project, Site 398, Leg 47B. *Initial Reports of the Deep Sea Drilling Project*, **47**, 493–496.

Letolle, R., Renard, M., Bourbon, M. & Filly, A. 1978. O^{18} and C^{13} isotopes in Leg 44 carbonates: a comparison with the Alpine series. *Initial Reports of the Deep Sea Drilling Project*, **44**, 567–573.

Letolle, R., Bergnaud-Grazzini, C. & Pierre, C. 1979. Oxygen and carbon isotopes from bulk carbonates and foraminiferal shells at DSDP Sites 400, 401, 402, 403 and 406. *Initial Reports of the Deep Sea Drilling Project*, **48**, 741–755.

Lipps, J.H. 1970. Plankton evolution. *Evolution*, **24**, 1–22.

Lohmann, G.P. 1978. Abyssal benthic foraminifera as hydrographic indicators in the western South Atlantic. *J. Foram. Res.*, **8**, 6–34.

Lohmann, G.P. 1981. Modern benthic foraminiferal biofacies: Rio Grande Rise. *EOS*, **62**, 903, (abstract).

Mankinen, E.A. & Dalrymple, G.B. 1979. Revised geomagnetic polarity time scale for the interval 0 to 5 m.y. *B.P.J. Geophys. Res.*, **84**, 615–626.

McDougall, I. 1977. *The Present Status of the Geomagnetic Polarity Time Scale*, Publ. No. 1288, Research School of Earth Sciences, Australian National University, Canberra, ACT, 34 pp.

Michael, H.V., Asaro. R., Alvarez, W., Alvarez, L.W. & Johnson, D.A. 1983. Abundance profiles of iridium and other elements near the Cretaceous/Tertiary boundary in Hole 516F of Deep Sea Drilling Project Leg 72. *Initial Reports of the Deep Sea Drilling Project*, **72**, 931–936.

Miller, K.G. & Curry, W.B. 1982. Eocene to Oligocene benthic foraminiferal isotopic record in the Bay of Biscay. *Nature*, **296**, 347–350.

Moore, T.C., van Andel, T.H., Sancetta, C. & Pisias, N. 1978. Cenozoic hiatuses in pelagic sediments. *Micropaleontol.*, **24**, 113–138.

Murphy, M.G. & Kennett, J.P. 1983. The Eocene/Oligocene boundary: new isotopic evidence from DSDP Leg 90, temperate southwest Pacific. *Geol. Soc. Am., Abstracts with Programs*, **15**, 649.

Muza, J.P., Williams, D.F. & Wise, S.W. 1983. Paleogene oxygen isotope record for Deep Sea Drilling Sites 511 and 512, Subantarctic South Atlantic Ocean: paleotemperatures, paleoceanographic changes, and the Eocene/Oligocene boundary event. *Initial Reports of the Deep Sea Drilling Project*, **71**, 409–422.

O'Neil, J.R., Clayton, R.N. & Mayeda, T.K. 1969. Oxygen isotope fractionation in divalent metal carbonates. *J. Chem. Phys.*, **51**, 5547–5558.

Perch-Nielsen, K., McKenzie, J. & He, Q. 1982. Biostratigraphy and isotope stratigraphy and the 'catastrophic' extinction of calcareous nannoplankton at the Cretaceous/Tertiary boundary. *Geol. Soc. Am. Spec. Paper*, **190**, 353–371.

Poore, R.Z. & Matthews, R.K. 1984. Late Eocene–Oligocene oxygen and carbon isotope record from South Atlantic Ocean, Deep Sea Drilling Project Site 522. *Initial reports of the Deep Sea Drilling Project*, **73**, 725–735.

Prell, W.L. 1982a. Oxygen and carbon isotope stratigraphy for the Quaternary of Hole 502B: Evidence for two modes of isotopic variability. *Initial Reports of the Deep Sea Drilling Project*, **68**, 455–464.

Prell, W.L. 1982b. A re-evaluation of the initial of Northern Hemisphere glaciation at 3.2 m.y. New isotopic evidence. *Geol. Soc. Am., Abstracts with Programs*, **14** (7), 592.

Prell, W.L. 1983. Oxygen and carbon isotope stratigraphy for the Quaternary of Hole 502B: evidence for two modes of isotopic variability. *Initial Reports of the Deep Sea Drilling Project*, **68**, 455–466.

Pujol, C. 1983. Cenozoic planktonic foraminiferal biostratigraphy of the southwest Atlantic (Rio Grande Rise). *Initial Reports of the Deep Sea Drilling Project*, **72**, 623–653.

Reid, J.L., Nowlin, W.S. & Patzert, W.C. 1976. On the characteristics and circulation of the Southwestern Atlantic Ocean. *J. Phys. Oceanogr.*, **7**, 62–91.

Rona, P.A. 1973. Worldwide unconformities in marine sediments related to eustatic changes of sea level. *Nature*, **244**, 25–26.

Sancetta, C. 1979. Paleogene Pacific microfossils and paleoceanography. *Mar. Micropaleontol.*, **4**, 363–398.

Savin, S.M. 1977. The history of the earth's surface temperature during the past 100 million years. *Ann. Rev. Earth Planet. Sci.*, **5**, 319–355.

Savin, S.M., Douglas, R.G. & Stehli, F.G. 1975. Tertiary marine paleotemperatures. *Geol. Soc. Am. Bull.*, **64**, 67–87.

Schnitker, D. 1974. West Atlantic abyssal circulation during the past 120 000 years. *Nature*, **248**, 385–387.

Scholle, P.A. & Arthur, M.A. 1980. Carbon isotope fluctuations in Cretaceous pelagic limestones: potential stratigraphic and petroleum exploration tool. *Am. Ass. Petrol. Geol. Bull.*, **64**, 67–87.

Shackleton, N.J. 1974. Attainment of isotopic equilibrium between ocean water and the benthonic foraminifera genus *Uvigerina*: isotopic changes in the ocean during the last glacial. *Cent. Nat. Rech. Sci. Colloques Internat.*, **219**, 203–209.

Shackleton, N.J. & Boersma, A. 1981. The climate of the Eocene ocean. *J. Geol. Soc. London*, **138**, 153–157.

Shackleton, N.J. & Kennett, J.P. 1975. Paleotemperature history of the Cenozoic and the initiation of Antarctic glaciation: Oxygen and carbon isotope analyses in DSDP Sites 277, 279 and 281. *Initial Reports of the Deep Sea Drilling Project*, **29**, 743–756.

Shackleton, N.J. & Opdyke, N.D. 1973. Oxygen isotope and paleomagnetic stratigraphy of equatorial Pacific core V28–238: Oxygen isotope temperatures and ice volumes on a 10^5 and 10^6 year scale. *Quat. Res.*, **3**, 39–55.

Shackleton, N.J. & Opdyke, N.D. 1976. Oxygen isotope and paleomagnetic stratigraphy of Pacific core V28–239, Late Pliocene to Latest Pleistocene. In: Investigations of Late Quaternary Paleoceanography and Paleoclimatology, ed. R.M. Cline & J.D. Hays, *Geol. Soc. Am. Mem.*, **145**, 449–464.

Shackleton, N.J. & Opdyke, N.D. 1977. Oxygen isotope and paleomagnetic evidence for early Northern Hemisphere glaciation. *Nature*, **270**, 216–219.

Shackleton, N.J., Wiseman, J.H.D. & Buckley, H.A. 1973. Non-equilibrium isotope fractionation between sea-water and planktonic foraminiferal tests. *Nature*, **242**, 177–179.

Shackleton, N.J., Backman, J., Kent, D. *et al.* 1982. The evolution of climate response to orbital forcing: Results over three million years from DSDP Site 552A. In: *International Symposium on Milankovitch and Climate: Understanding the Response to Orbital Forcing, Abstract Volume*, ed. A. Berger & J. Imbrie.

Shackleton, N.J., Hall, M.A. & Lang, S. 1983. Carbon isotope data in core V19–30 confirm reduced carbon dioxide concentration in the ice age atmosphere. *Nature*, **306**, 319–322.

Shackleton, N.J., Backman, J., Zimmerman, H., Kent, D.V., *et al.* 1984. Oxygen isotope calibration of the onset of ice-rafting and history of glaciation in the North Atlantic region. *Nature*, **307**, 620–623.

Smit, J. & Hertogen, J. 1980. An extraterrestrial event at the Cretaceous–Tertiary boundary. *Nature*, **285**, 198–200.

Streeter, S.S. 1973. Bottom water and benthonic foraminifera in the North Atlantic: Glacial–interglacial contrasts. *Quat. Res.*, **3**, 131–141.

Thiede, J. 1977. The subsidence of aseismic ridges: evidence from sediments on the Rio Grande Rise. *Am. Ass. Petrol. Geol. Bull.*, **61**, 929–940.

Thierstein, H. & Berger, H. 1978. Injection events in earth history, *Nature*, **276**, 461–464.

Thunell, R.C. 1979. Pliocene–Pleistocene paleotemperature and paleosalinity history of the Mediterranean Sea, results from DSDP Sites 125 and 132. *Mar. Micropaleontol.*, **4**, 173.

Thunell, R.C. 1981a. Cenozoic paleotemperature changes and planktonic foraminiferal speciation. *Nature*, **289**, 670–672.

Thunell, R.C. 1981b. Late Miocene–Early Pliocene planktonic foraminiferal biostratigraphy and paleoceanography of low-latitude marine sequences. *Mar. Micropaleontol.*, **6**, 71–90.

Thunell, R.C. & Corliss, B.H. 1983. Late Eocene and early Oligocene carbonate deposition in the deep sea. *Am. Ass. Petrol. Geol. Annual Convention*, (abstracts), p. 170.

Thunell, R.C. & Williams, D.F. 1983. The stepwise development of Pliocene–Pleistocene paleoclimatic and paleoceanographic conditions in the Mediterranean. *Utrecht Micropaleontol. Bull.*, **30**, 111–127.

van Andel, Tj.H. 1975. Mesozoic/Cenozoic calcite compensation depth and the global distribution of calcareous sediments. *Earth Planet. Sci. Lett.*, **26**, 187–194.

van Andel, Tj.H., Heath, G.R. & Moore, T.C. 1975. Cenozoic history and paleoceanography of the central equatorial Pacific Ocean. *Geol. Soc. Am. Mem.*, 134, 1–134.

van Andel, Tj.H., Thiede, J., Sclater, J.G. & Hay, W.W. 1977. Depositional history of the South Atlantic Ocean during the last 125 million years. *J. Geol.*, 85, 651–698.

van Donk, J. 1976. O^{18} record of the Atlantic Ocean for the entire Pleistocene epoch. In: Investigation of Late Quaternary paleoceanography and paleoclimatology, ed. R.M. Cline & J.D. Hays, *Geol. Soc. Am. Mem.*, 145, 147–163.

Vergnaud-Grazzini, C., Pierre, C. & Letolle, R. 1978. Paleoenvironment of the North-East Atlantic during the Cenozoic: oxygen and carbon isotope analyses at DSDP Sites 398, 400A and 401. *Oceanol. Acta*, 1, 381–390.

Vergnaud-Grazzini, C., Grably, M., Pujol, C. & Duprat, J. 1983. Oxygen isotope stratigraphy and paleoclimatology of southwestern Atlantic Quaternary sediments (Rio Grande Rise) at Deep Sea Drilling Project Site 517. *Initial Reports of the Deep Sea Drilling Project*, 72, 871–884.

Vincent, E. 1981. Neogene carbonate stratigraphy of Hess Rise (Central North Pacific) and paleoceanographic implications. *Initial Reports of the Deep Sea Drilling Project*, 62, 571–606.

Weiss, W. 1983. Upper Cretaceous planktonic foraminiferal biostratigraphy from the Rio Grande Rise: Site 516 of Leg 72, Deep Sea Drilling Project. *Initial Reports of the Deep Sea Drilling Project*, 72, 715–722.

Williams, D.F., Sommer, M.A. & Bender, M.L. 1977. Carbon isotopic compositions of Recent planktonic foraminifera of the Indian Ocean. *Earth Planet. Sci. Lett.*, 36, 391–403.

Williams, D.F., Healy-Williams, N., Leventer, A., Leonard, K., & Johnson, P. 1981. Stable carbon and oxygen isotope records from DSDP Leg 72, South Atlantic. *EOS, Trans. Am. Geophys. Union*, 62, 903.

Williams, D.F., Healy-Williams, N., Thunell, R.C. & Leventer, A. 1983. Detailed stable isotope and carbonate records from the late Maestrichtian–early Paleocene section of Site 515F (Leg 72) including the Cretaceous/Tertiary boundary. *Initial Reports of the Deep Sea Drilling Project*, 72, 921–929.

Woodruff, F., Savin, S.M. & Douglas, D.G. 1981. Miocene stable isotope record: A detailed deep Pacific Ocean study and its paleoclimatic implications. *Science*, 121, 665–668.

Worthington, L.W. 1970. The Norwegian Sea as a Mediterranean basin. *Deep Sea Res.*, 17, 77–84.

14

Abyssal teleconnections II.
Initiation of Antarctic Bottom Water flow
in the southwestern Atlantic

D. A. JOHNSON

Abstract

Three principal factors are potentially important in governing the
initiation of Antarctic Bottom Water (AABW) flow through the deep
ocean basins during the Paleogene, and subsequent fluctuations in its
intensity during the Neogene and Quaternary. These are: (a) the
beginning of circumpolar flow, leading to thermal isolation of the
Antarctic continent and decreasing surface water temperatures around
the Antarctic perimeter; (b) injection of high-salinity waters into the
Antarctic region at intermediate depths; and (c) rifting and subsidence of
sills and passages below critical threshold depths. Of these factors,
salinity has most probably been the limiting one. Formation of bottom
water in the northern Atlantic began near the Eocene/Oligocene
boundary via overflow from the Norwegian Sea through deepening sills
in the Iceland–Scotland Ridge. This overflow, or proto-North Atlantic
Deep Water (NADW), injected high-salinity water into the central and
southern Atlantic at intermediate depths, and was probably essential to
AABW production. A prominent seismic discontinuity (Horizon A of
Ewing) marks the initiation of AABW flow through the southwestern
Atlantic; drilling through this discontinuity in the Brazil Basin at DSDP
Site 515 shows thick, rapidly accumulating (> 20 m/Ma) bottom current
deposits beginning in the middle Oligocene (32 Ma), and continuing
through the Quaternary. The effects of AABW in the deep North Atlantic
were also well established by the middle Oligocene ($\sim$ 30 to 33 Ma), with
the advection of older and ^{13}C-depleted bottom waters as far north as the
Bay of Biscay. Beginning in the middle Miocene ($\sim$ 16 Ma), an alternative
source region developed in the northwestern Indian Ocean for supplying
high-salinity waters to high southern latitudes. The gradual closing of the
eastern Tethyan seaways during the Neogene dramatically altered the
evaporation/precipitation budget of southwestern Asia, allowing the
Arabian Sea and its marginal basins to become source areas for
southward-flowing hypersaline intermediate and deep waters. During the
climatic cycles of the late Neogene and Quaternary, the most recent

episodes of intense AABW flow appear to precede the strongest glacial maxima at 50 000 and 150 000 years BP. These episodes are not directly associated with maxima in NADW production which may occur at a 40 000 year period. Thus the production of high-salinity deep water in the northernmost Atlantic (i.e. NADW) is not necessarily the only precondition to the formation of AABW, provided that an alternative source of hypersaline water (e.g. the northwestern Indian Ocean) is available.

Introduction

The purpose of this discussion is to review our present state of ignorance regarding the initiation of Antarctic Bottom Water (AABW) flow in the deep basins of the world's oceans: for example, what is the chronology of major pulses of AABW production during the Cenozoic, and how precisely can we identify the limiting factors (i.e. the necessary conditions and the sufficient conditions) associated with AABW formation around the Antarctic perimeter?

I shall begin by proposing a fundamental assumption to provide a frame of reference for considering deep thermohaline convection and its imprint upon the geological record. This assumption is perhaps implicit in many of the strategies which paleoceanographers have applied, yet I believe it is sufficiently important to be re-stated at the outset. Suppose that one considers a particular paleoceanographic objective: in this instance, determining the chronology of AABW initiation during the Paleogene, and the limiting factors associated with its formation. The fundamental assumption is this: That the critical information in the geologic record pertaining to this particular objective is *not* randomly distributed in time and space, but is tightly clustered in key geographic regions and in relatively narrow time slices. Our task as paleoceanographers is to identify these critical clusters of information, and then focus a diverse array of geologic interpretative strategies upon these relatively narrow targets, even if it requires excluding the temporal and spatial surroundings of the chosen 'cluster'.

For those of us focusing our attention upon the *deep* ocean circulation, this approach of identifying and concentrating our attention upon restricted targets of information is a marked contrast to the strategies used by our colleagues in describing and interpreting surface ocean conditions during the past. The successful multidisciplinary efforts of the CLIMAP and CENOP projects, for example, have required a broad spatial coverage of samples in order to map global distribution patterns of paleontological indicators of surface ocean properties (e.g. CLIMAP Project Members, 1981). The analysis of widely dispersed core samples yields the most satisfactory data base from which one may contour desired properties of the surface oceans. The maps of these properties in turn may be used as boundary conditions in constructing models of the atmosphere–ocean climate system at designated time-slices of interest (e.g. Gates, 1977; Manabe & Hahn, 1977). While this map-making approach is appropriate for the surface oceans, I would argue that an identical strategy is not necessarily suitable for understanding deep ocean circulation.

Suppose that we could construct, for example, an accurate and detailed map of ocean surface temperature on a global scale for any desired time-slice. Moreover, suppose that the further refinement of our techniques allowed the construction of maps of *other* important properties in surface waters as well (e.g. salinity, oxygen, nutrients, etc.). Question: Would a complete set of such maps allow either a qualitative or a quantitative description of deep thermohaline circulation for that particular time-slice? The answer is almost certainly *No*. If the modern ocean is a representative analog for the ancient oceans, the geographic localities and the limiting oceanographic conditions required for deep convection to occur are dependent upon water properties down to and including the main pycnocline, not simply the surface water conditions (e.g. Warren, 1981a; Killworth, 1983). The relatively restricted localities where bottom waters are formed today do not appear anomalous in the distribution of surface water properties compared to the surrounding oceans. Most commonly it is the properties of subsurface waters and to some extent the geometrical boundary conditions imposed by the ocean floor which prove to be limiting conditions to deep convection in a given area. Moreover, there are a number of regions such as the North Pacific which are potentially suitable for bottom water formation, but where deep convection does *not* occur because a critical water property (e.g. high salinity) is not available via the possible teleconnective linkages to other oceanic areas (Warren, 1983).

If the mapping of surface ocean properties alone is an unsatisfactory approach to identifying deep convection, what is the best alternative? Using the modern oceans as an analog, deep thermohaline currents today are typically broad (several hundred kilometres) and sluggish, with mean flow velocities of only a few centimetres per second (Warren, 1981a), which is not sufficient for sediment erosion and transport. Time-series observations of the deep flow typically show the strongest tidal components in the total energy spectrum at tidal or eddy frequencies (Schmitz, 1978), the exceptions being in relatively restricted straits or sills where the flow is accelerated and rectified, and the mean flow field may overshadow the higher-frequency components of the flow (e.g. Schmitz & Hogg, 1983). It is in such restricted regions that we have focused our attention in geologic studies to describe and understand the history of deep thermohaline circulation because the geologic imprints are most striking in such locations. This paper will incorporate the results of studies in three critical passages through which modern AABW flow is substantial: the Vema Channel of the southwestern Atlantic, the Amirante Passage of the western Indian Ocean, and the Samoan Passage of the southwestern Pacific. In addition, results from select DSDP sites will be cited when such results have been obtained at appropriate 'clusters' in time and space pertinent to describing deep ocean circulation.

In this paper I will consider a particular event in the history of abyssal circulation, namely, the initiation of northward-flowing AABW through the deep basins of the world's oceans. We shall examine the evidence pertaining to the initial triggering of AABW flow during the Paleogene, and some evidence bearing on the chronology of significant fluctuations during the latest Cenozoic. We shall identify the probable essential conditions associated with AABW formation, and then indicate several geological arguments which suggest problems with this simplest scenario. I shall then outline an alternative model for AABW formation during the Neogene which includes a different set of mechanisms to those of the Paleogene when AABW flow began initially.

Modern AABW flow: oceanographic and geologic aspects

We shall consider the basins of the western Atlantic for describing and interpreting AABW flow, although it is clear that volumetrically equivalent quantities of AABW enter the deep Pacific and Indian Oceans as well (Warren, 1981a). The western Atlantic, unlike the Pacific Ocean, has both northern and southern sources of present-day bottom water, and hence the contrasts in the diagnostic water properties are especially striking in vertical sections. The first-order aspects of the deep circulation appear on cross-sections of salinity and silicate, extending northward from the Weddell Sea into the Argentine and Brazil Basins, thence into the western North Atlantic (Figs. 1–3). A sharp salinity minimum ($S < 34.5\,\%o$) centered near 1000 m depth denotes Antarctic Intermediate Water (AAIW), among the shallowest of the major thermohaline currents. This water mass obtains its properties in the surface waters of the Antarctic south of the Polar

Front, where it sinks and flows northward, extending as far as tropical latitudes in each of the ocean basins. Beneath AAIW is a broad salinity maximum centered between 2000 and 3500 m (Fig. 2) corresponding to southward-flowing North Atlantic Deep Water (NADW). Much of NADW can be traced to overflow from the Norwegian Sea (Steele *et al.*, 1962), although a substantial contribution to NADW may come from open-ocean convection in the Labrador Sea as well (Clarke & Gascard, 1982). NADW is of comparable density to Circumpolar Deep Water (Reid *et al.*, 1977), and the salinity maximum associated with NADW (Fig. 2) can be traced eastward within the zonal circumpolar flow into the southern Indian and Pacific Oceans (Reid & Lynn, 1971, Fig. 8). Beneath the southward-flowing NADW, AABW flows northward. The

Fig. 1. Location map for transect of GEOSECS stations in the central and southwestern Atlantic, after Bainbridge (1980). Dashed line denotes cross-sections of salinity and dissolved silica (Fig. 2 and 3, respectively). Solid line indicates position of salinity transect in western Weddell Sea (Fig. 4).

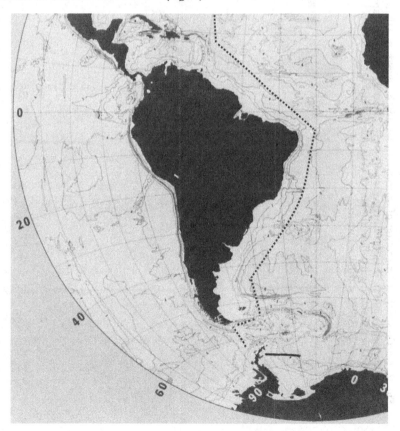

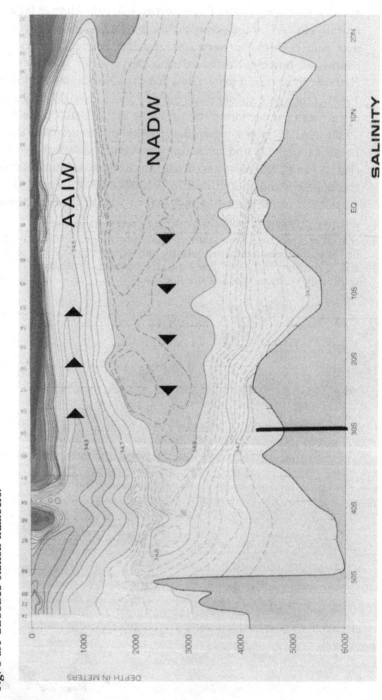

Fig. 2. North–south cross-section of salinity (in ‰) in the western Atlantic, after Bainbridge (1980). The salinity minimum near 1000 m denotes northward-flowing Antarctic Intermediate Water (AAIW), and the salinity maximum near 2500 m denotes southward-flowing North Atlantic Deep Water (NADW). The numbers along the top of this figure and Fig. 3 are GEOSECS station numbers.

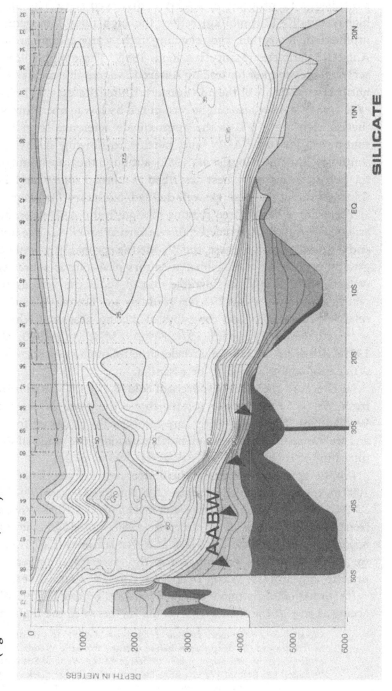

Fig. 3. Dissolved silica in the western Atlantic, after Bainbridge (1980). Highest values (> 120 μmol/kg) mark northward-flowing Antarctic Bottom Water (AABW), which presently extends to the latitudes of Bermuda and the eastern United States (e.g. Richardson *et al.*, 1981).

most diagnostic property of AABW is dissolved silica (Fig. 3), whose high values ($>120\,\mu$mol/kg) reflect the high rate of biogenic silica production, deposition, and recycling in the water column around the Antarctic perimeter (Mantyla & Reid, 1983). These high silica values serve as a key marker for tracing Antarctic-source abyssal waters as far north as the latitude of the northeastern United States (e.g. Richardson et al., 1981). AABW also can be recognized by low temperature values; the isotherm $\theta = 1.8\,°$C is the approximate reference level between southward-flowing NADW and northward-flowing AABW in the southwest Atlantic (Hogg et al., 1982), although the transition between the two water masses is best described as a broad zone nearly 1 km in thickness rather than a single well-defined 'boundary' (Johnson, 1982).

There are at least three regions of significant AABW production around the Antarctic perimeter: the western Weddell Sea, the Ross Sea, and the Adelie Coast; of these, the Weddell Sea probably contributes the greatest proportion to the global inventory of Antarctic-source bottom waters. There is still considerable debate about the various sources which contribute to AABW (or Weddell Sea Bottom Water*), their relative proportions, and the types of mixing processes which may occur (e.g. Foster & Carmack, 1976; Foster & Middleton, 1980; Warren, 1981a; Killworth, 1983). Nevertheless the following aspects of bottom water production appear to be essential.

(a) There are one or more sources of cold ($\theta < -1.5\,°$C) and relatively fresh ($S = 34.52$‰) surface waters above the shelf of the western Weddell Sea. These cold waters are associated with sea ice formation, and are transported offshore and northward along the Weddell shelf at a total flux of about $2 \times 10^6\,$m^3/s (Warren, 1981a, p. 20).

(b) The break in slope at the shelf edge, between depths of 500 and 1000 m, corresponds with a tongue of Warm Deep Water ($\theta = +0.4\,°$C; $S > 34.68$‰). This intermediate water mass (Fig. 4) is ultimately derived from NADW (Foster & Middleton, 1980), although the salinity maximum associated with NADW at lower latitudes (Fig. 2) becomes somewhat diluted by entrainment in the eastward circumpolar flow prior to appearing as Warm Deep Water in the Weddell Sea.

(c) Formation of bottom water (Weddell Sea Bottom Water, WSBW) occurs at or near the shelf break via mixing between the high-salinity

---

* We shall retain the more familiar designation 'AABW' in this discussion in referring to the deep thermohaline outflow from the Weddell Sea. More precisely, one might prefer to designate the Weddell Sea outflow as 'Weddell Sea Bottom Water', since its properties change markedly via mixing with the overlying water as it is advected northward (Reid et al., 1977).

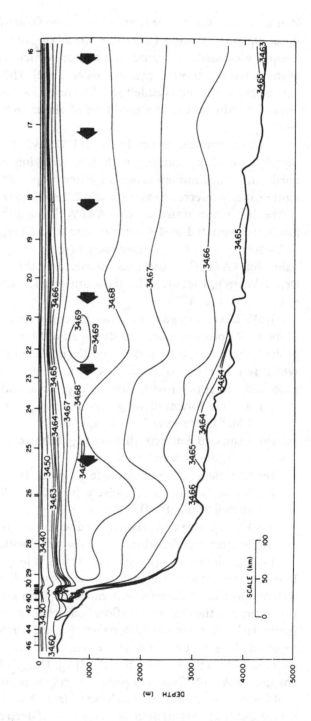

Fig. 4. Salinity transect in the western Weddell Sea, after Foster & Middleton (1980). The salinity maximum near 1000 m is ultimately derived from high-salinity water of northern origin (e.g. NADW) which becomes entrained in the circumpolar flow. The numbers along the top of this figure are the hydrographic stations of USCGC *Glacier*, IWSOE 76.

Warm Deep Water and the relatively cold shelf water(s). Due to the non-linear equation of state of seawater, the mixing of two water masses of comparable density will produce a mixture which is denser than either of the two original water masses (Foster, 1972). This 'cabbeling' effect may be critical in the Weddell Sea. The resulting outflow of WSBW to abyssal depths occurs at a total flux of about $16 \times 10^6 \,\mathrm{m}^3/\mathrm{s}$ (Warren, 1981a, p. 17).

(d) The properties of newly formed WSBW ($\theta < -0.7\,^\circ\mathrm{C}$) become rapidly altered via mixing with the overlying water as it moves northward. Such mixing produces water properties ($\theta = -0.5$ to $0\,^\circ\mathrm{C}$) most properly referred to as Antarctic Bottom Water (Reid et al., 1977).

The flux of northward-flowing AABW through the western Atlantic has been computed as $4 \times 10^6 \,\mathrm{m}^3/\mathrm{s}$ near $31\,^\circ\mathrm{S}$ (Hogg et al., 1982), and $\sim 2 \times 10^6 \,\mathrm{m}^3/\mathrm{s}$ near $4\,^\circ\mathrm{N}$ (Whitehead & Worthington, 1982). Comparable values for AABW flux have been estimated for the southwestern Indian Ocean (Warren, 1981b) and for the southwestern Pacific (Warren, 1973; Reid & Lonsdale, 1974).

AABW leaves its signature in the geological record as a consequence of its in situ properties and its strong advection. Among the possible geologic indicators, there is no single diagnostic 'fingerprint' for AABW which is preferable to another. A wide range of indices have been proposed and applied with varying degrees of success, including:

(a) Benthic foraminiferal assemblages (e.g. Bremer & Lohmann, 1982; Peterson & Lohmann, 1982).

(b) Displaced Antarctic diatoms (e.g. Johnson et al., 1977; Burckle, 1981; Jones & Johnson, 1984).

(c) Clay mineral assemblages (e.g. Jones, 1984).

(d) Stable isotope geochemistry (e.g. Curry & Lohmann, 1982; Hodell et al., 1983).

(e) Particle size analysis (e.g. Ledbetter, 1984).

(f) Magnetic grain fabric (e.g. Ellwood, 1984).

(g) Dissolution of calcareous microfossils (e.g. Thunell, 1982).

It is important to realize that each of these indices is monitoring a different aspect of the near-bottom water (Johnson, 1982). The water properties and the near-bottom flow field are not necessarily co-variant, particularly in geologically interesting areas such as channels and topographic elevations where second-order complexities may be introduced into the near-bottom flow (e.g. Hogg, 1983). Consequently changes in AABW flow in a particular region need not appear in the geologic record as simultaneous 'events' in each of the several potential geologic indices. At this point in our studies of deep circulation history, we should be aware of the importance of applying a number of geologic indices, and that concordance between the different indices should *not* necessarily be expected since each index is reflecting different and non-co-varing aspects of the near-bottom water.

Figs. 5 through 9 give examples of some of the geologic indicators of AABW flow in the southwestern Atlantic. Displaced Antarctic diatoms are common in surface sediment of the Argentine and Brazil Basins (Fig. 6), and have been traced as far north as the latitude of Bermuda (L. Burckle, personal communication). On a macroscopic scale, seismic profiling has revealed striking evidence of bottom current activity in the form of complex patterns of differential sediment accumulation (Kumar *et al.*, 1979), buried acoustic reflectors (Le Pichon *et al.*, 1971; Gamboa *et al.*, 1983), and surface bedforms generally associated with near-bottom flow (Ewing *et al.*, 1971; Damuth & Hayes, 1977). In the Vema Channel region, a strong acoustic reflector sequence occurs at a total depth of ~6.5 s, and often is expressed as a paleochannel lying several hundred meters below and laterally displaced from the modern channel axis (Fig. 7). This paleochannel is equivalent to seismic Reflector A (Horizon A of Ewing *et al.*, 1964), and may correspond with the beginnings of deep flow during the early Cenozoic (Gamboa *et al.*, 1983).

AABW flow is geologically significant not only in the region of flow acceleration through the narrow Vema Channel, but also in the southern Brazil Basin where the flow diverges and weakens. Strong AABW flow continues eastward along the base of the Rio Grande Rise, following the trend of the Rio Grande Fracture Zone (Gamboa & Rabinowitz, 1981). In this deep region, local flow acceleration has eroded down to intermediate and deep reflectors (Fig. 8) which have been sampled by standard piston coring (Johnson & Rasmussen, 1984). A thick wedge of sediment has accumulated at the Vema Channel exit in the southwestern Brazil Basin (Figs. 8 and 9), and represents the erosional debris entrained in AABW and re-deposited in the southwestern Brazil Basin as the flow decelerated. These patterns of sediment erosion and preferential accumulation in the southwestern Atlantic (Damuth & Hayes, 1977; Kumar *et al.*, 1979) are representative of what one encounters in comparable regions of strong AABW flow in the western Pacific (e.g. Hollister *et al.*, 1974; Lonsdale, 1981) and western Indian Ocean (e.g. Johnson & Damuth, 1979).

Initial formation of AABW in the Paleogene
During the 1970s, a considerable array of evidence from stable isotopic analyses (Savin *et al.*, 1975; Shackleton & Kennett, 1975) and faunal assemblages (Benson, 1975; Corliss, 1981) pointed to a major re-

structuring of the water column on a global scale near the Eocene/
Oligocene boundary, perhaps associated with the onset of cooling and
deep thermohaline convection in high latitudes. Recent DSDP cruises
have included among their objectives the further documentation of the
onset of deep convection, the various source areas, and the chronology
of bottom water formation from different sources. In the southwestern

Fig. 5. Bathymetry of the northern Vema Channel and southern
Brazil Basin. Solid lines denote seismic profiles shown in Figs. 7
and 8.

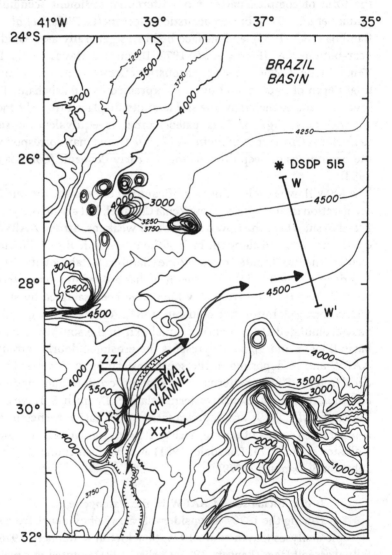

Fig. 6. Distribution of displaced Antarctic-source diatoms in surface sediments of the southwestern and central Atlantic (after Johnson *et al.*, 1977). This advected component can be used as a qualitative indicator of the spatial and temporal distribution of AABW (e.g. Burckle, 1981; Jones & Johnson, 1984).

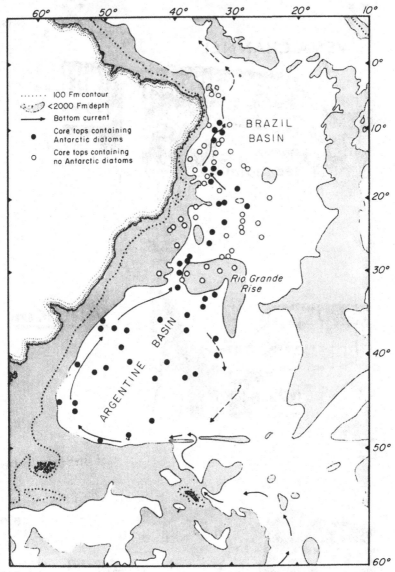

Fig. 7. Seismic reflection profiles across the Vema Channel axis
near 30° S (location on Fig. 5). The prominent acoustic reflector
near 6.5 s takes the form of a buried erosional channel, marking
the onset of strong bottom water flow through the Vema Channel
during the early Cenozoic (Gamboa *et al.*, 1983).

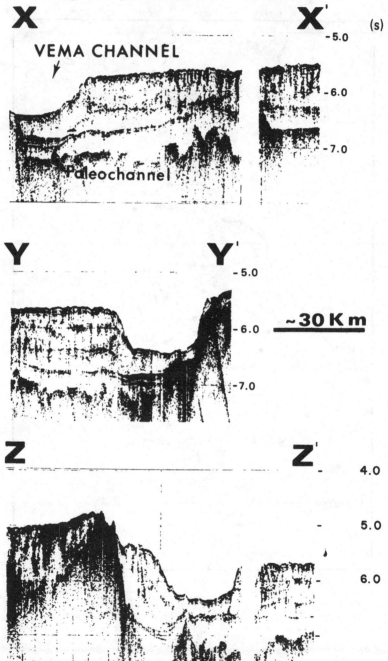

Fig. 8. Seismic profile extending from DSDP Site 515 across the Vema Channel axis at the base of the north flank of the Rio Grande Rise. Truncated and discontinuous reflectors near the channel indicate localized erosion and differential accumulation in response to bottom current flow through the channel.

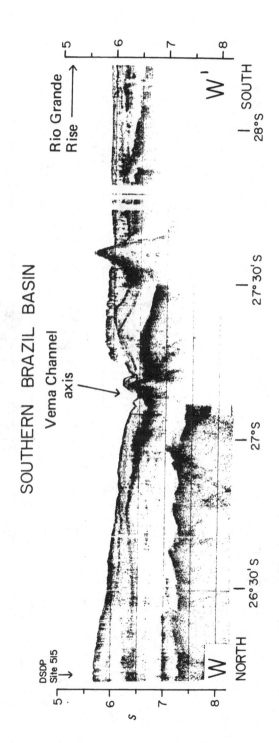

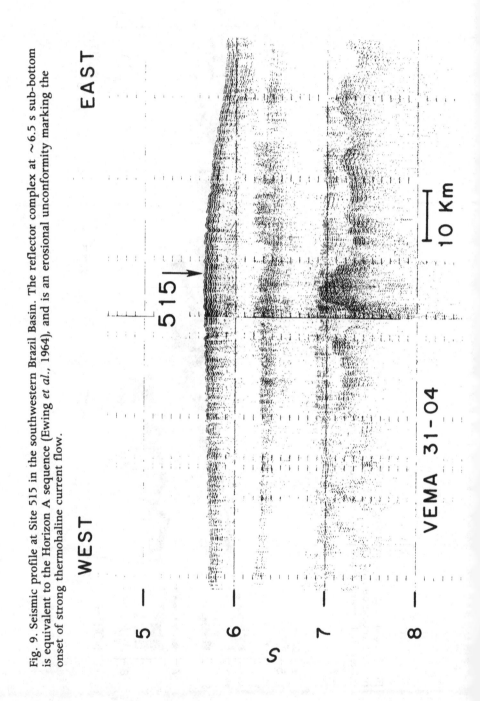

Fig. 9. Seismic profile at Site 515 in the southwestern Brazil Basin. The reflector complex at ~6.5 s sub-bottom is equivalent to the Horizon A sequence (Ewing et al., 1964), and is an erosional unconformity marking the onset of strong thermohaline current flow.

Atlantic, seismic stratigraphic studies show that the onset of large-scale sediment erosion and re-deposition in the early Cenozoic is marked by a prominent reflecting sequence which has been designated 'Reflector A' (e.g. Gamboa *et al.*, 1983). This seismic reflector complex is well developed in the Vema Channel and Brazil Basin (Figs. 7–9), and was the target of drilling at DSDP Site 515 in order to establish the age of the deep circulation event corresponding to Reflector A. One of the primary objectives of drilling at this site was to determine the relationship between the previously established faunal and isotopic shift occuring near the Eocene/Oligocene boundary (Shackleton & Kennett, 1975; Keigwin, 1980) and more direct evidence for the beginning of AABW advection in the deep western Atlantic (Barker, Carlson, Johnson *et al.*, 1983).

Drilling at Site 515 during DSDP Leg 72 penetrated the seismic discontinuity corresponding to Reflector A at a subbottom depth of 625 m (Fig. 10). Below the discontinuity is a pelagic limestone of early

Fig. 10. Lithology and stratigraphy of DSDP Site 515. The unconformity at ~620 m separates early Eocene pelagic limestones from current-deposited siliceous mudstones of post-middle Oligocene age.

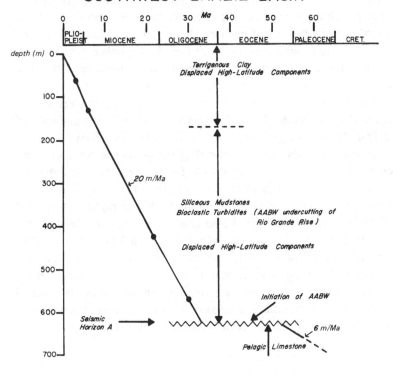

Eocene age. The calcareous microfossils in this unit are well preserved and a reliable magnetic stratigraphy was obtained. The material directly beneath the unconformity corresponds to zones NP 13 and P 9/P 8, and falls within the normal polarity interval of magnetic anomaly 23. Geochronological correlation establishes the age of this material as 54 Ma (Berggren et al., 1983). This early Eocene pelagic limestone unit shows a net accumulation rate of 6 m/Ma (Fig. 10) and no evidence of significant re-working. Directly overlying the unconformity is a thick sequence of rapidly accumulating (> 20 m/Ma) siliceous mudstones, with common silt and sand-sized detrital mineral grains and inter-mittent carbonate turbidites several centimetres thick which show laminations indicative of current re-working (Barker, Carlson, Johnson et al., 1983). The sediment directly overlying the unconformity corresponds to foraminiferal zone P 21 and the reversely magnetized interval below magnetic anomaly 12, or approximately 32 Ma (Berggren et al., 1983).

From the evidence at Site 515, we can conclude that sediment re-working under the influence of AABW had become established in the Brazil Basin by 32 Ma. Because of the 22 Ma hiatus encountered at Site 515, however, we must look elsewhere for independent evidence to place additional constraints on the age of AABW initiation.

Drilling on several DSDP legs in the deep North Atlantic has yielded evidence which appears to reflect not only the initiation of northern sources of bottom water, but perhaps the initial intrusion of southern sources of bottom water as well. Miller & Tucholke (1983) have determined that a widely distributed seismic discontinuity within the basins south of Iceland (Reflector R_4) represents an erosional unconformity near the Eocene/Oligocene boundary. They were able to trace this reflector around the perimeters of basins both east and west of the Mid-Atlantic Ridge, and concluded that overflow of deep thermohaline currents through the deepening Faeroe–Shetland and Faeroe Bank Channels was responsible for the widespread erosional unconformity. This change in abyssal circulation is also reflected in a major benthic foraminiferal turnover in both the northeastern and northwestern abyssal Atlantic (Miller et al., 1982; Miller, 1983). Stable isotopic evidence in the earliest Oligocene (~ 36.5 Ma) shows a 1‰ increase in benthic foraminifera $\delta^{18}O$ and a 0.6‰ increase in $\delta^{13}C$, which Miller et al. (1984) suggest as indicating a temperature drop, a decrease in age of bottom water, and an increase in intensity of abyssal circulation due to the initiation of overflow into the deep Atlantic from the Norwegian Sea. Thus, the seismic, stratigraphic, lithologic, paleontological, and geochemical indicators in the deep North Atlantic are consistent with northern sources of bottom water (proto-NADW) beginning near the Eocene/Oligocene boundary.

At somewhat higher stratigraphic levels in DSDP sites from the Bay of Biscay, Miller & Curry (1982) have established a middle Oligocene increase in the abundance of *Nuttalides umbonifera* (Fig. 11), a benthic foraminifera whose distribution is strongly correlated with that of AABW (Lohmann, 1978a). Concurrently with the benthic faunal shift, Miller & Curry (1982) note a 1 ‰ depletion in benthic $\delta^{13}C$, implying the replacement of relatively young bottom waters by older and more corrosive waters. The carbon isotope depletion spans the interval from

Fig. 11. Benthic foraminiferal assemblages in a portion of the Paleogene cores from DSDP Site 119 in the Bay of Biscay (after Miller & Curry, 1982). The benthic species *Nuttalides umbonifera* is closely associated with Antarctic Bottom Water (Lohmann, 1978a); its increase in abundance during the middle Oligocene may mark the initial entry of AABW into the northeastern Atlantic. The $\delta^{18}O$ and $\delta^{13}C$ values are given as ‰.

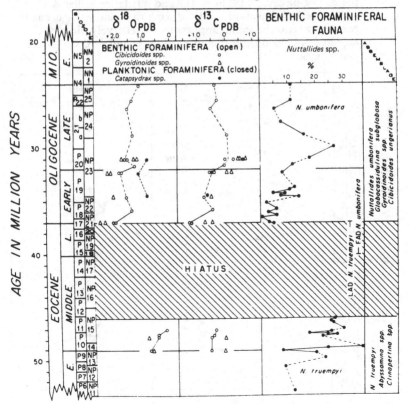

32.5 to 31 Ma at DSDP Sites 119 and 401 (Fig. 12). A comparable depletion of nearly 1‰ in $\delta^{13}C$ has also been documented at Site 549 in the Bay of Biscay, and spans at least the P 21 Zone, extending from 30 to 27 Ma (Miller *et al.*, 1984, Fig. 3). The available evidence thus is consistent with the gradual replacement of young, proto-NADW in the deep North Atlantic by an older, more corrosive proto-AABW during the early and middle Oligocene. The initial intrusion of AABW $(32 \pm 2\,Ma)$ thus appears to have followed the onset of NADW $(36 \pm 2\,Ma)$ by perhaps several million years. AABW flow through the Brazil Basin was established by 32 Ma, as reflected in the drilling at Site 515 (Fig. 10). These older bottom waters gradually filled the deeper basins of the North Atlantic and replaced the younger northern-source NADW, perhaps over an interval of several million years during the middle Oligocene.

We note that the presumed age of AABW initiation in the southwest Atlantic (32 Ma) is consistent with evidence from the southwest Pacific

Fig. 12. Oxygen and carbon isotopes in benthic foraminifera from the Bay of Biscay, DSDP Sites 119 and 401 (after Miller & Curry, 1982). Marked depletion in $\delta^{13}C$ during the middle Oligocene is consistent with the initial appearance of relatively old, ^{13}C-depleted AABW during this interval. $\delta^{18}O$ and $\delta^{13}C$ values are given as ‰.

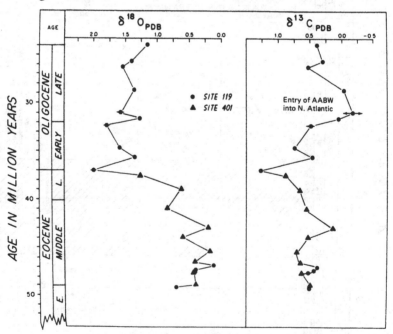

which points to the onset of strong AABW flow through the Samoan Passage by the early Oligocene (Johnson, 1974). Thus it may be that the initiation of AABW during the Oligocene affected the abyssal oceans on a global scale, particularly in those regions where topographic constraints accentuate the effects of AABW flow upon the geologic record.

While the stable isotopic evidence from the deep Atlantic is consistent with the scenario outlined above for AABW initiation, the multiplicity of factors controlling the isotopic geochemistry of microfossils requires caution at this early stage of interpretation. Shifts in foraminiferal $\delta^{13}C$ may in some instances reflect advection, but in other instances may be controlled by sea-level changes and the associated global re-distribution of isotopes of carbon between numerous reservoirs (e.g. Shackleton, 1977; Broecker, 1982). If indeed the early to middle Oligocene was marked by a major sea-level regression (Vail *et al.*, 1980), it is likely that the stable isotopic record during this time interval reflects first-order geochemical balances in addition to water column advection.

Proposed model: essential conditions for AABW production

I shall next outline a set of assumptions and essential conditions which we might examine as to their applicability in describing the production of AABW. I suggest that this scenario be viewed as a testable framework which will require critical examination and revision as further observations become available. The scenario is summarized in Table 1; the following discussion will amplify and clarify several aspects.

A.   Scenario for Paleogene AABW production

Assumption No. 1: The initiation of deep thermohaline flow in high latitudes is limited primarily by salinity, not by temperature. It is clear that in the modern oceans, salinity is the critical water property which allows deep convection to occur in the northern North Atlantic, around Antarctica, and as outflows from marginal seas in subtropical latitudes (Warren, 1981a, pp. 15–26). The lack of deep convection in the North Pacific is a direct result of the lack of water of sufficiently high salinity to mix with the cold surface water (Warren, 1983). If we examine paleoclimatic records from both oceanic and terrestrial sediments from the early Cenozoic, we observe a consistent pattern of warm and equable climates extending to middle and high latitudes during the Paleocene and Eocene, followed by several episodes of global climatic cooling during the Oligocene and the Neogene (e.g. Berggren & Hollister, 1974; Savin et al., 1975). One might initially propose that one or more of these episodes of surface cooling in high latitudes was the essential condition required to trigger deep convection and the initiation of the 'psychrosphere'. However, if the modern ocean is a suitable analog for the Paleogene oceans, there are no modern examples in which surface water is cooled to a certain threshold value, beyond which point deep convection is initiated. In the colder ranges of sea water temperature ($+2\,^{\circ}$C to $-2\,^{\circ}$C), isopycnals (σ_0) become nearly parallel to isotherms on temperature–salinity diagrams, indicating that a required change in density is much more readily achieved by a salinity perturbation than by a temperature perturbation.

Assumption No. 2: Because of the teleconnective linkages between deep ocean water masses and currents (Johnson, 1982), the initiation of deep convection in a particular region may be controlled not only by the ambient water properties but also by the injection of external water masses from relatively distant sources and of notably different character. Thus, in order to consider the limiting conditions to Weddell Sea convection, one may need to extend the limits of observations beyond the Weddell Sea itself.

Table 1. *Conditions governing AABW production during the Paleogene*

Assumptions

1. Initiation of deep thermohaline convection in high latitudes is limited primarily by *salinity*, not by temperature.
2. Thermohaline currents are highly *interactive*. Whether or not deep convection will occur in a given location may be limited by water properties and/or flow patterns very remote from the potential site of convection.
3. Role of geographic boundary conditions: flow is possible only when conditions exceed certain threshold values:
    (a) size and geometry of reservoirs;
    (b) sill depth of critical passages.

Sequence of events

1. Hypersaline water from the northernmost Atlantic (Norwegian Sea overflow) filled the interior basins of the deep Atlantic, eventually reaching the Antarctic perimeter (earliest Oligocene, *c.* 36 Ma).
2. Return flow of AABW began shortly thereafter, creating erosional unconformities and drift deposition in the southwest Atlantic, advecting older bottom waters into the northeast Atlantic (early to middle Oligocene, *c.* 32 Ma).

Assumption No. 3: In certain regions of modern bottom water formation, the sea-floor geometry plays an important role in restricting the amount of circulation and mixing between the relatively limited sites of deep convection and the adjacent oceanic areas. In some instances these topographic boundary conditions take the form of restricted or closed basins (e.g. the Red Sea and Mediterranean Sea), wheras elsewhere a marginal shelf may serve as an appropriate reservoir from which deep convection can begin (e.g. the western Weddell Sea). Ordinarily these boundary conditions will change on relatively long time scales ($10^6–10^7$ years) which greatly exceed the shorter-period changes which occur in bottom water production. Yet in some instances the changing tectonic configuration of a particular sill may be an essential condition in initiating or terminating deep convection, such as the evolving connection between the western Mediterranean and the open Atlantic during the Messinian stage of the late Miocene.

The sill depth of deep passages through which thermohaline currents flow must be considered as a possible constraint on advection. Pathways of modern bottom water are constrained by the distribution of mid-ocean ridges and aseismic rises which lack a sill of sufficient depth to allow bottom water passage. Most of the passages through which deep water *can* flow (e.g. the Vema Channel, Samoan Passage, and Amirante Passage) lie on oceanic crust of pre-Cenozoic age, and therefore the rate of crustal subsidence due to thermal cooling is relatively low (e.g. van Andel *et al.*, 1977, Fig. 4). In considering the Vema Channel, for example, the present sill depth is 4550 m (Johnson, 1984), and the underlying crustal age is approximately 95 Ma. Thus the paleodepth of this sill at the Eocene/Oligocene boundary would have been perhaps 4200 m, by back-tracking along the crustal subsidence curve. Inasmuch as northward flowing AABW within the Vema Channel presently extends up to depths of ~ 3650 m (Hogg *et al.*, 1982), the relatively slow subsidence of the channel during the past 40 Ma has probably not been an important threshold condition governing AABW flow through the western Atlantic.

Thus, it is clear that topographic constraints may provide suitable reservoirs for one or more water masses which supply deep thermohaline flow, and may also limit the available pathways of the resulting flow. Only in special cases, however, would changing boundary conditions be sufficient to produce an abrupt initiation or termination of deep current flow.

Let us now consider the initial formation of AABW during the Paleogene, and postulate the following sequence of events (see Table 1):

(a) Formation of proto-North Atlantic Deep Water (NADW) began near the Eocene–Oligocene boundary, c. 36 Ma, via overflow from the Norwegian Sea through the deepening channels across the Iceland–Scotland Ridge. This event is reflected as a relatively sharp transition in the benthic foraminiferal assemblages, stable isotope geochemistry, and seismic stratigraphy of the deep North Atlantic basins south of Iceland on both sides of the proto-Mid-Atlantic Ridge (see Miller & Curry, 1982; Miller et al., 1982; Miller & Tucholke, 1983).

(b) The development of proto-NADW advected young, high-salinity waters southward through the abyssal western Atlantic, creating extensive erosional surfaces and depositional bedforms along the continental margin within and above the Horizon A complex of seismic reflectors (Tucholke & Mountain, 1979).

(c) Proto-NADW filled the deep basins of the southwestern Atlantic, and continued southward across the Scotia Arc complex to the Antarctic perimeter. The Drake Passage was not yet opened (Barker & Burrell, 1977), and consequently the high-salinity of proto-NADW was not substantially diluted by circumpolar flow.

(d) The injection of high-salinity proto-NADW into the subsurface waters around Antarctica was the necessary trigger to initiate deep convection of proto-Antarctic Bottom Water (AABW). There is little information regarding the ambient surface water temperature around Antarctica at this time, because suitable calcareous material for faunal and stable isotopic analyses has proven to be very sparse in high southern latitudes (Haq et al., 1977). Moreover, there is still much uncertainty regarding the initial appearance of ice on and around Antarctica (e.g. Matthews & Poore, 1980; Mercer, 1983). Whatever the temperature of proto-AABW may have been, it was sufficiently dense to be advected northward beneath the less-dense proto-NADW, and was sufficiently strong to create erosional unconformities and associated seismic reflectors in the deep southwestern Atlantic (Ewing et al., 1964; Le Pichon et al., 1971; Gamboa et al., 1983). AABW flow during its early history carved one or more paleochannels within the Vema Gap (Fig. 7), and eroded into the Paleogene biogenic sediments of the Brazil Basin (Figs. 8 and 10).

(e) Transport of proto-AABW through the Brazil Basin was well established by 32 Ma (Fig. 10), although its initiation may have occurred somewhat earlier during the Oligocene. The intrusion of proto-AABW into the deep North Atlantic occurred gradually over several million years during the middle Oligocene, c. 32 to 27 Ma, based on evidence from benthic foraminiferal assemblages and stable isotope geochemistry

(Figs. 11 and 12). This older southern-source water replaced the younger proto-NADW in several of the deeper basins of the North Atlantic. There is no evidence that the overflow of proto-NADW was shut off at the time, although it may have weakened during the early Neogene, based on seismic stratigraphic evidence showing an upward transition from strong reflectors to drift deposits (Miller & Tucholke, 1983).

B.   *Possible problems with this scenario*
    In the preceding description, NADW production is postulated as a necessary precondition to AABW formation, and one would therefore predict that geologic indicators of the two water masses should show evidence of strong co-variation. If indeed the mixing time of the oceans is relatively short (10^3 years), then one would predict that an 'event' in NADW production should be reflected in a corresponding 'event' in AABW flow, with little or no observable lag between the two (i.e. on geologic time scale of 10^5 to 10^6 years or longer). Let us consider some evidence bearing on deep circulation during the late Pleistocene/ Holocene, since it is possible to achieve high-resolution stratigraphic control over this time frame. A wide range of paleontological and geochemical evidence from the deep Atlantic has suggested that NADW production during the last glacial maximum was severely reduced relative to the present (e.g. Streeter & Shackleton, 1979; Schnitker, 1980; Boyle & Keigwin, 1982), and may have been terminated entirely (e.g. Streeter *et al.*, 1982). More recently, time-series analyses of geologic indices of NADW production have identified what may be significant periodicities in deep convection during the late Pleistocene/ Holocene. Boyle (1983) has determined that the cadmium geochemistry of certain species of benthic foraminifera is strongly correlated with their isotope geochemistry ($\delta^{13}C$). Because $\delta^{13}C$ is a potential index of the relative 'age' of deep water (Kroopnick, 1980; Graham *et al.*, 1981; Curry & Lohmann, 1982), variations in the Cd/Ca ratio of benthic foraminifera may therefore be a useful index of NADW production rates. Fig. 13 shows the results of geochemical analyses of a late Pleistocene/Holocene core from the east flank of the Vema Channel, at a depth corresponding to the salinity maximum associated with modern NADW. The Cd/Ca ratio in the benthic foraminifera *Planulina wuellerstorfi* is strongly co-variant with $\delta^{13}C$ in the same species (Boyle, 1984). High values of Cd/Ca are associated with more negative values of $\delta^{13}C$, and thus imply minima in NADW production (Boyle, 1984, Figs. 1 and 2).
    The Cd/Ca ratio is low near the core top, reaches a maximum near the

last glacial, and undergoes several high-frequency fluctuations over the past 300 000 years. Spectral analysis shows significant variance occurring at the 41 000 year period (Fig. 13). Because the 41 000 year tilt cycle strongly controls summer insolation in the high latitudes (north of ~65°N) (Berger, 1976; Ruddiman & McIntyre, 1981), there may be a

Fig. 13. Cadmium geochemistry of benthic foraminifera from a late Pleistocene/Holocene core from the east flank of the Vema Channel (after Boyle, 1983). Spectral analysis shows a strong peak at 41 000 years, suggesting orbital forcing of conditions favourable to NADW production.

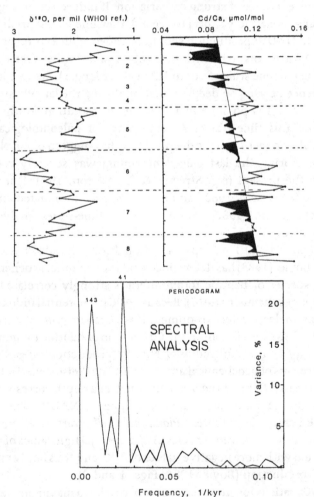

direct relation between NADW production and the solar radiation balance at the latitudes of bottom water formation.

Evidence bearing on late Pleistocene/Holocene fluctuations in AABW production is relatively limited. A diverse set of potential geologic indices have been proposed (see previous discussion); yet their application has most commonly been limited by the absence of high-resolution stratigraphic control, an essential condition for reliable time-series analysis. In the deeper regions where AABW is dominant, calcite dissolution commonly removes most of the planktonic and benthic foraminifera, which are essential to stratigraphic control. Consequently there is a relatively narrow depth interval which contains a significant portion of Antarctic source waters, and where calcareous microfossils remain useful for stratigraphy. Initial studies of Pleistocene benthic foraminifera in the Vema Channel by Lohmann (1978b) suggested possible long-period ($> 10^5$ years) fluctuations in AABW production over the Brunhes interval of the late Pleistocene. However, sample spacing and stratigraphic resolution in this study was not sufficient to identify shorter-period fluctuations (e.g. the orbital forcings of 23 000 and 41 000 years). Jones & Johnson (1984) recently examined displaced Antarctic diatoms in a precisely dated core from the east flank of the Vema Channel at a depth (4148 m) which is sufficiently deep to be well within AABW ($\theta = +0.7\,°C$), yet contains sufficient $CaCO_3$ for reliable stratigraphic control. Counts of displaced diatoms in closely spaced samples over the past 200 000 years show remarkably high concentrations (up to 700 000 specimens per gram) within some samples. For three of the dominant diatom species, maxima in their abundance appear to occur immediately preceding the strong glacial maximum at the stage 7/6 transition and the stage 3/2 transition (Fig. 14). Diatom maxima do not appear at the transition into stage 4, and there is no evidence for high-frequency fluctuations (e.g. 41 000 years), even though sample spacing in this study was quite close, corresponding to every 2000 to 4000 years (Jones & Johnson, 1984). Thus, we see no indications that the 41 000 year NADW cycle (Fig. 13) shows parallel fluctuations in AABW intensity.

A second line of evidence suggests a lack of co-variance between NADW and AABW flow during the Quaternary. The advection of Antarctic diatoms through the Vema Channel below depths of ~4000 m persisted during glacial maxima as well as during interglacials (Jones & Johnson, 1984), even though NADW was substantially (or entirely) shut off. An alternative source must have continued to supply saline waters to the Antarctic if AABW persisted concurrently with cessation of NADW.

If further studies of the late Pleistocene confirm the lack of a close correspondence between AABW flow and that of NADW, one possible explanation is that the response time of the deep oceans is significantly longer than 10^3 years, and that the injection of high-salinity NADW into the deep north Atlantic requires a substantial period of time to influence the circum-Antarctic region and produce a high-salinity 'wedge' in the subsurface at the shelf edge (e.g. Fig. 4). If indeed there was a measurable lag (10^6 years) between the first overflow of proto-NADW and the onset of AABW production in the early Oligocene, then there would be some support for this notion. Unfortunately the relatively poor biostratigraphic resolution in the early Oligocene is such that a more precise chronology for the early Oligocene 'events' will be difficult to achieve. An alternative explanation for the non-parallel histories of AABW and NADW is that there has been more than one source of high-salinity water supplying the circum-Antarctic region with the excess salinity necessary to initiate deep convection.

Fig. 14. Time-series record of displaced Antarctic diatom species in a late Pleistocene–Holocene core from the flank of the Vema Channel (after Jones & Johnson, 1984). Diatom maxima (up to 700 000 specimens per gram) are associated with strong deglaciations at the stage 7/6 and 3/2 transitions, approximately 150 000 and 50 000 years. BP.

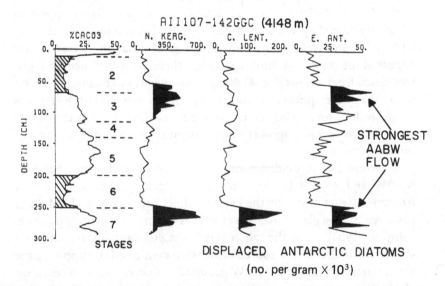

Western Indian Ocean: possible role in AABW generation

The highest salinities in the world ocean today are in the northwestern Indian Ocean and the associated marginal seas: the Red Sea, the Gulf of Aden, and the Persian Gulf (Wyrtki, 1971). The evaporation/precipitation budget in these subtropical latitudes, together with the topographic constraints in the partially enclosed basins bordering the Arabian Sea, leads to thermohaline convection of both intermediate and deep waters towards the south. Let us examine the essential aspects of the intermediate and deep flow in the western Indian Ocean to determine its possible role in influencing the circum-Antarctic waters. A more detailed discussion is presented in recent papers by Wyrtki (1973) and Warren (1974, 1981b).

The zonal hydrographic sections in the southern Indian Ocean at 23°S and 18°S (Warren, 1974, 1981b) are critical in identifying the 'permanent' (i.e non-seasonal) aspects of the western Indian Ocean hydrography which may influence the global circulation. In the upper kilometer of the water column the general sense of the flow is northward, with a geostrophic volume transport of the order of $20 \times 10^6 \, \text{m}^3/\text{s}$ (Warren, 1981b, p. 771). This upper water includes several layers, the shallowest of which is a salinity maximum near 250 m, designated as 'subtropical surface water' (Wyrtki, 1973). This shallow layer originates in the surface waters between 25°S and 35°S where evaporation greatly exceeds precipitation, and it extends northward as a subsurface layer to near 10°S where it loses its signature in the low-salinity water entering from the Pacific in the south Equatorial Current. The northward-flowing upper layer includes the salinity minimum associated with Antarctic Intermediate Water (AAIW), centred between 600 and 900 m. This salinity minimum extends to $\sim 10°$S in the western Indian Ocean, where it intersects hypersaline Arabian Sea waters of comparable density and loses its signature. In the depth interval below AAIW between 1000 and 2000 m, the meridional transport is difficult to determine because the zonal density differences are very weak, and transport shear is thus very low (Warren, 1981b, p. 772).

Below a depth of ~ 4000 m in the Madagascar and Mascarene Basins the abyssal flow is also toward the north (Warren, 1974). In sections east of Madagascar, the near-bottom isotherms slope downward toward the east with temperatures as low as 0.6°C and high oxygen values $O_2 > 4.75 \, \text{ml/l}$), clearly indicating an Antarctic source region. AABW extends through fracture zones in the southwest Indian Ridge (Warren, 1978), and through the Amirante Passage near 9°S below depths of

4000 m (Johnson & Damuth, 1979). The northward transport of AABW near 18°S has been computed to be 12×10^6 m³/s (Warren, 1981b, Table 1).

Between the northward-flowing surface water (0–1000 m) and AABW (>4000 m) lies a broad salinity maximum, centered near 2500 m, with concentrations exceeding 34.47‰ (Figs. 15 and 16). The salinity distribution alone does not allow one to determine the sense of the flow; one could plausibly associate the salinity maximum with either an extension of North Atlantic Deep Water which became entrained in circumpolar flow (e.g. Reid & Lynn, 1971) or with southward-flowing hypersaline outflow from the Arabian Sea (Warren,

Fig. 15. Bathymetry of the western Indian Ocean. The dashed line denotes hydrographic transect at 23°S (Fig. 16), and the box outlines the Amirante Passage region (Fig. 17). The heavy arrows show the principal pathway of Antarctic Bottom Water.

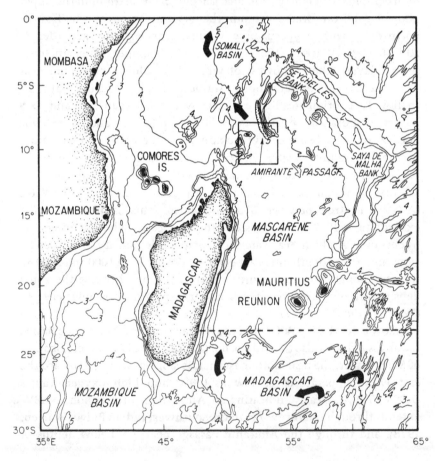

1981b). Mass balance considerations alone would favour southward flow at intermediate depths (2000–4000 m), inasmuch as northward flow is clearly required in the overlying surface layer and in AABW beneath. Moreover, the distribution of other properties associated with the salinity maximum strongly suggest southward flow as well. In particular, the deep oxygen minimum near 2500 m and very high values of dissolved silica (> 120 μmol/kg) in the depths below 2500 m (Fig. 16) are inconsistent with a North Atlantic origin, and strongly suggest southward flow from the highly productive hypersaline surface waters of the Arabian Sea. Warren (1981b, Table 1) has estimated the southward transport in this high-salinity layer ($\sim$ 2 to 4 km depth) to be $7 \times 10^6 \, \text{m}^3/\text{s}$.

In summary, the present-day tectonic and climatic conditions in the northwestern Indian Ocean produce among the saltiest waters in the world's oceans within and around the Arabian Sea. A significant fraction of this high-salinity, high-silica surface water sinks and flows southward at intermediate depths (2–4 km) through the Mascarene and Madagascar Basins, and continues southward to merge with the eastward circumpolar flow near the subtropical convergence at 40°S. The northwestern Indian Ocean therefore must be considered potentially as important as the Norwegian Sea overflow in supplying high-salinity waters to the circum-Antarctic, and thus for the generation of AABW.

Fig. 16. Salinity (‰) and dissolved silicate (μmol/kg) in the western Indian Ocean near 23°S, after Warren (1974).

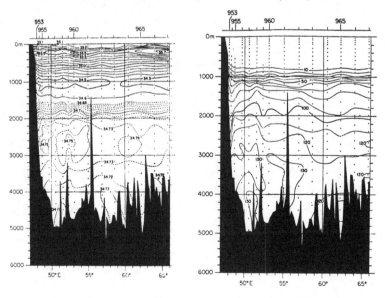

Some initial geologic evidence exists which allows us to speculate about the timing of the events which caused the northwestern Indian Ocean to begin to supply hypersaline deep waters to the southern hemisphere. A suite of cores was obtained in the Amirante Passage region near 9°S (Fig. 17), where AABW flows between the Mascarene and Somali Basins (Johnson & Damuth, 1979). The cores span a depth range from 3300 to 4500 m (Fig. 18), which includes AABW (> 4000 m) and the transition zone to the overlying high-salinity deep water. In a previous report, Johnson *et al.* (1983) identified a prominent erosional unconformity in each of the shallower cores from the Amirante Passage which suggested a marked increase in deep flow above 4000 m at around 16 Ma (Fig. 18). They contrasted this evidence for erosion in the western Indian Ocean with evidence of the depositional continuity in Neogene cores at comparable depths within NADW in the southwestern

Fig. 17. Bathymetry of the Amirante Passage, and locations of piston cores.

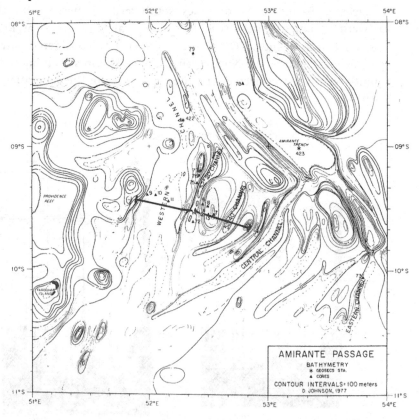

Atlantic. They concluded that if these erosional episodes reflected NADW pulsations, one would expect to encounter even more striking evidence for erosion farther 'upstream' in NADW, but such is not the case. Johnson *et al.* (1983) concluded that another water mass was responsible for the Amirante Passage erosion above 4000 m, and suggested an intensification of circumpolar flow in the middle Miocene. Another plausible explanation, not discussed by Johnson *et al.* (1983), is that the erosional hiatus above 4000 m (Fig. 18) reflects the onset of southward thermohaline flow from the Arabian Sea. It is clear that the closure of the eastern Tethyan seaways during the Neogene has produced a dramatic change in the evaporation/precipitation budget of southwestern Asia, the Arabian Sea, and the adjoining marginal basins. It may be that the 16 Ma 'event' implied by the Amirante Passage cores

Fig. 18. Lithology and stratigraphy of Amirante Passage cores (after Johnson *et al.*, 1983). A marked erosional unconformity at mid-depths (3000–4000 m) begins during the middle Miocene, *c.* 16 Ma. This event may mark the initiation of the hypersaline outflow from the northern Indian Ocean in response to the closure of the eastern Tethys and the production of high-salinity surface waters in the northwestern Indian Ocean. DWBA = Deep Western Boundary Current.

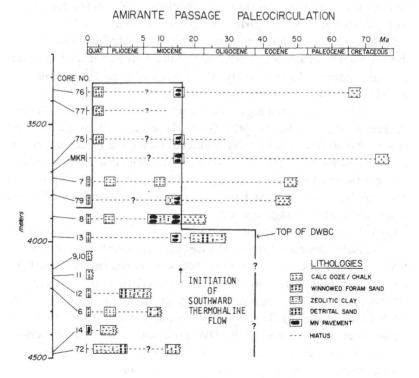

(Fig. 18) is direct evidence of the closure of the eastern Tethys beyond a certain point, leading to extensive deep convection of hypersaline waters from the northwestern Indian Ocean. The Amirante cores suggest that pulsations of deep southward flow have continued intermittently until the Recent; if so, these episodes may play an important, and as yet undetermined, role in AABW production.

Conclusions

(1) The initiation of Antarctic Bottom Water (AABW) flow through the basins of the southwestern Atlantic began during the early Oligocene, c. 32 ± 2 Ma. This event is marked by a seismic discontinuity in the Vema Channel and Brazil Basin (Horizon A of Ewing et al., 1964), and the deposition of a thick wedge of bottom current drift deposits at the Vema Channel exit (DSDP Site 515) by the time of magnetic anomaly 12, 32 Ma. Benthic foraminiferal assemblages and stable isotope geochemistry at sites farther downstream in the northeastern Atlantic suggest the gradual intrusion of older, more corrosive, ^{13}C-depleted bottom waters (proto-AABW) over several million years during the middle Oligocene, c. 32 to 27 Ma. This southern-source bottom water filled the deeper basins, replacing North Atlantic Deep Water (NADW), whose initial overflow through the Greenland–Iceland–Scotland ridge had occurred during the earliest Oligocene, c. 36 Ma.

(2) As is true in the modern oceans, deep thermohaline convection in the Paleogene oceans was probably salinity-limited rather than temperature-limited. Consequently the essential condition governing initial formation of AABW around the Antarctic perimeter was probably the injection of high-salinity waters (i.e. proto-NADW) from the north at intermediate depths. Gradual cooling of the surface waters around Antarctica undoubtedly occurred during the Paleogene as seafloor spreading produced geographic and thermal isolation of Antarctica, yet the most dramatic cooling associated with full development of circumpolar flow may not have occurred until the earliest Miocene when the Drake Passage opened.

(3) While the formation of AABW during the Paleogene may have been closely associated with Norwegian Sea overflow (i.e. proto-NADW), such a close association between northern and southern sources of bottom water has not been characteristic of the Quaternary. NADW production was severely reduced during glacial maxima, while AABW flow persisted. Pleistocene cores suggest a strong 41 000 year signal in geologic indices of NADW production, whereas indices of AABW indicate major pulses at longer periods (e.g. during onsets of glacial maxima at the stage 7/6 and 3/2 transitions).

(4) The northwestern Indian Ocean and its marginal seas are a major source area today for hypersaline, silica-rich waters flowing southward at intermediate depths (2–4 km). This southward flow merges with the eastward-flowing circumpolar current, and thereby serves as an additional possible source of hypersaline waters to fuel deep convective overturn around the Antarctic perimeter. The total cessation of bottom water production (NADW) in the far northern Atlantic, therefore, need not lead to a cessation of AABW production, provided that an alternative source is available for injecting hypersaline water into the circumpolar flow.

(5) The gradual closure of the eastern Tethyan land masses and the collision of India with southern Asia during the Neogene dramatically altered the evaporation/precipitation budget in the northwestern Indian Ocean, allowing hypersaline surface waters to be formed in the Arabian Sea and associated marginal seas. Deep convection and southward advection of this hypersaline water at intermediate depths (2–4 km) apparently began in the early to middle Miocene, *c.* 16 Ma, and has continued intermittently to the present day.

Acknowledgements

I thank K.J. Hsü and the other organizers of the First International Conference on Paleoceanography (July 1983) for the invitation to participate in the special symposium on South Atlantic Paleoceanography, and to contribute a paper in this volume. Partial support for preparation of this paper was provided by Working Group No. 7 of the International Lithosphere Program. The author's research on the paleoceanography of the southwestern Atlantic is supported under NSF Grant OCE 80-25208. I thank C.W. Baker, G.A. Jones, D. Martinson, and L. Keigwin for discussions and critique of the manuscript. This is the Woods Hole Oceanographic Insitution Contribution No. 5489.

REFERENCES

Bainbridge, A.E. 1980. *GEOSECS Atlantic Expedition: Volume 2, Sections and Profiles.* National Science Foundation, Washington, DC, 196 pp.
Barker, P.F. & Burrell, J. 1977. The opening of Drake Passage. *Mar. Geol.,* **25**, 15–34.
Barker, P.F., Carlson, R.L., Johnson, D.A. *et al.* 1983. *Initial Reports of the Deep Sea Drilling Project,* **72**, US Govt Printing Office, Washington, DC, 1024 pp.
Benson, R.H. 1975. the origin of the psychrosphere as recorded in changes of deep-sea ostracode assemblages. *Lethaia,* **8**, 69–83.
Berger, A. 1976. Obliquity and precession for the last 5 000 000 years. *Astron. Astrophys.,* **51**, 127–135.

Berggren, W. A. & Hollister, C. D. 1974. Paleogeography, paleobiogeography and the history of circulation in the Atlantic Ocean. In: W. W. Hay (ed.), Studies in Paleo-oceanography, Soc. Econ. Pal. Min., Spec. Publ., 20, 126–186.

Berggren, W. A., Hamilton, N., Johnson, D. A., Pujol, C., Weiss, W., Cepek, P. & Gombos, A. 1983. Magnetobiostratigraphy of Deep Sea Drilling Project Leg 72, Sites 515–518, Rio Grande Rise (South Atlantic). Initial Reports of the Deep Sea Drilling Project, 72, US Govt Printing Office, Washington, DC, pp. 939–948.

Boyle, E. A. 1984. Abyssal hydrography and the 41 Kyr tilt cycle. In: T. Takahashi & J. Hansen (eds.), Climate Sensitivity: Response to Solar Insolation and CO_2, American Geophysical Union, Washington, DC, pp. 360–368.

Boyle, E. A. & Keigwin, L. D. 1982. Deep circulation of the North Atlantic over the last 200 000 years: geochemical evidence. Science, 218, 748–787.

Bremer, M. L. & Lohmann, G. P. 1982. Evidence for primary control of the distribution of certain Atlantic Ocean benthonic foraminifera by degree of carbonate saturation. Deep-Sea Res., 29, 987–998.

Broecker, W. S. 1982. Ocean chemistry during glacial time. Geochim. Cosmochim. Acta, 46, 1689–1705.

Burckle, L. H. 1981. Displaced Antarctic diatoms in the Amirante Passage. Mar. Geol., 39, M39–M43.

CLIMAP Project Members, 1981. Seasonal reconstruction of the earth's surface at the last glacial maximum. Geol. Soc. America Map Chart Series, MC-36.

Corliss, B. H. 1981. Deep-sea benthonic foraminiferal faunal turnover near the Eocene/Oligocene boundary. Mar. Micropaleontol., 6, 367–384.

Curry, W. B. & Lohmann, G. P. 1982. Carbon isotopic changes in benthic foraminifera from the western South Atlantic: reconstruction of glacial abyssal circulation patterns. Quat. Res., 18, 218–235.

Damuth, J. E. & Hayes, D. E. 1977. Echo character of the East Brazilian continental margin and its relationship to sedimentary processes. Mar. Geol., 24, M73–M95.

Ellwood, B. B. 1984. Magnetic fabric and remanence analyses of cores from the U.S. continental rise and the Vema Channel. Mar. Geol., 58, 151–164.

Ewing, M., Ludwig, W. J. & Ewing, J. 1964. Sediment distribution in the oceans: the Argentine Basin. J. Geophys. Res., 10, 2003–2032.

Ewing, M., Eittreim, S. L., Ewing, J. I. & Le Pichon, X. 1971. Sediment transport and distribution in the Argentine Basin. 3. Nepheloid layer and processes of sedimentation. In: L. H. Ahrens, F. Press, S. K. Runcorn & H. C. Urey (eds.), Physics and Chemistry of the Earth, vol. 8, Pergamon Press, New York, pp. 51–77.

Foster, T. D. 1972. An analysis of the cabbeling instability in sea water. J. Phys. Oceanogr., 2, 294–301.

Foster, T. D. & Carmack, E. C. 1976. Frontal zone mixing and Antarctic Bottom Water formation in the southern Weddell Sea. Deep-Sea Res., 23, 301–317.

Foster, T. D. & Middleton, J. H. 1980. Bottom water formation in the western Weddell Sea. Deep-Sea Res., 27A, 367–381.

Gamboa, L. A. P. & Rabinowitz, P. D. 1981. The Rio Grande Fracture Zone in the western South Atlantic and its tectonic implications. Earth Planet. Sci. Lett., 52, 410–418.

Gamboa, L. A. P., Buffler, R. T. & Barker, P. F. 1983. Seismic stratigraphy and geologic history of the Rio Grande Gap and southern Brazil Basin. Initial Reports of the Deep Sea Drilling Project, 72, US Govt Printing Office, Washington, DC, pp. 481–497.

Gates, W. L. 1977. The numerical simulation of ice-age climate with a global general circulation model. J. Atmosph. Sci., 33, 1844–1873.

Graham, D. W., Corliss, B. H., Bender, M. L. & Keigwin, L. D. 1981. Carbon and oxygen isotopic disequilibria of Recent deep-sea benthic foraminifera. Mar. Micropaleontol., 6, 483–497.

Haq, B. U., Premoli-Silva, I. & Lohmann, G. P. 1977. Calcareous plankton paleobiogeographic evidence for major climatic fluctuations in the early Cenozoic Atlantic Ocean. J. Geophys. Res., 82, 3861–3876.

Hodell, D. A., Kennett, J. P. & Leonard, K. A. 1983. Climatically-induced changes in vertical water mass structure of the Vema Channel during the Pliocene: evidence from DSDP Sites 516A, 517 and 518. *Initial Reports of the Deep Sea Drilling Project*, **72**, US Govt Printing Office, Washington, DC, pp. 907–919.

Hogg, N.G. 1983. Hydraulic control and flow separation in a multi-layered fluid with applications to the Vema Channel. *J. Phys. Oceanogr.*, **13**, 695–708.

Hogg, N.G., Biscaye, P., Gardner, W. & Schmitz, W.J., Jr 1982. On the transport and modification of Antarctic bottom water in the Vema Channel. *J. Mar. Res.*, **40** (supplement), 231–263.

Hollister, C.D., Johnson, D.A. & Lonsdale, P.F. 1974. Current-controlled abyssal sedimentation: Samoan Passage, equatorial west Pacific. *J. Geol.*, **82**, 275–300.

Johnson, D.A. 1974. Deep Pacific circulation: intensification during the early Cenozoic. *Mar. Geol.*, **17**, 71–78.

Johnson, D.A. 1982. Abyssal teleconnections: interactive dynamics of the deep ocean circulation. *Palaeogeogr. Palaeoclimatol. Palaeoecol.*, **38**, 93–128.

Johnson, D.A. 1984. The Vema Channel: physiography, structure, and sediment–current interactions. *Mar. Geol.*, **58**, 1–34.

Johnson, D.A. & Damuth, J.E. 1979. Deep thermohaline flow and current-controlled sedimentation in the Amirante Passage: western Indian Ocean. *Mar. Geol.*, **33**, 1–44.

Johnson, D.A. & Rasmussen, K.A. 1984. Late Cenozoic turbidite and contourite deposition in the southern Brazil Basin. *Mar. Geol.*, **58**, 225–262.

Johnson, D.A., Ledbetter, M.T. & Burckle, L.H. 1977. Vema Channel paleoceanography: Pleistocene dissolution cycles and episodic bottom water flow. Mar. Geol., **23**, 1–33.

Johnson, D.A., Ledbetter. M.T. & Damuth, J.E. 1983. Neogene sedimentation and erosion in the Armirante Passage, western Indian Ocean. *Deep-Sea Res.*, **30**, 195–219.

Jones, G.A. 1984. Advective transport of clay minerals in the region of the Rio Grande Rise: insights into processes of sedimentation in the deep sea. *Mar. Geol.*, **58**, 187–212.

Jones, G.A. & Johnson, D.A. 1984. Displaced Antarctic diatoms in Vema Channel sediments: late Pleistocene/Holocene fluctuations in AABW flow. *Mar. Geol.*, **58**, 165–186.

Keigwin, L.D., Jr 1980. Paleoceanographic change in the Pacific at the Eocene–Oligocene boundary. *Nature*, **287**, 722–725.

Killworth, P.D. 1983. Deep convection in the world ocean. *Rev. Geophys. Space Phys.*, **21**, 1–26.

Kroopnick, P. 1980. The distribution of C^{13} in the Atlantic Ocean. *Earth Planet. Sci. Lett.*, **49**, 469–484.

Kumar, N., Leyden, R., Carvalho, J. & Francisconi, O. 1979. Sediment isopach map: Brazilian continental margin. *Am. Ass. Petrol. Geol.*, Tulsa, map series.

Ledbetter, M.T. 1984. Bottom-current speed in the Vema Channel recorded by particle size of sediment in fine-fraction analyses. *Mar. Geol.*, **58**, 137–149.

Le Pichon, X., Ewing, M. & Truchan, M. 1971. Sediment transport and distribution in the Argentine Basin. 2. Antarctic bottom current passage into the Brazil Basin. In: L.H. Ahrens, F. Press, S.K. Runcorn & H.C. Urey (eds.), *Physics and Chemistry of the Earth*, vol. 8, Pergamon Press, New York, pp. 31–48.

Lohmann, G.P. 1978a. Abyssal benthonic foraminifera as hydrographic indicators in the western South Atlantic Ocean. *J. Foram. Res.*, **8**, 6–34.

Lohmann, G.P. 1978b. Response of the deep sea to ice ages. *Oceanus*, **21**, 58–64.

Lonsdale, P.F. 1981. Drifts and ponds of reworked pelagic sediment in part of the southwest Pacific. *Mar. Geol.*, **43**, 153–193.

Manabe, S. & Hahn, D.G. 1977. Simulation of the tropical climate of an ice age. *J. Geophys. Res.*, **82**, 2889–2911.

Mantyla, A.W. & Reid, J.L. 1983. Abyssal characteristics of the World Ocean waters. *Deep-Sea Res.*, **30**, 805–833.

Matthews, R.K. & Poore, R.Z. 1980. Tertiary $\delta^{18}O$ record and glacioeustatic sea-level fluctuations. *Geology*, **8**, 501–504.

Mercer, J.H. 1983. Cenozoic glaciation in the southern hemisphere. *Ann. Rev. Earth Planet. Sci.*, **11**, 99–132.

Miller, K.G. 1983. Eocene–Oligocene paleoceanography of the deep Bay of Biscay:

benthic foraminiferal evidence. *Mar. Micropaleontol.*, **7**, 403–440.

Miller, K.G. & Curry, W.B. 1982. Eocene to Oligocene benthic foraminiferal isotopic record in the Bay of Biscay. *Nature*, **296**, 347–350.

Miller, K.G. & Tucholke, B.E. 1983. Development of Cenozoic abyssal circulation south of the Greenland–Scotland Ridge. In: M. Bott, M. Talwani, J. Thiede & S. Saxov (eds.), *Structure and Development of the Greenland–Scotland Ridge*. Plenum Press, New York, pp. 549–589.

Miller, K.G., Gradstein, F.M. & Berggren, W.A. 1982. Late Cretaceous to early Tertiary agglutimated benthic foraminifera in the Labrador Sea. *Micro-paleontol.*, **28**, 1–30.

Miller, K.G., Curry, W.B. & Osterman, D.R. 1984. Late Paleogene (Eocene to Oligocene) benthic foraminiferal paleoceanography of the Goban Spur region, DSDP Leg 80. *Initial Reports of the Deep Sea Drilling Project*, **80**, US Govt Printing Office, Washington, DC, 1984, pp. 505–538.

Peterson, L.C. & Lohmann, G.P. 1982. Major change in Atlantic deep and bottom waters 700 000 yr ago: benthonic foraminiferal evidence from the South Atlantic. *Quat. Res.*, **17**, 26–38.

Reid, J.L. & Lonsdale, P.F. 1974. On the flow of water through the Samoan Passage. *J. Phys. Oceanogr.*, **4**, 58–73.

Reid, J.L. & Lynn, R.J. 1971. On the influence of the Norwegian–Greenland and Weddell seas upon the bottom waters of the Indian and Pacific Oceans. *Deep-Sea Res.*, **18**, 1063–1088.

Reid, J.L., Nowlin, W.D., Jr & Patzert, W.C. 1977. On the characteristics and circulation of the southwestern Atlantic Ocean. *J. Phys. Oceanogr.*, **7**, 62–91.

Richardson, M.J., Wimbush, M. & Mayer, L. 1981. Exceptionally strong near-bottom flows on the continental rise of Nova Scotia. *Science*, **213**, 887–888.

Ruddiman, W.F. & McIntyre, A. 1981. Oceanic mechanisms for amplification of the 23 000-year ice-volume cycle. *Science*, **212**, 617–627.

Savin, S.M., Douglas, R.G. & Stehli, F.G. 1975. Tertiary marine paleotemperatures. *Geol. Soc. Am. Bull.*, **86**, 1479–1510.

Schmitz, W.J., Jr 1978. Observations of the vertical distribution of low frequency kinetic energy in the Western North Atlantic. *J. Mar. Res.*, **36**, 295–310.

Schmitz, W.J., Jr & Hogg, N.G. 1983. Exploratory observations of abyssal currents in the South Atlantic near Vema Channel. *J. Mar. Res.*, **41**, 487–510.

Schnitker, D. 1980. Quaternary deep-sea benthic foraminifers and bottom water masses. *Ann. Rev. Earth Planet. Sci.*, **8**, 343–370.

Shackleton, N.J. 1977. Tropical rainforest history and the equatorial Pacific carbonate dissolution cycles. In: N. Andersen & A. Malahoff (eds.), *The Fate of Fossil Fuel CO_2 in the Oceans*, Plenum Press, New York, pp. 401–428.

Shackleton, N.J. & Kennett, J.P. 1975. Paleotemperature history of the Cenozoic and the initiation of Antarctic glaciation: oxygen and carbon isotope analyses at DSDP Sites 277, 279, and 281. *Initial Reports of the Deep Sea Drilling Project*, **29**, 743–755, US Govt Printing Office, Washington, DC.

Steele, J.H., Barrett, J.R. & Worthington, L.V. 1962. Deep currents south of Iceland. *Deep-Sea Res.*, **9**, 465–474.

Streeter, S.S. & Shackleton, N.J. 1979. Paleocirculation of the deep North Atlantic: 150000-year record of benthic foraminifera and oxygen-18. *Science*, **203**, 168–171.

Streeter, S.S., Belanger, P.E., Kellogg, T.B. & Duplessy, J.C. 1982. Late Pleistocene paleo-oceanography of the Norwegian–Greenland Sea: benthic foraminiferal evidence. *Quat. Res.*, **18**, 72–90.

Thunell, R.C. 1982. Carbonate dissolution and abyssal hydrography in the Atlantic Ocean. *Mar. Geol.*, **47**, 165–180.

Tucholke, B.E. & Mountain, G.S. 1979. Seismic stratigraphy, lithostratigraphy and paleosedimentation patterns in the North American basin. In: M. Talwani, W. Hay & W.B.F. Ryan (eds.), *Deep Drilling Results in the Atlantic Ocean: Continental Margins and Paleoenvironment*, American Geophysical Union, Washington, pp. 58–86.

Vail, P.R., Mitchum, R.M., Jr, Shipley, T.H. & Buffler, R.T. 1980. Unconformities of the North Atlantic. *Phil. Trans. R. Soc. London A*, **294**, 137–155.

van Andel, Tj.H., Thiede, J., Sclater, J.G. & Hay, W.W. 1977. Depositional history of the South Atlantic Ocean during the last 125 million years. *J. Geol.*, **85**, 651–698.

Warren, B.A. 1973. Transpacific hydrographic sections at Lats. 43°S and 28°S: The SCORPIO Expedition – II. Deep water. *Deep-Sea Res.*, **20**, 9–38.

Warren, B.A. 1974. Deep flow in the Madagascar and Mascarene basins. *Deep-Sea Res.*, **21**, 1–21.

Warren, B.A. 1978. Bottom water transport through the Southwest Indian Ridge *Deep-Sea Res.*, **25**, 315–321.

Warren, B.A., 1981a. Deep circulation in the world ocean. In: B.A. Warren & C. Wunsch (eds.), *Evolution of Physical Oceanography*, MIT Press, Cambridge, Mass., pp. 6–41.

Warren, B.A. 1981b. Transindian hydrographic section at Lat. 18°S: Property distributions and circulation in the South Indian Ocean. *Deep-Sea Res.*, **28A**, 759–788.

Warren, B.A. 1983. Why is no deep water formed in the North Pacific? *J. Mar. Res.*, **41**, 327–347.

Whitehead, J.A. & Worthington, L.V. 1982. The flux and mixing rates of Antarctic Bottom Water within the North Atlantic. *J. Geophys. Res.*, **87**, 7903–7924.

Wyrtki, K. 1971. *Oceanographic Atlas of the International Indian Ocean Expedition*. National Science Foundation, Washington, DC, 531 pp.

Wyrtki, K. 1973. Physical oceanography of the Indian Ocean. In: B. Zeitzschel (ed.), *The Biology of the Indian Ocean*, Springer-Verlag, New York, pp. 18–36.

15

Cenozoic evolution of polar water masses, southwest Atlantic Ocean

S. W. WISE, A. M. GOMBOS and J. P. MUZA

Abstract

Although clustered in a relatively small geographic area on or near the Falkland Plateau in the southwest Atlantic, Deep Sea Drilling Project Leg 71 sites, together with those drilled by Leg 36, provide evidence of a succession of Cenozoic Southern Ocean paleoenvironmental events. These include strong regional erosion during the early Paleocene, the initiation of the psychrosphere during the early Oligocene, the opening of Drake Passage (late Oligocene or early Miocene, timing not well constrained), the build-up of massive biosiliceous/carbonate drifts on submarine promontories while the East Antarctic Ice Cap formed during the middle Miocene, the development of late Miocene Antarctic glaciation and the birth of the West Antarctic Ice Sheet, and climatic-induced migrations of the Polar Front during the Pliocene/Pleistocene.

Surface water productivities were high and temperatures were relatively warm at DSDP Site 511 on the eastern Falkland Plateau during the middle to late Eocene, but surface and bottom waters cooled markedly during the early Oligocene. Sharp changes in the radiolarian faunas and other microfossil groups all suggest a major expansion of the Antarctic water mass during the early Oligocene. Comparison of faunal and floral diversities with oxygen isotopic values and the absence of ice rafting at the site strongly suggest the presence of an ice cap on the East Antarctic craton, but not one which reached sea level to produce fringing ice shelves.

The opening of Drake Passage and the onset of the Antarctic Circumpolar Current are evidenced by the paucity of early to early middle Miocene sediments on the Falkland Plateau whereas the development of the East Antarctic Ice Cap (14 to 10 Ma) is reflected in the rapid accumulation of biogenic sediments. Late Miocene glaciation and the development of the West Antarctic Ice Sheet is evidenced by the initiation of ice rafting at these latitudes which followed an episode of strong bottom current erosion which ended at about 9.5–9.0 Ma. By the end of the Miocene, the Polar Front had migrated well north of the

Falkland Plateau to the vicinity of Site 513. The pattern of climatic deterioration during the Pliocene/Pleistocene is revealed by fluctuations in the position of the Polar Front Zone which is recorded at Site 514 by at least eight alternations between cold and warm water radiolarian faunas. The last six oscillations, beginning at about 2.5 Ma, may coincide with pulses of the Northern Hemisphere glaciations.

Introduction

The four Deep Sea Drilling Project Sites (511, 512, 513, and 514) drilled in the high latitude South Atlantic during *Glomar Challenger* Leg 71 (Ludwig, Krasheninnikov *et al.*, 1983) doubled the number of pre-Quaternary sections available for study from that region (Fig. 1). Although clustered in a relatively small area due to logistical constraints encountered during drilling at these latitudes,* the sites now provide coverage of some deep basinal areas (Sites 328, 513, and 514) as well as of a major intermediate depth plateau (Sites 327, 329, 330, 511, and 512) where sediments containing carbonate have accumulated over most of the past 150 Ma.

Despite the fact that major gaps still exist in this South Atlantic record, the Leg 71 sections provide a Cenozoic record sufficiently complete to allow documentation of a number of major ocean-wide paleoceanographic events. For instance, those at Sites 513 and 514 provide a record of climatically induced fluctuations of the Polar Front over the past 4 Ma, whereas those at Sites 511 and 512 provide a stable isotope record which encompasses the Eocene/Oligocene boundary and the development of the psychrosphere. The latter event was first defined from the study of sections from the Pacific sector of the Southern Ocean (Shackleton & Kennett, 1975; Kennett & Shackleton, 1976), and until now, no comparable sections have been available from the Atlantic sector for direct comparison. As far as the terminal Cretaceous/Tertiary boundary event is concerned, the K/T section in this study area is incomplete. Instead, the new record obtained during Leg 71 reveals further evidence of severe and prolonged erosion which may indicate an early Tertiary oceanic event of major regional importance.

The purpose of this paper is to highlight the paleoceanographic

---

* Due largely to weather and ice conditions, *Challenger* was able to drill or core only 10.7 days during her 55 day Leg 71 cruise; on Leg 36 she managed to drill or core 20 days during that 53 day cruise (still a low average for DSDP cruises in general). The fact that the DSDP *Initial Reports* volumes for both cruises easily exceeded their 1000 page limits, however, speaks well for the scientific value of the material recovered.

Fig. 1. Location of DSDP Leg 71 (large numbers) and Leg 36 drill sites (small numbers).

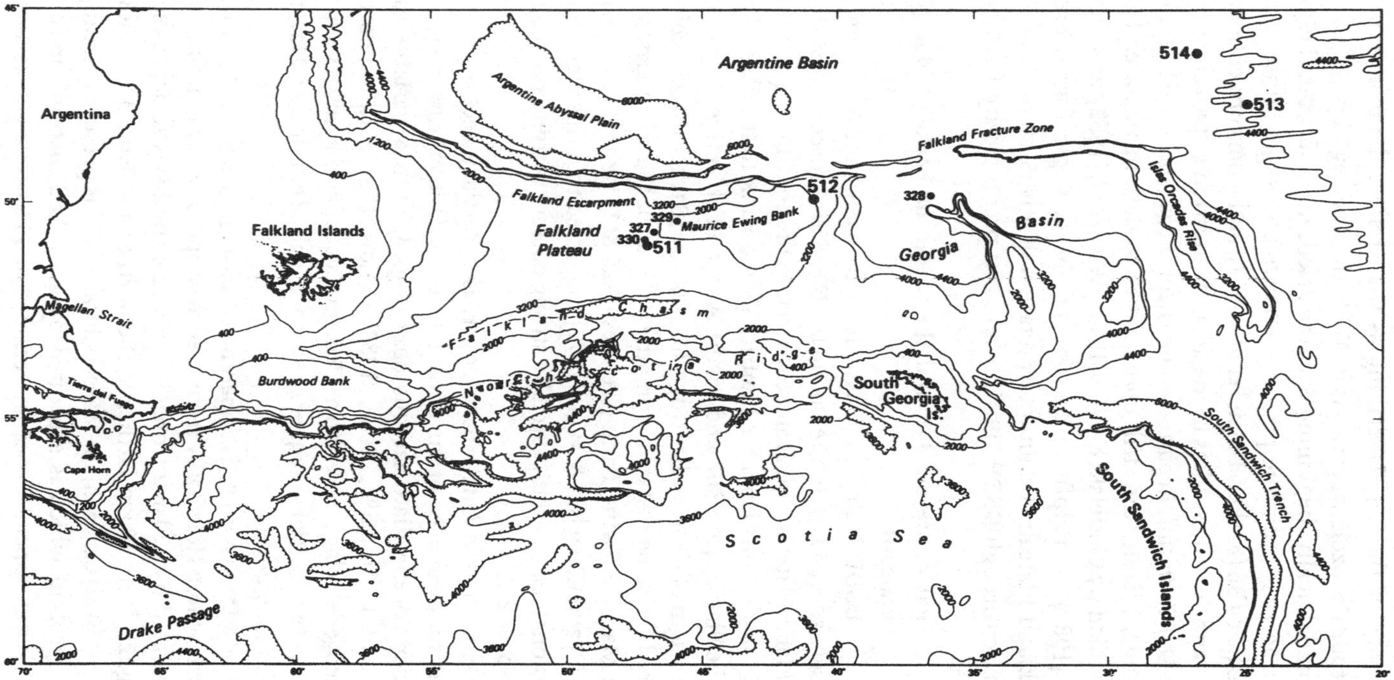

findings of the Leg 71 drilling as they reflect the evolution of high latitude Cenozoic water masses. To do this the Leg 71 results are related chronologically, beginning with the Paleocene, and set in the context of data accrued from the previous *Glomar Challenger* cruise to the area, DSDP Leg 36 (Barker, Dalziel *et al.*, 1977), and from conventional piston core and seismic studies in the area. In addition to pertinent reports in the DSDP *Initial Reports* volumes for Legs 36 and 71, synopses of seismic, piston, and drill core studies to date in this area have been provided by Ciesielski & Wise (1977), Wise *et al.* (1982), and Ciesielski *et al.* (1982). A helpful discussion of the geologic framework of the Falkland Plateau and an interpretation of a N–S multichannel seismic profile through DSDP Site 511 has been presented by Ludwig (1983).

Paleocene erosion and the origin of Barker Ridge (Falkland Plateau)

DSDP Hole 511 was drilled to complement a suite of three closely spaced Leg 36 sites along the NE–SW transect across the eastern Falkland Plateau (Fig. 2) and to ascertain the continuity (or lack thereof) of geologic units from the Maurice Ewing Bank to the basinal province of the Plateau (the Falkland Basin). A second objective was to determine the origin of the peculiar ridge-like feature (Barker Ridge) which can be traced for some distance along the contour of the southeastern flank of the Maurice Ewing Bank (see Ciesielski & Wise, 1977, Fig. 4). Barker (1977) and Ciesielski & Wise (1977) suggested that this was probably an erosional feature sculptured by strong current activity during the early Paleocene.

Below a thin Quaternary cover, the Tertiary section in Hole 511 consists primarily of an unusually thick (186 m) lower Oligocene to upper Eocene diatomaceous ooze (Fig. 3) with a variable carbonate content (1 to 26%, Bode, 1983). Below that a thin poorly dated Paleogene pelagic clay and a 9.5 m interval of non-recovery separate the recovered Tertiary sequence from a subjacent lower Maestrichtian calcareous ooze.

The Maestrichtian is apparently missing entirely only 10 km to the north where Hole 330A recovered an upper Eocene diatomaceous ooze separated by a 20 m uncored interval from middle Cretaceous sediments (Fig. 2). Only 10 km farther to the northeast at Hole 327A, however, a thick (50 m) Campanian–Maestrichtian chalk is present and capped by a hardground which is separated by a marked disconformity from an overlying upper Paleocene section. The Paleocene consists of 22 m of zeolitic clay devoid of fossils except for palynormorphs. The zeolitic clay

Fig. 2. Geologic section through DSDP Leg 36 and 71 sites and selected piston core localities on the Maurice Ewing Bank of the eastern Falkland Plateau (updated from Ciesielski & Wise, 1977, Fig. 7).

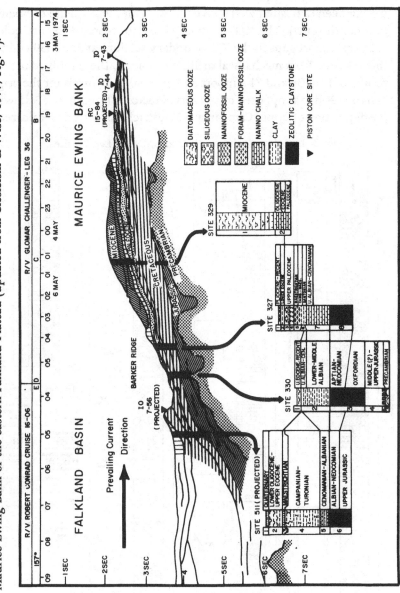

grades upwards into 38 m of diatom ooze which becomes slightly calcareous (up to 13% $CaCO_3$ according to Cameron, 1977) toward the top. At Site 329, some 450 m higher up on the Ewing Bank along the transect, a roughly equivalent section is highly calcareous (78% $CaCO_3$), having been deposited well above the late Paleocene calcite compensation depth (CCD). The K/T boundary was not reached in that hole.

Despite short gaps in the K/T boundary sections at Sites 330 and 511 that resulted from mechanical drilling problems, the combined drilling results of Legs 36 and 71 support an erosional hypothesis for the origin of Barker Ridge as opposed to any sort of tectonic hypothesis. Following the deposition of upper Campanian–Maestrichtian siliceous–calcareous

Fig. 3. Lithologic columnar sections, DSDP Leg 71 sites.

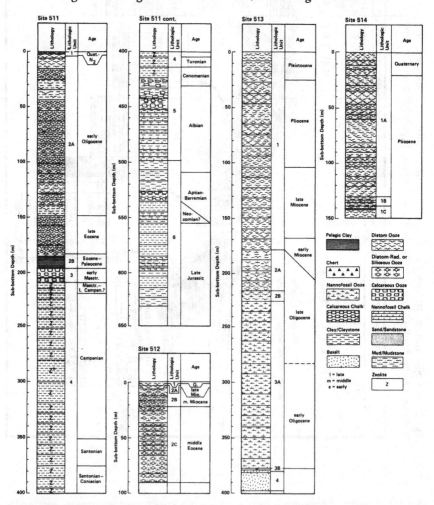

oozes (which are thicker at Site 327 than at 511 due to a higher elevation above the CCD), strong currents developed during the Danian breached basinward-dipping units along a broad front, cutting deeply down to mid-Cretaceous units in the vicinity of Site 330 to form a cuesta-like ridge. The Maestrichtian oozes may have been more subject to erosion along the flanks of the bank where sediment rich in opal-CT may have been absent. This is suggested by the fact that porcellanites sampled along the higher elevations of the plateau at Site 327 and at piston core locality IO 7–44 (Fig. 2) were not cored at Site 511 where only zeolites were present. A second reason for the considerable amount of section missing at Site 330 is the fact that the Cenomanian to lower Campanian sediments were apparently relatively thin there. Only 10 m thick at Site 327, this interval thickens rapidly toward the Falkland Basin where it exceeds 200 m at Site 511 (Fig. 2). Containing very little carbonate, the thickness of these clays was apparently not governed so much by the CCD, which was quite high at this time, as by basin topography. The unit is essentially a basin fill sequence.

The record at Site 327 indicates that the main erosional event(s?) was confined to the Danian, with upwelling conditions setting in during late Paleocene time to produce the rich siliceous and calcareous oozes cored at Sites 327 and 329. Interestingly, an upper Paleocene diatom flora, rather similar to but slightly younger than that first described by Gombos (1977) from the Falkland Plateau, has subsequently been found by that author at DSDP Site 524 in the northern Cape Basin (Gombos, 1984). This represents the only other locality in which Paleocene diatoms have been reported from the South Atlantic proper. Thus upwelling conditions may have been quite widespread in this region during the late Paleocene.

On the Falkland Plateau, current activity was sufficient to maintain some scour zones free of sediment well into the Paleocene and beyond. Currents flowing over the Barker Ridge apparently produced vortices which kept the trough behind the ridge swept free of sediment. Ciesielski & Wise (1977) reported that Paleocene sediments high on the Ewing Bank at piston core locality RC 15–84 (Fig. 2) contained reworked upper Cretaceous coccoliths. In addition, Gombos (Appendix 1) has observed the Upper Cretaceous silicoflagellate *Lyramula furcula* in the upper Paleocene of *Islas Orcadas* Core 16–49. A coccolith (*Nannoconus truitti*) known only from the Aptian–Albian in this region was noted reworked in the lower Eocene of Site 327 (Ciesielski & Wise, 1977). By late Eocene times, however, upwelling, enhanced productivity, and reduced current velocities were sufficient to leave a drape of pelagic sediments over the entire area, particularly in the Falkland Basin where the thick upper Eocene–lower Oligocene sequence laps against the Barker Ridge. There it was better protected from subsequent erosion than at sites exposed higher on the Bank. Nevertheless, sediments of this age have been dated in the trough behind the ridge and along promontories such as on the ridge itself (*Islas Orcadas* Core 7–56) and on the Bank at *Islas Orcadas* station 7–43 (Fig. 2).

Certainly the most vexing question concerning the early Tertiary erosion in this area is the origin of the currents involved. Were they strictly local or was there some form of circumpolar flow involved? It is tempting to speculate on the latter possibility in view of the apparent severity of the event. Circumpolar flow operating at depths of 2000 m as postulated for these sites during the Paleocene (Barker *et al.*, 1977a, Fig. 9) would suggest either a major breach in the then extant South American–Antarctic isthmus or at least an intermediate depth seaway across the Antarctic continent itself.

Evidence of the possible existence of seaways across at least portions of pre-glacial Antarctica has recently been gleaned from high elevation glacial deposits along the crest of the present Transantarctic Mountain chain which contain marine microfossils dating back to the Paleogene (Harwood *et al.*, 1983; Webb *et al.*, 1984). These fossils were presumably excavated from Tertiary marine deposits in basins on the East Antarctic craton by late Neogene glaciers. This evidence supports earlier speculations that major seaways and perhaps some rudimentary (shallow?) form of circum-Antarctic flow existed through the intracratonic basins or through a seaway linking the Weddell and Ross Seas during Paleogene times. Such speculations had previously been based largely on the detection by geophysical means of major depressions or subglacial basins well below sea level beneath the present-day ice cap (see map and discussion by Webb *et al.*, 1984). The Transantarctic Mountains presumably did not begin to ascend rapidly until the Neogene (Gleadow *et al.*, 1982).

Unfortunately, the tectonic history of Antarctica and of the isthmus that presumably linked that continent with South America until the Oligocene is still poorly understood. Some authors believe that the Antarctic Peninsula and West Antarctica are composed of a group of microplates which have had a complex Tertiary history (e.g. Dalziel, 1982). Others believe that the configuration of the Peninsula and its relationship to East Antarctica during the Paleogene was the same as it is today (see discussion by Barron *et al.*, 1981). There is, therefore, no consensus as to the relationship of the Peninsula to the Falkland Plateau and many rather different alignments have been suggested in recent years (compare the reconstructions of Smith & Hallam (1970), Barker *et al.* (1977a), de Wit (1977), Jones & Plafker (1977), Sliter (1977), Barron *et al.* (1981), and Dalziel (1982). Many of these reconstructions, however, suggest the presence of an isthmus between Antarctica and South America during the Paleogene.

The configuration of the southern continents by Barron *et al.* (1981) shows the known extent of emerged and submerged lands. Their reconstruction for the Danian (60 Ma) seems to permit little possibility for a deep or intermediate water circumpolar flow, particularly in the region of the Falkland Plateau. If that is true, then either shallow seaway connections for circumpolar flow or purely local currents would have to be cited as the cause for the Paleogene erosion of the Falkland Plateau. At the time of writing, the prevailing uncertainties and lack of data on the geologic history of Antarctica strongly limit our ability to understand the Paleogene depositional and erosional history of the Plateau.

Lower to middle Eocene

DSDP Leg 36 drilling revealed little about Eocene sedimentation. Less than a metre of lower Eocene chalk had been recovered at Site 329 and equivalent sediments at Site 327 were carbonate-poor zeolitic clays less than 10 m thick. Piston cores taken at higher elevations along the northern margin of the Maurice Ewing Bank, however, had indicated the presence of a long (*c.* 400 m) Paleocene–Oligocene section replete with well preserved calcareous and siliceous microfossils. Limited seismic information prevented drilling of the optimal site high on the Bank during DSDP Leg 71, but a short break in the adverse weather and current conditions did permit the acquisition of an 89 m middle Eocene and Miocene section via hydraulic piston coring (HPC) at Site 512 (Figs. 1 and 3). This middle Eocene section coupled with the long upper Eocene–lower Oligocene sequence at Site 511 provided the first Paleogene material from this sector of the Atlantic suitable for stable isotope analysis. Preservation of siliceous and calcareous microfossils at both sites was good to excellent.

The magnetic polarity signature for the 56 m middle Eocene section at Site 512 (Fig. 4) fits well the reversal sequence for Anomalies 19 and part of 18 (Ledbetter, 1983). The interval is well dated biostratigraphically and falls within the calcareous nannofossil *Discoaster bifax* Subzone (CP14a) of Okada & Bukry (1980) and the *Globigerapsis index* foraminiferal Zone of New Zealand (Krasheninnikov & Basov, 1983;

Wise, 1983). The diversities of these microfossil assemblages are reasonably high for these latitudes, equivalent to those reported from the New Zealand area (Appendix 2, Fig. A). Dissolution among the nannofossils, however, prevented identification of the smaller reticulofenestrids. The siliceous faunas are also diverse, are well preserved, and include many high latitude provincial taxa, a number of which remain undescribed. A local high latitude zonation has been erected for the silicoflagellates based on newly described taxa (Shaw & Ciesielski, 1983), but formal zones have not been established for the diatoms and radiolarians due to the lack of comparable sections in this region. The silicoflagellate assemblage is characterized by *Dictyocha grandis*, a form which has only been found at one other locality, *Eltanin* Core Site E54-3 on the Kerguelen Plateau which lies at a comparable latitude in the Indian Ocean (Mostajo, Gombos, and Wise, unpublished data).

Although the interval in question is well defined by paleomagnetics and biostratigraphy, the ages assigned to this middle Eocene section using the different time scales available vary considerably with a duration ranging from 3.1 to 1.0 Ma for the interval (Fig. 4). The choice of time scale can make a considerable difference in calculating sedimentation rates. The maximum value is given by the La Brecque *et al.* (1977) scale which has no internal calibration points. The minimum value is given by Okada & Bukry (1980) whose assigned ages for their coccolith zones are taken from Bukry (1973). Ledbetter (1983) followed Ness *et al.* (1980) in arriving at the value of 2.8 Ma for the interval. The scale of Lowrie & Alvarez (1981), on the other hand, is calibrated by tie

Fig. 4. Biostratigraphy and paleomagnetic correlation of the middle Eocene interval of DSDP Hole 512.

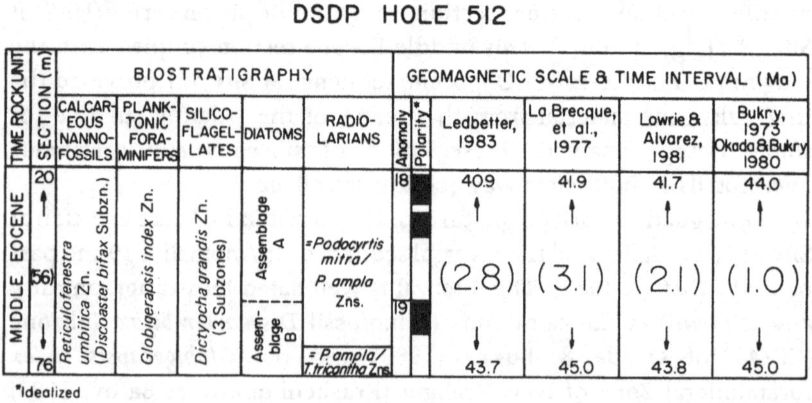

DSDP HOLE 512

*Idealized

points derived from detailed biostratigraphic and paleomagnetic studies of the Contessa sections near Gubbio, Italy (Lowrie *et al.*, 1982). Their results, however, rest more heavily on the foraminiferal record than on the less well preserved calcareous nannofossils.

The oxygen isotope curves for benthic and planktonic foraminifers from the middle Eocene of Site 512 (Fig. 5) fluctuate within a narrow

Fig. 5. Oxygen and carbon isotope data from planktonic and benthic foraminifers for the middle Eocene interval of Hole 512 (from Muza *et al.*, 1983, Fig. 5). Stratigraphic divisions are not based on thickness or absolute age. Samples are plotted arbitrarily at constant divisions (core sections). $\delta^{18}O$ and $\delta^{13}C$ are reported as deviations per mil from PDB. Paleotemperatures are calculated using Shackleton's (1974) paleotemperature equation and a δ_w of $-1.2‰$, his correction factor for a pre-Miocene 'ice-free world'. Dashed sections indicate gaps in isotopic record because of missing data.

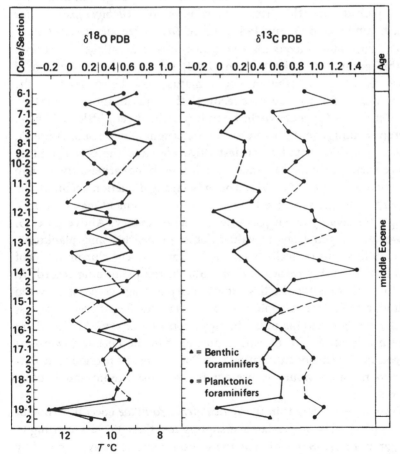

range. However, a depletion in ^{18}O at the base of the section may indicate a short warming or possible diagenetic effects. If the isotope values are converted to absolute temperature values using the formula applied for an ice-free Paleogene ocean at these latitudes by Shackleton & Kennett (1975), then the paleotemperatures for both benthic and planktonic foraminifers range between 8.5 and 12 °C. Little temperature contrast is indicated between bottom and surface waters at this intermediate water depth (about 1600 m) over the Plateau during the mid-Eocene.

Eocene/Oligocene boundary event (Site 511)

Although the intervening middle–upper Eocene record has not been recovered on the Falkland Plateau, little change is noted between the oxygen isotopic values discussed above and those derived for the upper Eocene at Site 511 (Fig. 6). There the Eocene/Oligocene boundary was picked on the last occurrences of *Globigerapsis index* (Krasheninnikov & Basov, 1983) and of *Discoaster saipanensis* (Wise, 1983). The simultaneous disappearance of five dinoflagellate species at this point (Goodman & Ford, 1983) supports the suggestion of Basov *et al.* (1983) that a slight hiatus may be present at the boundary. Unfortunately, core recovery through this interval was low, and Core 511–16 just above the boundary yielded very little carbonate, thereby making age and stable isotope determinations difficult. Core 15, however, is well dated as earliest Oligocene, and the corresponding oxygen isotope values fall generally in line with the earlier trend seen in the middle Eocene of Site 512. The only strong deviation occurs in the upper two sections of Core 17 where an ^{18}O enrichment may indicate a temporary cooling event just prior to the boundary. Above Core 15, however, oxygen isotopic values for both benthic and planktonic foraminifers shift radically, showing by Core 9 an enrichment in ^{18}O of 3‰ in the benthic record and of 1.5‰ in the planktonic record. In addition, there is for the first time a sharp average contrast between bottom and surface isotopic values of over 1‰. These trends persist with some fluctuations until the upper portion of Core 511–3 where they nearly return to the values prevalent in the Eocene–lowermost Oligocene. The major fluctuation is in Core 6 where a pronounced shift toward negative values is present in both planktonic and benthic curves.

Core 511–3 falls within the *Reticulofenestra hillae* coccolith Zone of Bukry (CP 16c) which is dated at slightly less than 35 Ma according to Berggren *et al.* (1985). Although the section at Site 511 was not cored by

Fig. 6. Correlation of the stable isotope record (oxygen and carbon data from planktonic and benthic foraminifers), radiolarian assemblages, and inferred water masses for the upper Eocene–lower Oligocene interval of Site 511. Isotopic and paleotemperature curves are from Muza *et al.*, (1983, Fig. 4) and are plotted in the same manner as those in Fig. 5. As noted in the text, however, these isotopic paleotemperature estimates are considered to be too low and are taken as evidence for the existence of a moderate-sized early Oligocene ice sheet on Antarctica. The radiolarian assemblages are those determined by Weaver (1983) for core catcher samples only: A = Antarctic; SA = sub-Antarctic; CT = cool temperate. The curve showing the migration of water masses across Site 511 is based on both the isotope and radiolarian assemblage data.

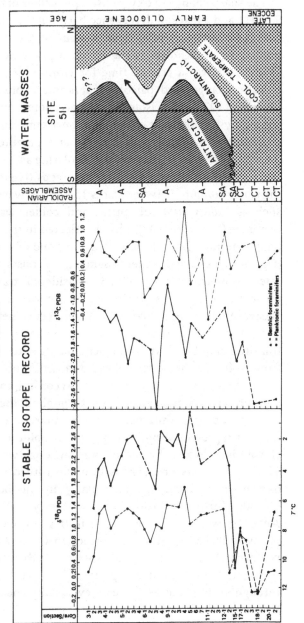

HPC, and detailed paleomagnetics are not available, the time represented by the lower Oligocene is estimated to be of the order of about 2 Ma.

The sharp change in oxygen isotopic values in the lower Oligocene at Site 511 and a marked divergence between the planktonic and benthic curves reflect the initial development of the present-day system of bottom waters known as the psychrosphere (Kennett & Shackleton, 1976). This was a critical time for worldwide oceanic development, climatic change, and paleobiogeographic evolution and distribution (Kennett, 1978). The isotopic changes noted on site 511 are accompanied by concomitant changes in the assemblage compositions of the attendant radiolarians and benthic and planktonic foraminifers. Of these, the changes in the overall radiolarian assemblage with respect to species dominance seem to be the most sensitive to changes throughout the water column. Between Cores 15 and 12, warm-water radiolarians, such as *Calocycletta*, disappear and colder water genera begin to dominate (Weaver, 1983). This is reflected in the replacement of a cool temperate high latitude assemblage by one of sub-Antarctic affinities (Fig. 6). This cooler water assemblage persists through Core 11, an interval which Weaver (1983) considered transitional, suggesting upwelling conditions during a progressive climatic deterioration. Weaver further suggested that a convergence between cool temperate and sub-Antarctic water masses probably lay over the site during this transitional interval. With further climatic deterioration, however, the front shifted to the north and eventually the sub-Antarctic water mass was displaced by an Antarctic one. According to the radiolarian data, the Antarctic water mass persisted throughout the deposition of Cores 9 to 3 except for a shift back to sub-Antarctic conditions during the interval represented by Core 5. At that point a slight amelioration of climate is indicated by the reappearance of two warm-water genera. This corresponds precisely with a short-lived warming trend in the oxygen isotope curves, particularly in the benthic record which provides further indication of a southerly shift of the warmer sub-Antarctic water mass over the site (Fig. 6). Overall, Weaver (1983) believed that the radiolarian record in Cores 9 to 3 represented the northern elements of a broad Antarctic–sub-Antarctic biogeographic province that expanded significantly during the early Oligocene throughout the Southern Ocean, eventually encompassing over 10° of latitude in the circum-Antarctic.

The foraminiferal assemblages at Sites 511 and 512 can be roughly correlated with the radiolarian and isotopic data. Benthic species diversity drops from 10–15 species during the middle Eocene to 5–10 species per sample during the late Eocene and early Oligocene, presumably as a result of a drop in temperature (Basov & Krasheninnikov, 1983). Planktonic species diversity decreases from the middle Eocene to the late Eocene to a level which is maintained through the early Oligocene (Krasheninnikov & Basov, 1983; see also Appendix 2, Figs. A and B). There, however, warm-water genera such as *Globigerapsis* and *Acarinia* have disappeared and have been replaced by more cosmopolitan forms. Coccolith diversities at Sites 511, 512, and 513, already somewhat limited at these latitudes, fall from 31 species in the mid-Eocene to 24 in the late Eocene, then to 15–18 above the cooling event in the early Oligocene, finally stabilizing at about a dozen species for the remainder of the Oligocene (Fig. 7; Appendix 2, Figs. A–C).

A most significant change among the coccoliths is the progressive extinctions of key marker species in the Oligocene, a phenomenon which has resulted in a zonation for these latitudes based exclusively on last occurrences (Wise, 1983). The warm-water sphenolith group, which evolved rapidly in the low latitudes to provide a number of first occurrence datums for the Oligocene zonation there, is not represented on the Falkland Plateau except by the ubiquitous *Sphenolithus moriformis*.

As noted previously, Shackleton & Kennett (1975) interpreted the dramatic cooling in the vicinity of the Eocene/Oligocene boundary as a critical stage in the development of the psychrosphere. Their pioneering study was done on DSDP Site 277 on the Campbell Plateau at latitude 52°14'S (compared to latitude 51°00'S for our Site 511). The present water/sediment depth to the Eocene/Oligocene boundary at Site 277 is 1522 m whereas that at Site 511 is 2737 m. Shackleton & Kennett (1975) dated the cooling event as having occurred during the early Oligocene. In a more detailed study of the same section, Keigwin (1980) dated this event as having occurred across the Eocene/Oligocene boundary. Although sediment recovery at Site 511 in Cores 12 to 20 was poor (20%), sedimentation rates were high during the early Oligocene compared to the Campbell Plateau (77 m/Ma versus 5 m/Ma), thus biostratigraphic resolution for dating the event is good. Assuming that the cooling was synchronous at all latitudes, then our results support the early Oligocene date for the event as suggested by Shackleton & Kennett (1975).

As far as the question of worldwide synchroneity of this event is concerned, studies of the late 1970s have been rather ambiguous (Fig. 8), mainly because of the sparsity of data points in rotary cored sections which, by their nature, are not ideal for making such a determination.

Fig. 7. Correlation of the sedimentary record from the DSDP Leg 36 and 71 drill sites (as reported in the appropriate DSDP *Initial Reports*) and from selected piston cores taken on the Falkland Plateau (Appendix 1). The right-hand column provides a chronology of major regional tectonic, paleoceanographic, or paleoclimatic events. For comparison purposes all data are correlated against the calcareous nannofossil zonation of Wise (1983), the diatom zonations of Weaver & Gombos (1981), and Ciesielski (1983), Gombos & Ciesielski (1983), or according to the ages given by the authors cited. Species diversity data are taken from Gombos (1983), Krasheninnikov & Basov (1983), and Wise (1983) and are averaged per diatom zone. More detailed diversity curves are given in Appendix 2. References cited are: (1) La Brecque & Rabinowitz (1977); (2) Barker & Burrell (1977); (3) Barker & Burrell (1982); (4) La Brecque *et al.* (1983); (5) Haq (1980); (6) Savin *et al.* (1981); and (7) Ciesielski & Weaver (1983). Specific cores are numbered for DSDP Site 329; for Site 513, RW = reworked microfossils and * = oldest ice-rafted clast.

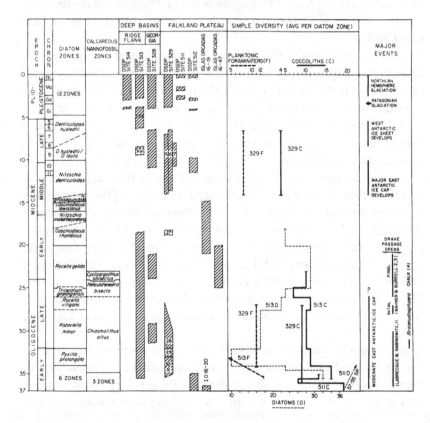

More recent studies on HPC sections (Oberhänsli & Toumarkine, this volume chapter 9), however, seem to support an early Oligocene timing for the event.

Muza *et al.* (1983), however, cautioned that the isotopic expression and possibly the timing of the event can be affected by local conditions, particularly by the character and evolution of the overlying water masses. This was suggested by the strong shifts in water masses and fronts of regional character seen at Site 511 (Fig. 6), and is further indicated by a comparison between that site and Site 277.

Although the timing of the event over the Falkland and Campbell Plateaus seems to be closely synchronous, there is a distinct difference in the contrast between benthic and planktonic foraminiferal oxygen isotope records at the two sites. Whereas Muza *et al.* (1983) see a sharp contrast over the Falkland Plateau of about 1‰, Keigwin (1980) detected no significant increase in contrast at Site 277. This indicated to him that the cooling affected the entire water column there equally. Perhaps this could have been due to the more intense upwelling and mixing of the water column which might occur if this shallow site was positioned close to a front between adjacent water masses. As the Eocene/Oligocene boundary at Site 277 lies at a present depth of only 1522 m, mixing to its paleodepth during the Oligocene is conceivable.

Fig. 8. Comparison of benthic foraminiferal oxygen isotope records ($\delta^{18}O$ in ‰ PDB) for the Eocene–Oligocene of Site 511 (Falkland Plateau) with published records from the following areas (from Schnitker, 1980, Fig. 5): North Atlantic (Vergnaud-Grazzini *et al.*, 1978); South Atlantic (Boersma & Shackleton, 1977); subtropical Pacific (Savin, *et al.*, 1975); and the subantarctic Pacific (Kennett & Shackleton, 1976). The record from the Falkland Plateau indicates that the psychrosphere developed during the earliest Oligocene.

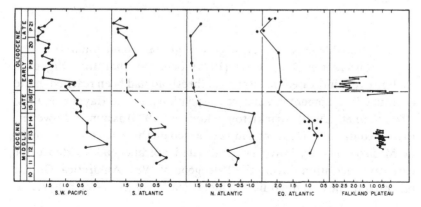

Barker & Burrell (1982) have commented on the fact that plateaus and rises in the Southern Ocean tend to create their own oceanography, causing upwelling which would promote homogenization within the water column. Site 511, on the other hand, had reached a depth of over 2500 m by the Oligocene (Barker *et al.*, 1977a) and, being significantly deeper than Site 277, its benthic record would not have been noticeably affected by mixing along the contact zone between surface water masses. Thus a truer indication of the temperature difference between deep surface and benthic records may be provided by the deeper site.

An alternative explanation is that the production of cold bottom waters during the early Oligocene only affected the deeper ocean water masses to any appreciable extent. One can assume that the early Oligocene ocean was initially rather stratified. The initiation of bottom water production off Antarctica would have tended to fill the ocean basins with cold dense water from the bottom upwards. The upper water column extending down to 1500m or so may not have been so drastically affected. At no time would thermohaline circulation have been as vigorous as that of today.

Interestingly, Keigwin (1980) found significant contrast between the benthic and planktonic foraminiferal isotopic records at Site 292 which is also located at a fairly deep water depth (2943 m) some 16°N of the Equator on the Benham Rise in the Philippine Sea. The tropical plankton, little affected by polar events, showed only a minor (0.3‰) ^{18}O enrichment whereas the benthos recorded a decrease of nearly 1‰, directly comparable to that of Site 511. Thus the cooling of deep waters was far more noticeable throughout the Pacific than that of the surface waters. Again, the geographic and paleobathymetric position of the site coupled with the physical oceanographic setting seems to have been important in determining the exact expression of the Oligocene cooling event.

Probable existence of Oligocene glacial ice on Antarctica
Shackleton & Kennett (1975) believed that the Oligocene cooling event indicated that a critical threshold had been reached in the eventual development of Antarctic glaciation (see also Hayes & Frakes, 1975; Kennett, 1978). Shackleton & Kennett (1975) assumed, however, that no continental ice sheet had yet formed on the Antarctic continent. Le Masurier (1972), however, postulated volcanogenic evidence for extensive glaciation during the Paleogene of West Antarctica. On the other hand, the first unequivocal identification of ice-rafted material in

Antarctic marine sequences is from drill cores close to the continent in the Ross Sea and is late Oligocene in age (Hayes & Frakes, 1975). Most workers do agree that there was an extensive ice cap on East Antarctica by 14 Ma (Kennett, 1978; Kerr, 1982) and that a stable ice sheet first formed on West Antarctica by late Miocene times (Kennett, 1978). For purposes of calculating paleotemperatures, therefore, Shackleton & Kennett (1975) assumed an essentially ice-free world prior to the mid-Miocene and applied a correction to the paleotemperature equation of Shackleton (1974) of 0.9‰ to account for the isotopically light water now locked up in present-day ice sheets. This compares with corrections of 0.3‰ suggested by Emiliani (1954) and of 0.5‰ by Craig (1965). Shackleton & Kennett (1975) then interpreted the abrupt early Oligocene temperature drop as representative of the formation of sea ice around the Antarctic continent and the initiation of the formation of bottom waters close to freezing. Their assumption, however, that no significant continental ice was present on Antarctica and that the enrichment in their oxygen isotope values was purely a function of temperature, has recently been challenged by a number of workers (Matthews & Poore, 1980; Miller & Fairbanks, 1983b; Keigwin & Keller, 1984; Poore & Matthews, 1984; Shackleton *et al.* 1984a).

Muza *et al.* (1983) applied the same correction factor as did Shackleton & Kennett (1975) so that the results from the Falkland Plateau could be closely compared with those from the Campbell Plateau. The resulting paleotemperature values are scaled along the bottom of Fig. 6. Muza *et al.* (1983) noted that the values calculated for the early Oligocene cooling event are comparable with those of the present day at Site 511 (as are those for Site 277). This would imply a temperature/circulation regime similar to that of the present day.

The above deduction presents a number of problems, particularly that the sedimentary and microfossil records of the lower Oligocene at both sites are quite different from those of the present day. At Site 511, a significant component of ice-rafted debris has been present in the sediments only since the late Miocene. Secondly, the diversity of Oligocene calcareous microfossils is far higher in the lower Oligocene than in the late Neogene or at present. The diversity of indigenous calcareous nannofossils identified in the light microscope, for instance, drops from 12–18 in the lower Oligocene to 7 in the mid-Miocene, and stands at about 5 in the Holocene (Fig. 7; Appendix 2, Figs. A–C). Clearly, climatic conditions were not as severe during the early Oligocene as today, a fact well recognized by Shackleton & Kennett (1975) and by Muza *et al.* (1983).

The question arises, then, as to why the isotopic paleotemperature values derived from the early Oligocene are so low. Assuming no fundamental error in the methods used, we conclude, as do the various authors cited above, that there must have been a significant volume of ice present on Antarctica. The unanswered questions are: (1) where was the ice located, (2) why is there no record of ice-rafting in lower Oligocene deep sea marine sediments surrounding the continent, (3) why did the diversities of planktonic organisms remain relatively high over the Falkland Plateau and at other localities some distance from the continent, and (4) what was the extent of the ice development?

Frakes (1983) suggested that the presently known geologic record is simply inadequate to reveal the extent of ice-rafting during the Paleogene, and this is certainly true for the relatively inaccessible interior and margins of the continent. Limited exploration has revealed evidence supportive of local sea level glaciation as early as 45–55 Ma in the McMurdo Sound area (Webb, 1983) and of more widespread late Oligocene–early Miocene ice-rafting in the Ross Sea proper (Hayes & Frakes, 1975). The deep marine records for the early Oligocene at the latitude of the Falkland and Campbell Plateaus, however, are now adequately known, and, as mentioned above, these reveal no record of ice-rafting or of greatly restricted faunal and floral diversities. The implication, then, is that if significant ice did exist on the Antarctic continent during the early Oligocene as suggested by the isotopic data, it was not of the magnitude (volume) of the present-day ice, nor did it reach sea level to any appreciable extent to form ice shelves from which large icebergs could be generated to cause widespread ice-rafting to lower latitudes. We are left then with the prospect of a moderate-sized ice cap more or less 'hidden' on the interior of the Antarctic continent. If that were the case, ice-rafting should have been negligible and the effect on surface plankton in the oceans beyond the continent would not have been as severe. Sea ice, however, could have been present in the Ross (and Weddell) Sea to cause bottom water formation as Shackleton & Kennett (1975) suggested, thereby influencing the benthic as well as planktonic isotopic record of the deep sea. These isotopic values, therefore, would record both a temperature and an ice-volume effect.

Is it reasonable to assume a significant accumulation of ice on the interior of Antarctica rather than at sea level along the continental margins? We believe so. G. H. Denton (1983, personal communication; see also Denton et al., 1971, Kvasov & Verbitsky, 1981) has pointed out that ice sheets of both hemispheres today are not centred on and probably were not initiated in mountain ranges. Instead, they occupy the lowlands. Ice cap formation in Antarctica during the Oligocene, therefore, may well have begun at comparatively low elevations (2500–3000 m, Denton *et al.* 1971) on plateaus of the East Antarctic craton. The important requirement in addition to low temperatures would have been the presence of moisture sources to nourish the ice mass. Such sources would likely have been the interior seas such as those postulated by Zinsmeister (1978) and Webb *et al.*, 1984). If the ice sheet did reach sea level during the early Oligocene, it may well have done so along the interior seaways where any ice-rafting would have been rather localized. This is not to suggest that mountain glaciers did not form as well. They would also have been more localized and volumetrically less important, however, compared to the cratonic ice sheet.

The speculative nature of the above scenario is indicated by the lack of consensus on a number of points made. Oberhänsli *et al.* (1984; this volume, chapter 9) argue against the presence of Oligocene ice. They cite isotope data from the mid South Atlantic similar to that of Poore & Matthews (1984) and Shackleton *et al.* (1984a) who argue the affirmative case. Frakes (1983) postulates that the build-up of Oligocene ice occurred in West Antarctica rather than East Antarctica due to more mountainous topography there and a more convenient source of precipitation.

If we are correct, however, in assuming that the strong enrichment in ^{18}O noted for the early Oligocene in Fig. 6 must be attributed to both temperature and ice-volume effects (rather than to just temperature alone), then how much of a correction should be applied for ice volume? This is difficult to calculate without further independent evidence. Only then can meaningful isotopic paleotemperature values be assigned. For the present, however, it is clear that those indicated in our Fig. 6 are too low since they are not corrected for ice volume. Interestingly, if the corrections of 0.3–0.5‰ suggested by Emiliani (1954) and Craig (1965) were applied for the early Oligocene, the resulting paleotemperature values would be about 3 to 4 °C warmer, more in line with the paleontological evidence. In agreement with that estimate, Keigwin & Keller (1984) suggested an ice-volume effect of about 0.4‰ based on the benthic-planktonic foraminiferal covariance in $\delta^{18}O$ at two equatorial Pacific DSDP sites.

It is also interesting to note that the carbon isotopic curves generally parallel those for the oxygen isotopes (Fig. 6). The strong enrichment of ^{13}C values during cold phases suggests to us the influx of younger waters. Kroopnick (1974) has shown that modern ocean deep waters formed in the North Atlantic initially have relatively positive ^{13}C compositions due to biological utilization of light carbon in the surface water. As the water mass ages, the organic matter is gradually oxidized and the light carbon is returned to the water as dissolved carbon. The strong enrichment in organic carbon shown in Fig. 6, therefore, is consistent with the faster circulation, overturn, upwelling, and flushing of older water masses which one would expect with the development of a psychrosphere and the initiation of bottom water formation off nearby Antarctica. Miller & Fairbanks (1983a) noted a similar carbon isotope signal at several other Atlantic and Pacific localities, and considered this to be a worldwide phenomenon.

None of the above interpretations of the stable isotope or paleontologic record should imply, however, that the early Oligocene conditions persisted unabated until the next threshold in climatic deterioration was reached during the middle Miocene. On the contrary, an amelioration of climate during the early Miocene is a widely recognized global phenomenon and one well documented in the sub-Antarctic by oxygen isotope data and the diversity of planktonic foraminifers (Kennett, 1978). This could imply that any ice cap present on the Antarctic continent during the Oligocene may well have been destroyed by or during early Miocene times, and that any correction factor applied to early Oligocene isotopic paleotemperatures would probably not be applicable for the early Miocene. Unfortunately, as shown in the next section, our present sedimentary record from the Falkland Plateau for upper Oligocene–lower Miocene events is not adequate to document such a supposition.

Opening of Drake Passage

The acquisition of a continuously cored and reasonably complete Oligocene–lower Miocene record on the Falkland Plateau would be of high interest because this interval encompasses the predicted time of opening of Drake Passage, the effects of which presumably should be recorded in the sediments of the Plateau. Based on somewhat different interpretations of the magnetic lineations and topography within the present-day Drake Passage and western Scotia Sea, two different dates have been predicted for the establishment of deep water circulation through Drake Passage. As summarized on Fig. 7, La Brecque & Rabinowitz (1977) postulated a date of early Oligocene (Anomaly 10–13) which would allow this event to be linked with the *Braarudosphaera* coccolith blooms in the South Atlantic during Chron 10.25R time (La Brecque *et al.*, 1983). Barker & Burrell (1977), however, believed the first coherent spreading began during the late Oligocene (about Anomaly 8 time, well after the *Braarudosphaera* event). They argued that spreading led initially to the establishment of a shallow water connection, and that a complete barrier to circumpolar flow consisting of two ridges along the Shackleton Fracture Zone remained until about 23.5 ± 2.5 Ma, the approximate time of the Oligocene–Miocene boundary. The gradual subsidence of these two ridges and the establishment of the Antarctic Circumpolar Current would account for the apparent hiatus noted in the lower Miocene of Site 329 on the Falkland Plateau (Barker & Burrell, 1977, 1982).

On the face of it, a late Oligocene/early Miocene opening of Drake Passage as suggested by Barker & Burrell (1977) seems attractive in view of the reasonably continuous sedimentation pattern during the early Oligocene at Site 329 (no reworking noted) and the absence of uppermost Oligocene sediments (the *Reticulofenestra bisecta* coccolith Zone is not represented). A pronounced change in style of sedimentation and a marked change in sedimentation rate is evident in the thick Miocene sequence at Site 329 (Fig. 2) which is characterized by common reworked Oligocene microfossils (particularly coccoliths) (Ciesielski & Wise, 1977).

Either date proposed for the opening of Drake Passage could account for the strong erosion that truncates the Oligocene at Site 511. Considering that this locality is up-current from Site 329, this would be a logical source area for many of the Oligocene microfossils reworked into the Miocene of Site 329.

The problem of a lag effect in establishing a deep water passage following the initiation of sea-floor spreading as discussed by Barker & Burrell (1977) seems applicable to the La Brecque model as well; thus, one tends to seek more detail in the sedimentary record for discriminating the event. Unfortunately, the existing core coverage for the Oligocene/ early Miocene of the Falkland Plateau is not as complete as it might be. The uppermost Oligocene is missing at both Sites 511 and 512. The longest drilled section, that at Site 329, provided four continuous Oligocene cores (329-28 to 329-31), but recovery was poor as was the preservation of microfossils in the well lithified chalk matrix. In addition, the sequence is separated from the overlying Miocene and the underlying Eocene by 20 m coring gaps. The only upper Oligocene core appears to be Core 28 in which the lower Oligocene diatom *Pyxilla prolongata* is absent but the upper Oligocene coccolith marker *Chiasmolithus altus* is present (Fig. 7). Exactly where this core falls within the interval between 26 and 32 Ma cannot be specified. The next overlying core, Core 329–27, is dated at about 18.5 Ma (Haq, 1980), but it too is separated by a 20 m coring gap from the next overlying Miocene core.

The few piston cores from the Plateau from this interval (Fig. 7) do little at present to clarify the situation. Because of the small amount of section sampled, these are rather difficult to date precisely, even when the new zonations introduced for the Leg 71 sections are employed (Fig. 7). With this scattered core coverage, one cannot pinpoint a well defined hiatus indicative of Drake Passage opening. More helpful for such a study would be a HPC section through the Miocene/Oligocene section on the northern crest of the Maurice Ewing Bank. There the small amount of overlying section insures good preservation of siliceous and calcareous microfossils, and piston cores (S. W. Wise, unpublished) indicate that a relatively complete section can be expected. HPC cores would also permit correlation of the high latitude microfossil zones for the Oligocene/lower Miocene with the paleomagnetic time scale, something which has not yet been achieved in this area.

Middle Miocene

As summarized by Kennett (1978), the middle Miocene represents a crucial stage in the evolution of global climates in that the East Antarctic Ice Cap began to develop between 14 and 10 Ma (Savin et al., 1975; Shackleton & Kennett, 1975). The formation of an ice cap on East Antarctica is considered to be a direct consequence of the thermal isolation of the continent following the opening of Drake Passage and the onset of the Antarctic Circumpolar Current (ACC) Ciesielski et al. (1982) argued that no ice sheet was present in West Antarctica at this time, therefore no true Antarctic Bottom Water was being formed, although the production of North Atlantic Deep Water had begun (Schnitker, 1980).

It was also within this interval (between 14 and 11 Ma, depending on the author cited; see Fig. 7) that the rapid accumulation of calcareous–biosiliceous oozes occurred on the Falkland Plateau. This enormous build-up (over 300 m; Cores 329–25 to 329–2) took the form of a large drift deposit that heavily blanketed those portions of the Plateau that lay above the middle Miocene CCD. Fed by exceptionally high biological productivity in the surface waters, the drift was shaped by the increasingly strong ACC which, as mentioned previously, was actively eroding Oligocene and possibly older strata to the west. The low diversity coccolith assemblages of the drift were soon dominated by *Reticulofenestra perplexa* (= *Dictyococcites antarcticus*; see Wise, 1983), a form whose appearance at these latitudes (Haq, 1980) correlates with the build-up of the ice cap on Antarctica. Haq (1980, Fig. 12) also plotted oscillations in the abundance of this species which he believed may indicate variations in the stability of the Antarctic ice cap.

High biological productivity and sediment accumulation rates contributed to the entrapment of organic matter within the sediment, which led to the formation of authigenic pyrite and possibly gypsum. These authigenic minerals were discussed by Muza & Wise (1983) who considered the possibility that all the gypsum formed *in situ*. Re-examination of these cores after their complete desiccation as a result of nearly ten years of storage in the DSDP repository, however, has revealed gypsum crystals on the split core surfaces at some but not all of these intervals where gypsum was previously reported. This indicates that at least some and possibly all of the gypsum identified by Muza & Wise (1983) is an artifact of storage diagenesis (Wise, 1983, unpublished), and no direct correlation can be drawn between its occurrence and paleoproductivity.

Late Miocene glaciations

The late Miocene to Recent history of the DSDP Leg 71 sites has been reviewed in detail by Ciesielski & Weaver (1983) and will only be outlined here rather briefly. Progressing from the middle Miocene refrigeration event, the next stepwise deterioration in Cenozoic climate resulted in the late Miocene glaciations of Antarctica, the evidence for which is varied and widely seen both in the deep sea and land records of the Southern Ocean area and beyond (see summaries by Kennett, 1978; Wise, 1981; Ciesielski *et al.* 1982). As noted in Fig. 9, by this time the annulus of biosiliceous ooze deposited beneath the frigid surface waters surrounding the Antarctic continent had expanded well north of 50°S, ice-rafted clasts were being deposited at Site 513, and significant amounts of sand-sized ice-rafted detritus (IRD) were accumulating in the vicinity of the Falkland Plateau. There, as elsewhere, the sedimentary record was interrupted by episodes of current erosion (Ciesielski & Wise, 1977; Wise *et al.*, 1982). As in the Pacific sector, a sudden shift in the position of the proto(?) Polar Front occurred as warmer waters containing predominantly calcareous plankton were driven northward by the expanding Antarctic water mass (Kemp *et al.*, 1975; Ciesielski & Wise, 1977; see summary by Wise, 1981). The greatest change in the benthic foraminiferal oxygen isotope record of the Southern Ocean is an enrichment that occurred in sediments now dated as latest Miocene (Kapitean Stage of New Zealand; Messinian Stage of Europe; see van

Couvering et al., 1976; Weaver, 1976). From this Shackleton & Kennett (1975) inferred an expansion of the Antarctic ice sheet to a size 50% greater than present day. This figure has been challenged by Woodruff et al. (1981) who argued that the maximum late Miocene ice volume was probably similar to that of the present day. Mercer & Sutter (1982) estimated the late Miocene Antarctic ice volume to have been ½ or ⅓ that of the late Pleistocene, and suggested that any additional ice was present not in Antarctica, but in the Northern Hemisphere. Nevertheless, late Miocene Antarctic glaciation seems to have coincided with the grounding of shelf ice on the margins of Antarctica (Hayes & Frakes, 1975) and with a sea level drop recorded in the Kapitean of New Zealand (Kennett, 1965) as well as in the Messinian off the Spanish coast (Berggren & Haq, 1976). Elsewhere on land, Mercer (1979) has found evidence that the Patagonian glaciers of Argentina extended beyond the Andean mountain front, apparently for the first time, between 7.0 and 5.2 Ma.

The sequencing of these various late Miocene events related to the late Miocene glaciations of Antarctica and their ramifications are discussed in detail by Ciesielski et al. (1982). These authors take the accumulated evidence as support for the contentions of Drewry (1978) and Mercer (1978) that the West Antarctic Ice Sheet probably did not form before the late Miocene. This then led to the first production of

Fig. 9. Distribution of IRD and biogenic silica in the Southern Ocean during the late Paleogene and Neogene (from Ciesielski & Weaver, 1983, Fig. 10). Vertical lines represent the temporal distribution of DSDP sedimentary sequences. Arrows indicate the first occurrences of ice-rafted clasts. Bold portions of vertical columns represent an order of magnitude increase in sand-sized IRD. The dashed line signifies the lower limit of significant sand-sized IRD. The area between the curved lines represents the temporal and geographic distribution of biogenic siliceous sediment.

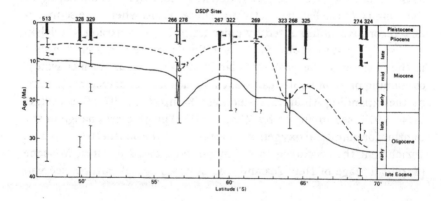

true Antarctic Bottom Water (AABW), thereby producing a major change in the world abyssal circulation.

The Leg 71 core record adds further to our understanding of these events. As related by Ciesielski & Weaver (1974), in the southeast Argentine Basin, strong bottom current erosion created a regional upper lower–lower upper Miocene disconformity. As sedimentation resumed at Site 513 about 9.5 to 9.0 Ma (Fig. 7), reworked microfossils and abundant rip-up clasts were incorporated in the basal sediments immediately above the disconformity, an indication that active erosion continued elsewhere in the region. Shortly thereafter (about 8.7 Ma, late Chron 9), the first ice-rafting began, judging from the occurrence of an ice-rafted clast in Section 513-10-1. This clast, apparently in place, is the oldest ice-rafted material yet known from the lower latitudes of the Southern Ocean. This, however, was soon followed by evidence in Hole 513 of a second episode of erosion or non-deposition that interrupted sedimentation between 8.5 and 6.5 Ma (Chron 8/7 time). Sedimentation then resumed about 6.5 Ma and continued uninterrupted until about 5.35 Ma (earliest Chron 6 to latest Chron 5). These sediments contain the oldest sand-sized IRD at Site 513 (about 5.57–5.35 Ma) which occurs in modest amounts compared to Pleistocene records for the area. These occurrences are highly significant because they provide one of the few records of sedimentation for this interval in the region. A last episode of bottom current erosion is inferred for the uppermost Chronozone 5 (between Cores 513A-4 and 513A-5) based on comparison with the record of *Islas Orcadas* piston core 7-2 from the Georgia Basin to the south.

These erosional events, so well documented at Site 513 where they are coupled with evidence of ice-rafting, suggest that there were several episodes of glaciation on the Antarctic continent during the late Miocene (assuming that the episodes of erosion represent the formation of AABW beneath ice shelves on the Antarctic continent). Estimates of the time and duration of late Miocene glacial events have previously been made from the sedimentary record of the Falkland Plateau (Ciesielski & Wise, 1977; Ciesielski et al., 1982), but the record there is far more condensed and difficult to decipher than at Site 513. The strong erosion of the Plateau during the late Miocene, as described by Ciesielski & Wise (1977), was responsible for the final sculpting of the sedimentary cover over the Maurice Ewing Bank and resulted in the ultimate topographic expression seen in Fig. 2. Ciesielski & Wise (1977) estimated that several tens of metres of sediment were eroded from the Bank at this time (see Wise, 1981, Fig. 9).

The mixtures and alternations of predominantly calcareous and biosiliceous oozes deposited over the Falkland Plateau at Sites 329 and 512 until about 8.5 Ma (through Chron 9) beg the existence of any true equivalent of the modern day Polar Front in this region throughout the late middle to early late Miocene. Predominantly biosiliceous sediment was restricted to the south at about 56–57 °S latitude (Fig. 9). Compared with the Oligocene, however, diversity of calcareous plankton over the Plateau had fallen sharply during the Miocene (Fig. 7; Appendix 2, Fig. D). Nonetheless, extensive ice shelves or ice tongues were not present along the Antarctic margin according to Ciesielski & Weaver (1983). Instead West Antarctica consisted of an archipelago and an open West Antarctic Sea (Ciesielski et al., 1982, Fig. 15). Most IRD was deposited close to the continent by small bergs carved from tidewater glaciers and small fringing ice shelves. During Chron 9 time, however, extensive ice shelves apparently formed in the West Antarctic Sea, eventually coalescing and thickening to form a grounded West Antarctic Ice Sheet. Cold Antarctic surface waters then expanded northward, and large bergs from the newly formed ice shelves began to carry glacial detritus to the lower latitudes of the Southern Ocean. The loss of carbonate in the youngest Miocene sediments indicates that the Polar Front and the associated biosiliceous sedimentary facies developed or migrated northward close to their present position over the Falkland Plateau. True Antarctic waters overlay the region and, as mentioned above, the first IRD was deposited at about 8.7 Ma.

The record of three late Miocene hiatuses at Site 513 discussed previously, indicates several distinct episodes of AABW formation during the late Miocene, suggesting that the West Antarctic Ice Sheet was relatively unstable, becoming ungrounded at various times, thereby affecting the production of AABW and the velocity of the flow of the ACC through Drake Passage (see discussion of the latter by Wise, 1981). The repeated and prolonged episodes of ungrounding may explain the paucity of IRD in upper Miocene sediments compared to those of the Plio/Pleistocene (Ciesielski & Weaver, 1983). The late Miocene glaciations were followed by a warming that extended into the early Pliocene (Hays & Opdyke, 1967; Bandy et al., 1971; Ciesielski & Weaver, 1974; Weaver, 1976). Keany (1978) dated a warm sub-Antarctic assemblage in the Antarctic and high sub-Antarctic as early Gilbert (5.26–4.7 Ma). Ciesielski & Weaver (1974) postulated early Pliocene paleotemperatures which should have caused at least a partial deglaciation (ungrounding) of the West Antarctic Ice Sheet (Ciesielski et al., 1982). By 4.35 Ma, however, the frequency and amount of IRD accumulation at the Leg 71 sites had increased significantly, suggesting a significant increase in the stability of the West Antarctic Ice Sheet.

Late Neogene history of the Polar Front (Sites 513 and 514)

Ciesielski & Wise (1977) noted that the Polar Front Zone (PFZ) had migrated well north of the Falkland Plateau by the end of the Miocene, but had no documentation as to how far it had moved before migrating back to its present location over the Plateau during the Holocene (Wise *et al.*, 1982, Fig. 16.5). The primary objective for DSDP Sites 513 and 514 was to obtain sections which might record the position and movements of the Polar Front during the late Neogene. An attempt to drill a site some 1250 km (775 miles) due east of Site 512 was frustrated by gale force winds and a field of icebergs which was being driven past the ship at speeds up to 3 knots. The actual selection of Site 513 some 290 km (180 miles) to the northwest, therefore, was a compromise between prevailing weather and ice conditions and the desire to monitor the southernmost fluctuations of the PFZ through time.

The final position of the site at 47°31'S rather than at the original location was fortuitous in that the latter would have been much too far south to have provided data relevant to the problem. Examination of the radiolarian fauna in Hole 513 showed no evidence of southerly shifts of the Polar Front during the Pliocene and Quaternary. Instead, only one significant transition was noted in the upper Miocene between Cores 513A-6 and 513A-5. There most warm-water radiolarians disappeared, radiolarian abundances were greatly reduced, and deeper-living forms became progressively more dominant. In Core 513A-5 the diatom *Thalassiothrix* spp. reached very high concentrations which, together with the deep-living radiolarians, suggests increased upwelling and very high productivity. In addition, the first major zone of IRD accumulation was detected between Sections 513A-5-6 and 513A-6-3 (Bornhold, 1983). All of the above suggested that by the end of the Miocene the Polar Front had migrated northward to a position at or very near Site 513, a location at least a degree or two north of its position today (Weaver, 1983).

To monitor the Pliocene/Pleistocene history of the Polar Front, Site 514 was then selected at 46°S, some 210 km (130 miles) northwest of Site 513 and 400 km (250 miles) north of the present position of the Polar Front. There a 150 m section down to the lower Pliocene was recovered via HPC before weather conditions forced the abandonment of this last site on the Leg. Excellent recovery and the strategic position of the site, however, permitted major trends in the movements of the PFZ to be monitored (Fig. 10) by means of qualitative estimates of downhole variations in radiolarian biofacies. These determinations were performed by Weaver (1983) on core catcher samples only and were first reported by Ludwig *et al.* (1980). Age control for the section was provided by radiolarian (Weaver, 1983) and diatom (Ciesielski, 1983) biostratigraphy coupled with paleomagnetics (Salloway, 1983). In addition, as summarized in Fig. 10, the occurrence in the section of hiatuses IRD (Bornhold, 1983), and calcareous microfossils, particularly foraminifers (Krasheninnikov & Basov, 1983), provided further information on the paleoenvironmental changes that accompanied the shifts in the position of the Polar Front. A somewhat finer subdivision of the radiolarian biofacies of the PFZ is given by Ciesielski & Weaver (1983, Fig. 6) who numbered the northerly shifts (Fig. 10) and whose interpretation of the section is summarized below with few modifications.

Having migrated as far north as Hole 513 by Chron 5 time, the PFZ maintained a position between Holes 513 and 514 from mid-Chron 5 through the late Gilbert Chron, and then assumed a position just south of Hole 514 by 4.7 Ma. At this point IRD of Zone I began to accumulate at

Fig. 10. Fluctuations of the Polar Front Zone at Site 514 (after Ludwig et al., 1980, Fig. 3; Krasheninnikov & Basov, 1983, Fig. 6; Ciesielski & Weaver, 1983, Fig. 6 and Table 1).

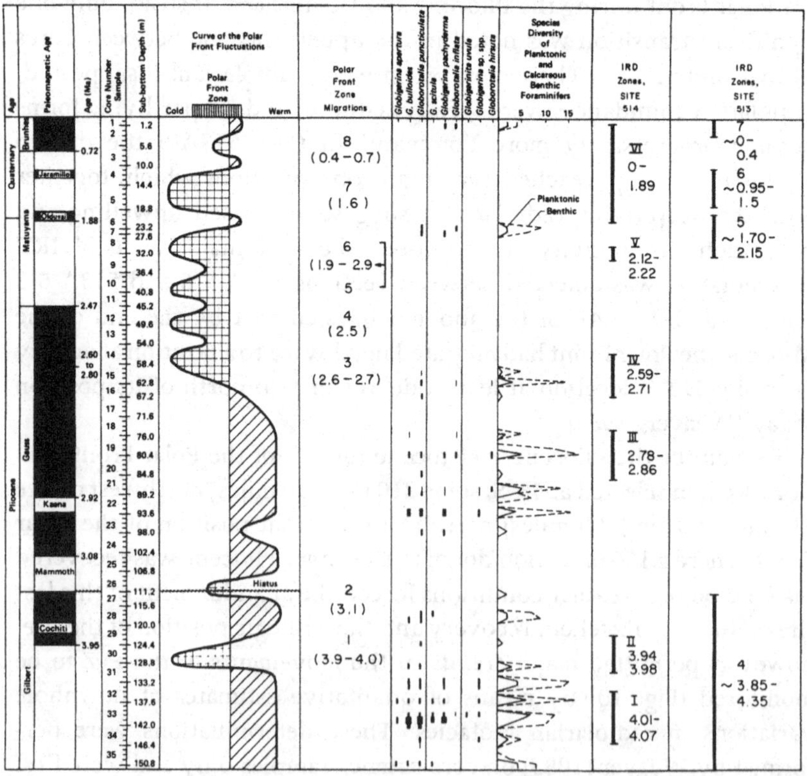

the site (Fig. 10). The rate and the frequency of major episodes of IRD accumulation at both Sites 513 and 514 subsequent to 4.35 Ma suggest the increased stability of the West Antarctic Ice Sheet noted previously.

At Site 514, the first major northward shift of the PFZ at 3.9–4.0 Ma (Fig. 10) was of considerable magnitude, causing a displacement of a radiolarian assemblage containing numerous warm-water species by one composed of cool-water species endemic to waters at or south of the Polar Front. It was accompanied by the deposition of IRD Zone II, in which was found the oldest ice-rafted clast noted at this site. This first major shift seems to reflect a regional climatic cooling in the sub-Antarctic and Antarctic which began toward the end of the Gilbert and terminated with a major northward shift in the ephemeral sea ice front around Antarctica and a major episode of glaciation in Argentina Patagonia at 3.59 Ma (Weaver, 1973; Mercer *et al.*, 1975; Mercer, 1976).

The second northward advance of the PFZ occurred during or shortly after the development of a 0.67 Ma hiatus which spans the uppermost Gilbert/lower Gauss (base of Core 514-26). Despite the missing section and some difficulty in dating precisely the hiatus, it is apparent that this disconformity marks an event of considerable importance. A significant increase in current velocities is inferred by the absence of sediments of this age at all Leg 71 sites (Fig. 7) and at many localities in the Southern Ocean (Ciesielski *et al.*, 1982; Ledbetter & Ciesielski, 1982; Osborn *et al.*, 1983). Sites on the Mid-Atlantic Ridge (Sites 513 and 514) probably reflect erosion by the Antarctic Bottom Water while those on the Falkland Plateau (Sites 511 and 512) were affected by the Circumpolar Deep Water component of the Antarctic Circumpolar Current. Above the hiatus foraminifers disappear from the record for several cores (Fig. 10) although the radiolarians eventually indicate a shift of the PFZ back to the south during the Mammoth.

The hiatus seems to encompass and probably reflects the well known cooling event at 3.2 Ma. On the basis of isotopic records, Shackleton & Opdyke (1977) suggested that this cooling represented the first Northern Hemisphere glaciation, an interpretation followed by Ciesielski & Weaver (1983) in their interpretation of the PFZ migrations. Ledbetter *et al.* (1978) had reported intensified abyssal circulation through the Vema Channel during this interval (about 3.0 to 3.4 Ma). Sedimentologic evidence from the North Atlantic (Backman, 1979) and new interpretation of the isotopic records from mid and high latitude regions (Prell, 1982; Shackleton *et al.*, 1984b), however, suggest that the initiation of Northern Hemisphere glaciation probably began near 2.5 Ma. The event near 3.2 Ma is believed instead to represent a fundamental change in intermediate and deep water circulation patterns related to global cooling and/or the closure of the Central American Seaway (Hodell et al., 1983a,b). Further work by Backman et al. (1983) suggests that the 3.25 event only affected surface plankton of the northern latitudes in a temporary sense, with a return to high *Discoaster* abundances noted thereafter until about 2.4 (or 2.5) Ma. This interpretation seems to be well reflected in the record at Site 514 (Fig. 10) which shows a return of foraminifers at that locality and an increase in ice-rafting to the south at Site 513 (Bornhold's IRD Zone 4) as the PFZ assumed a more southerly position.

The next major shift of the Pliocene PFZ at 2.7–2.6 Ma (Cores 514–18 to 514–14; Migration No. 3 in Fig. 10) heralded a significant change in style and frequency in the oscillations of the PFZ, and just precedes the date set for the onset of Northern Hemisphere glaciations of about 2.4 to 2.5 Ma (Backman, 1979; Prell, 1982). This marks the passing of the critical climatic threshold discussed by Kennett (1978) which led to the development of the present-day global glacial regime. This shift, plus the other five that followed in rapid succession (Fig. 10) may represent climatic coolings in the Southern Hemisphere which may be synchronous on a global scale with Northern Hemisphere glacial pulses (Ciesielski & Weaver, 1983). Foraminifera all but disappear during these Plio/Pleistocene oscillations except during two interglacials, one at the Plio/Pleistocene boundary and the other at the top of the Quaternary (probably Holocene).

Of particular interest is the fact that five samples from the Matuyama at Hole 514 exhibit IRD accumulation rates greater than any equivalent samples from Hole 513 (Bornhold, 1983; Ciesielski & Weaver, 1983). Bornhold's IRD Zone V in the uppermost Pliocene at Site 514 is not even represented at Site 513. The opposite relationship holds for Brunhes age samples which generally have lower IRD accumulation rates at Site 514 than those from Hole 513, thus indicating a shift of the PFZ from the more northerly Site 514 which it occupied during the Matuyama to the more southerly Site 513 where it remained for most or all of the Brunhes. Studies of Southern Ocean bottom current activity also indicate maximum velocities during the Matuyama Chron as opposed to the Brunhes (Ledbetter & Ciesielski, 1982; Osborn et al., 1983). The implication that climatic conditions in the Southern Ocean have been less severe during the past 700 000 years as compared with the preceeding 2 million years, however, is somewhat difficult to evaluate from the record at Site 514 due to the rather short Brunhes sequence preserved there (less than 6 m).

Conclusions

To date, deep sea drilling in the high latitude South Atlantic has been concentrated on the Falkland Plateau and its environs. This has been dictated primarily by logistics and the lack of other features (particularly to the north up to 30 °S) which could yield an appreciable carbonate record in this region. Although valuable carbonate sections representing every Cenozoic epoch have been sampled, the acquisition of a continuously cored Tertiary section well suited for stable isotope and high resolution calcareous/biosiliceous stratigraphy has remained an elusive target. In addition, the history of the opening of Drake Passage as reflected in regional sedimentation patterns remains somewhat obscure. A section to fill these sampling gaps could best be obtained by a series of slightly overlapping HPC holes down the exposed 400 m section on the northern flank of the Maurice Ewing Bank (roughly on or parallel to a line between *Islas Orcadas* Core Stations 16–47 and 16–50). The acquisition of such a section would establish the standard by which less extensive sections obtainable elsewhere in the South Atlantic sector of the Southern Ocean (e.g. Islas Orcadas Rise, Maud Rise) could be compared and placed into context. This, in conjunction with additional basinal HPC holes near the Polar Front Zone, would do much to elucidate the Cenozoic paleoceanography of the region.

Acknowledgements

Our special appreciation is extended to the Shipboard Scientific Staff of DSDP Leg 71 and their collaborators whose results published in the DSDP Leg 71 *Initial Reports* are summarized here. Mary E. Parker assisted in preparing the manuscript which was typed by Sharon Reeves and Jacie Blankenship and critically reviewed by James P. Kennett (University of Rhode Island). Sherwood W. Wise and Jay P. Muza were supported by National Science Foundation grant DPP 80-20382.

APPENDIX 1

Islas Orcadas piston cores shown in Fig. 7 were dated according to the diatom zonation of Weaver & Gombos (1981) or Gombos & Ciesielski (1983) as shown in the table. Lithologic descriptions, locations, and water depths for these cores are provided by Kaharoeddin *et al.* (1982).

Core	Interval (cm)	Age	Biostratigraphic zonal assignment
IO 16–19	180–181	early middle Miocene	*Coscinodiscus lewisianus* Zone
	240–241	early middle Miocene	*Coscinodiscus lewisianus* Zone
	310–311	early Miocene	probably upper *Rocella gelida* Zone
IO 16-20	129–130	early Oligocene	either *Melosira architecturalis* or *Asterolampra insignis* Zone
IO 16-47	70–71	early Miocene	*Rocella gelida* Zone
	100–101	early Miocene	*Rocella gelida* Zone
IO 16-49	29–30	Pliocene	*Nitzschia interfrigidaria* Zone
	84–85	late Paleocene, with reworked Upper Cretaceous silicoflagellates (*Lyramula furcula*)	

Age data for the other cores plotted in Fig. 7 were taken from the appropriate DSDP *Initial Reports* except in the case of DSDP Site 329 where somewhat different age determinations have been published subsequently (see figure caption for Fig. 7). The widest discrepancies are for Cores 28 and 29 which Wise & Wind (1977) dated as Oligocene and Savin *et al.* (1981) dated as early Miocene (19.26 to 22.37 Ma). The ages indicated for these and the subjacent Oligocene cores in Fig. 7 are based on their diatom and/or coccolith assemblages matched against the new zonations published in the Leg 71 *Initial Reports* (Gombos & Ciesielski, 1983; Wise, 1983).

APPENDIX 2

Simple species diversities averaged per core section from four DSDP Holes are given in Fig. A to D for planktonic microfossils as indicated. The data were taken from the appropriate DSDP *Initial Reports* as follows: diatoms, Gombos & Ciesielski (1983); planktonic foraminifers, Krasheninnikov & Basov (1983); coccoliths from Hole 329, Wise & Wind (1977); and coccoliths from Holes 511, 512, and 513A, Wise (1983).

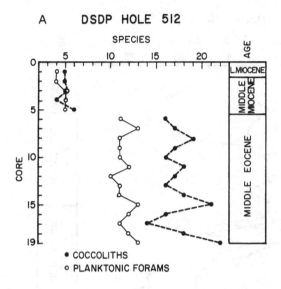

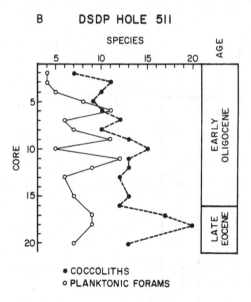

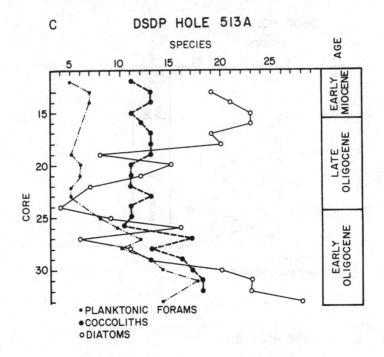

C   DSDP HOLE 513A

• PLANKTONIC FORAMS
● COCCOLITHS
○ DIATOMS

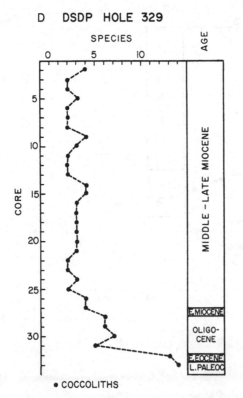

D   DSDP HOLE 329

● COCCOLITHS

REFERENCES

Backman, J. 1979. Pliocene biostratigraphy of DSDP Sites 111 and 116 from the North Atlantic Ocean and the age of Northern Hemisphere glaciation. *Stockholm Contrib. Geol.*, **32**, 115–137.

Backman, J., Shackleton, N.J. & Pestiaux, P. 1983. Data on Pliocene paleoclimates from the North Atlantic Ocean. In: *First International Conference on Paleoceanography, Zürich, 1983*, (abstracts), p. 10.

Bandy, O.L., Casey, R.E. & Wright, R.G. 1971. Late Neogene planktonic zonation, magnetic reversals, and radiometric dates, Antarctic to the tropics. In: J.L. Reid (ed.), Antarctic Oceanography 1. *Am. Geophys. Union Antarct. Res. Ser.*, **15**, 1–26.

Barker, P.F. 1977. Correlations between sites on the eastern Falkland Plateau by means of seismic reflection profiles, Leg 36, DSDP. In: P.F. Barker, I.W.D. Dalziel *et al.*, *Init. Repts. DSDP*, **36**, Washington (US Govt Printing Office), pp. 971–990.

Barker, P.F. & Burrell, J. 1977. The opening of the Drake Passage. *Mar. Geol.*, **25**, 15–34.

Barker, P.F. & Burrell, J. 1982. The influence upon Southern Ocean circulation sedimentation, and climate of the opening of Drake Passage. In: C. Craddock (ed.), *Antarctic Geoscience*. Univ. of Wisconsin Press, pp. 377–385.

Barker, P., Dalziel, I.W.D., Dinkleman, M.G., Elliot, D.H., Gombos, A.M., Lonardi, A., Plafker, G., Tarney, J., Thompson, R.W., Tjalsma, R.C., von der Borch, C.C., Wise, S.W., Harris, W. & Sliter, W.V. 1977a. Evolution of the Southwestern Atlantic Ocean Basin: Results of Leg 36, Deep Sea Drilling Project. In: P.F. Barker, I.W.D. Dalziel *et al.*, *Init. Repts. DSDP*, **36**, Washington (US Govt Printing Office), pp. 993–1014.

Barker, P.F., Dalziel, I.W.D., *et al.* 1977b. *Init. Repts. DSDP*, **36**, Washington (US Govt Printing Office).

Barron, E.J., Harrison, C.G.A., Sloan, J.L. & Hay, W.W. 1981. Paleogeography, 180 million years ago to the present. *Eclog. Geol. Helv.*, **74**, 443–470.

Basov, I.A. & Krasheninnikov, V.A. 1983. Benthic foraminifers in Mesozoic and Cenozoic sediments of the southwestern Atlantic as an indicator of paleoenvironment, Deep Sea Drilling Project Leg 71. In: W.J. Ludwig, V.A. Krasheninnikov *et al.*, *Init. Repts. DSDP*, **71**, part 2, Washington (US Govt Printing Office), 739–787.

Basov, I.A., Ciesielski, P.F., Krasheninnikov., V.A., Weaver, F.M. & Wise, S.W. 1983. Biostratigraphic and paleontological synthesis: Deep Sea Drilling Project Leg 71, Falkland Plateau and Argentine Basin. In: W.J. Ludwig, V.A. Krasheninnikov *et al.*, *Init. Repts. DSDP*, **71**, part 1, Washington (US Govt Printing Office), pp. 445–460.

Berggren, W.A. & Haq, B.U. 1976. The Andalusian Stage (Late Miocene): biostratigraphy, biochronology and paleoecology. *Palaeogeogr. Palaeoclimatol. Palaeoecol.*, **20**, 67–129.

Berggren, W.A., Kent, D.V., Flynn, J.J. & van Couvering, J.A. 1985. Paleogene geochronology and chromostratigraphy. In: N.J. Snelling (ed.), *Geochronology and the Geologic Time Scale*. Spec. Publ. Geol. Soc. London (in press).

Bode, G.W. 1983. Carbon and carbon analyses. In: W.J. Ludwig, V.A. Krasheninnikov *et al.*, *Init. Repts. DSDP*, **71**, part 2, Washington (US Govt Printing Office), pp. 1185–1187.

Boersma, A. & Shackleton, N. 1977. Tertiary oxygen and caробon isotope stratigraphy, Site 357 (mid latitude South Atlantic). In: P.R. Supko, K. Perch-Nielsen *et al.*, *Int. Repts DSDP*, **39**, Washington (US Govt Printing Office), pp. 911–924.

Bornhold, B.D. 1983. Ice-rafted debris in sediments from Leg 71, Southwest Atlantic Ocean. In: W.J. Ludwig, V.A. Krasheninnikov *et al.*, *Init. Repts. DSDP*, **71**, part 1, Washington (US Govt Printing Office), pp. 307–316.

Bukry, D. 1973. Coccolith stratigraphy, Eastern Equatorial Pacific, Leg 16, Deep Sea Drilling Project. In: Tj.H. van Andel, G.R. Heath *et al.*, *Init. Repts. DSDP*, **16**, Washington (US Govt Printing Office), pp. 653–771.

Cameron, D.H. 1977. Grain-size and carbon/carbonate analyses, Leg 36. In: P.F. Barker, I.W.D. Dalziel *et al.*, *Init. Repts. DSDP*, **36**, Washington (US Govt Printing Office), pp. 1047–1050.

Ciesielski, P.F. 1983. The Neogene and Quaternary diatom biostratigraphy of sub-Antarctic sediments, Deep Sea Drilling Project Leg 71. In: W.J. Ludwig, V.A. Krasheninnikov et al., Init Repts. DSDP, 71, part 2, Washington (US Govt Printing Office), pp. 635–665.

Ciesielski, P.F. & Weaver, F.M. 1974. Early temperature changes in the Antarctic Seas, Geology, 2, 511–515.

Ciesielski, P.F. & Weaver, F.M. 1983. Neogene and Quaternary paleoenvironmental history of Deep Sea Drilling Project Leg 71 sediments, Southwest Atlantic Ocean. In: W.J. Ludwig, V.A. Krasheninnikov et al., Init. Repts. DSDP, 71, part 1, Washington (US Govt Printing Office), pp. 461–477.

Ciesielski, P.F. & Wise, S.W. 1977. Geologic history of the Maurice Ewing Bank of the Falkland Plateau (Southwest Atlantic sector of the Southern Ocean) based on piston and drill cores. Mar. Geol., 25, 175–207.

Ciesielski, P.F., Ledbetter, M.T. & Ellwood, B.B. 1982. The development of Antarctic glaciation and the Neogene paleoenvironment of the Maurice Ewing Bank. Mar. Geol., 46, 1–51.

Craig, H. 1965. The measurement of oxygen isotope paleotemperatures. In: E. Tongiorgi (ed.), Stable Isotopes in Oceanographic Studies and Paleotemperatures. Pisa, Consiglio Nazionale dele Ricerche, pp. 161–182.

Dalziel, I.W.D. 1982. The early (pre-Middle Jurassic) history of the Scotia Arc Region: a review and progress report. In: C. Craddock (ed.), Antarctic Geoscience. Univ. of Wisconsin Press, pp. 111–126.

de Wit, M.J. 1977. The evolution of the Scotia Arc as a key to the reconstruction of southwest Gondwanaland. Tectonophys., 37, 53–81.

Denton, G.H., Armstrong, R.L. & Stuiver, M. 1971. The late Cenozoic glacial history of Antarctica. In: K.K. Turekian, The Late Cenozoic Glacial Ages, New Haven, Yale University Press, pp. 267–306.

Drewry, D.J. 1978. Aspects of the early evolution of West Antarctic ice. In: E.M. van Zinderen-Bakker (ed.), Antarctic Glacial History and World Paleoenvironments. Balkema, Rotterdam, pp. 25–32.

Emiliani, C. 1954. Depth habitats of some species of pelagic foraminifera as indicated by oxygen isotope ratios. Am. J. Sci., 252, 149–158.

Frakes, L.A. 1983. Problems in Antarctic marine geology: a review. In: R.L. Oliver, P.R. James & J.B. Jago, (eds.), Antarctic Earth Science, Cambridge Univ. Press, pp. 375–378.

Gleadow, A.J.W., McKelvey, B.C. & Ferguson, K.V. 1982. Uplift history of the Transantarctic Mountains in the Dry Valleys area, southern Victoria Land, from apatite fission track ages. Melbourne Univ. Prog. in Antarctic Studies Rept. No. 35, Melbourne, Australia.

Gombos, A.M. 1977. Paleogene and Neogene diatoms from the Falkland Plateau and Malvinas Outer Basin: Leg 36, Deep Sea Drilling Project. In: P.F. Barker, I.W.D. Dalziel et al., Init. Repts. DSDP, 36, Washington (US Govt Printing Office), pp. 575–687.

Gombos, A.M. 1984. Late Paleocene diatoms in the Cape Basin. In: K.J. Hsü, J.L. La Brecque et al., Init. Repts. DSDP, 73, Washington (US Govt Printing Office), pp. 495–511.

Gombos, A.M. & Ciesielski, P.F. 1983. Late Eocene to early Miocene diatoms from the southwest Atlantic. In: W.J. Ludwig, V.A. Krasheninnikov et al., Init. Repts. DSDP, 71, part 2, Washington (US Govt Printing Office), pp. 583–634.

Goodman, D.K. & Ford, L.N., Jr 1983. Preliminary dinoflagellate biostratigraphy for the middle Eocene to lower Oligocene from the Southwest Atlantic Ocean. In: W.J. Ludwig, V.A. Krasheninnikov et al., Init. Repts. DSDP, 71, part 2, Washington (US Govt Printing Office), pp. 859–877.

Haq, B.U. 1980. Biogeographic history of Miocene calcareous nannoplankton and paleoceanography of the Atlantic Ocean. Micropaleontol., 26, 414–443.

Harwood, D.M., Webb, P.N., McKelvey, B.C., Mercer, J.H. & Stott, L.D. 1983. Late Neogene and Paleogene diatoms in high elevation deposits of the Transantarctic Mountains, Antarctica. Geol. Soc. Am., Abstracts with Programs, 15, 592.

Hayes, D.E. & Frakes, L.A. 1975. General synthesis, Deep Sea Drilling Project Leg 28. In: D.E. Hayes, L.A. Frakes *et al.*, *Init. Repts. DSDP*, **28**, Washington (US Govt Printing Office), pp. 919–942.

Hays, J.D. & Opdyke, N.D. 1967. Antarctic radiolaria, magnetic reversals, and climate changes. *Science*, **158**, 1001–1011.

Hodell, D.A., Kennett, J.P. & Leonard, K.A. 1983a. Climatically induced changes in vertical water mass structure of the Vema Channel during the Pliocene: Evidence from Deep Sea Drilling Project Holes 516A, 517, and 518. In: P.F. Barker, R.L. Carlson, D.A. Johnson *et al.*, *Init. Repts. DSDP*, **72**, Washington (US Govt Printing Office), pp. 907–919.

Hodell, D.A., Williams, D.F. & Kennett, J.P. 1983b. Fundamental change in deep and intermediate water mass circulation patterns in the Western South Atlantic at 3.2 Ma. *Geol. Soc. Am., Abstracts with Programs*, **15**, 596.

Jones, D.L. & Plafker, G. 1977. Mesozoic megafossils from DSDP Hole 327A and Site 330 on the eastern Falkland Plateau. In: P.F. Barker, I.W.D. Dalziel *et al.*, *Init. Repts. DSDP*, **36**, Washington (US Govt Printing Office), pp. 845–855.

Kaharoeddin, F.A., Graves, R.S., Bergen, J.A., Eggers, M.R., Harwood, D.M., Humphreys, C.L., Goldstein, E.H., Jones, S.C. & Watkins, D.K. 1982. ARA ISLAS ORCADAS Cruise 1678 sediment descriptions. *Florida State University Sedimentol. Res. Contr.*, **50**, 1–172.

Keany, J. 1978. Paleoclimatic trends in early and middle Pliocene deep-sea sediments of the Antarctic. *Mar. Micropaleontol.*, **3**, 35–49.

Keigwin, L.D., Jr 1980. Oxygen and carbon isotope analyses from Eocene/Oligocene boundary at DSDP Site 277. *Nature*, **287**, 722–725.

Keigwin, L. & Keller, G. 1984. Middle Oligocene cooling from equatorial Pacific DSDP Site 77B. *Geology*, **12**, 16–19.

Kemp, E.M., Frakes, L.A. & Hayes, D.E. 1975. Paleoclimatic significance of diachronous biogenic facies, Leg 28, Deep Sea Drilling Project. *Init. Repts. DSDP*, **28**, Washington (US Govt Printing Office) pp. 909–917.

Kennett, J.P. 1965. Faunal succession in two upper Miocene–lower Pliocene sections, Marlborough, New Zealand. *Trans. R. Soc. N.Z.*, **3**, 197–213.

Kennett, J.P. 1978. The development of planktonic biogeography in the Southern Ocean during the Cenozoic. *Mar Micropaleontol.*, **3**, 301–345.

Kennett, J. 1982. *Marine Geology*. Englewood Cliffs, N.J., Prentice-Hall.

Kennett, J.P. & Shackleton, N.S. 1976. Oxygen isotopic evidence for the development of the psychrosphere 38 m. yr. ago. *Nature*, **260**, 513–515.

Kerr, R.A. 1982. New evidence fuels Antarctic ice debate. *Science*, **216**, 973–974,

Krasheninnikov, V.A. & Basov, I.A. 1983. Cenozoic planktonic foraminifers of the Falkland Plateau and Argentine Basin, Deep Sea Drilling Project Leg 71. In: W.J. Ludwig, V.A. Krasheninnikov *et al.*, *Init. Repts. DSDP*, **71**, part 2, Washington (US Govt Printing Office), pp. 821–858.

Kvasov, D.D. & Verbitzky, M.Y. 1981. Causes of Antarctic glaciation in the Cenozoic. *Quat. Res.*, **15**, 1–17.

Kroopnick, P. 1974. The dissolved O_2–CO_2–^{13}C system in the eastern equatorial Pacific. *Deep-Sea Res.*, **21**, 211–227.

La Brecque, J.L. & Rabinowitz, P.D. 1977. Magnetic anomalies bordering the continental margin of Argentina. *Am. Ass. Petrol. Geol., Spec. Map Ser.*, Tulsa.

La Brecque, J.L., Kent, D.V. & Cande, S.C. 1977. Revised magnetic polarity time scale for Late Cretaceous and Cenozoic time. *Geology*, **5**, 330–335.

La Brecque, J.L., Hsü, K.J., Carman, M.F., Karpoff, A.M., McKenzie, J.A., Percival, S.F., Petersen, N.P., Pisciotto, K.A., Schreiber, E., Tauxe, L., Tucker, P., Weissert, H.J. & Wright, R. 1983. DSDP Leg 73: Contributions to Paleogene stratigraphy in nomenclature, chronology and sedimentation rates. *Palaeogeogr. Palaeoclimatol. Palaeoecol.*, **42**, 91–125.

Ledbetter, M.T. 1983. Magnetostratigraphy of middle–upper Miocene and upper middle Eocene sections in Hole 512. In: W.J. Ludwig, V.A. Krasheninnikov *et al.*, *Init. Repts., DSDP*, **71**, part 2, Washington (US Govt Printing Office), pp. 1093–1096.

Ledbetter, M.T. & Ciesielski, P.F. 1982. Bottom current erosion in the South

Atlantic sector of the Southern Ocean. *Mar Geol.*, **46**, 329–341.

Ledbetter, M.T., Williams, D.F. & Ellwood, B.B. 1978. Late Pliocene climate and southwest Atlantic abyssal circulation. *Nature*, **272**, 237–239.

Le Masurier, W.E. 1972. Volcanic record of Cenozoic glacial history of Marie Byrd Land. In: R.J. Adie (ed.), *Antarctic Geology and Geophysics*, Oslo (Universitetsforlaget), pp. 251–260.

Lowrie, W. & Alvarez, W. 1981. One-hundred million years of geomagnetic polarity history. *Geology*, **9**, 392–397.

Lowrie, W., Alvarez, W., Napoleone, G., Perch-Nielsen, K., Toumarkine, M. & Premoli-Silva, I. 1982. Paleogene magnetic reversal stratigraphy in Umbrian pelagic carbonate rocks: the Contessa sections. *Geol. Soc. Am. Bull.*, **93**, 414–432.

Ludwig, W.J. 1983. Geologic framework of the Falkland Plateau. In: W.J. Ludwig, V.A. Krasheninnikov *et al.*, *Init. Repts. DSDP*, **71**, part 1, Washington (US Govt Printing Office), pp. 281–293.

Ludwig, W.J., Krasheninnikov, V.A., Basov, I.A., Bayer, U., Bloemendal, J., Bornhold, B., Ciesielski, P., Goldstein, E.H., Robert, C., Salloway, J.C., Usher, J.L., von der Dick, H., Weaver, F.M. & Wise, S.W. 1980. Tertiary and Cretaceous paleoenvironments in the southwest Atlantic Ocean: preliminary results of Deep Sea Drilling Project Leg 71. *Geol. Soc. Am. Bull.*, **91**, 655–664.

Ludwig, W.J., Krasheninnikov, V.A. *et al.* 1983. *Init. Repts. DSDP*, **71**, Washington (US Govt Printing Office).

Matthews, R.K. & Poore, R.Z. 1980. Tertiary ^{18}O record and glacio-eustatic sea-level fluctuations. *Geology*, **8**, 501–504.

Mercer, J.H. 1976. Glacial history of southernmost South America. *Quat. Res.*, **6**, 125–166.

Mercer, J.H. 1978. Glacial development and temperature trends in the Antarctic and in South America. In: E.M. van Zenderen-Bakker (ed.), *Antarctic Glacial History and World Paleoenvironments*. Balkema, Rotterdam, pp. 73–93.

Mercer, J.H. 1979. Late Miocene to earliest Pliocene glaciation in southern Argentina. *Geol. Soc. Am., Abstracts with Programs*, **11**, 478.

Mercer, J.H., Fleck, R.J., Mankinen, E.A. & Sander, W. 1975. Southern Patagonia: glacial events between 4MY and 1MY. In: R.P. Suggate, & M.M. Cresswell (eds.), *Quaternary Studies. Soc. N. Z. Bull.*, **13**, 223–230.

Mercer, J.H. & Sutter, J.F. 1982 Late Miocene–earliest Pliocene glaciation in Southern Argentina: Implications for a global ice-sheet history. *Palaeogeogr. Palaeoclimatol. Palaeoecol.*, **38**, 185–206.

Miller, K.G. & Fairbanks, R.G. 1983a. Abyssal circulation in the North Atlantic during the early to middle Miocene. *Geol. Soc. Am., Abstracts with Programs*, **15**, 645.

Miller, K.G & Fairbanks, R.G. 1983b. Evidence for Oligocene–middle Miocene abyssal circulation changes in the Western North Atlantic. *Nature*, **306**, 250–253.

Muza, J.P. & Wise, S.W. 1983. An authigenic gypsum, pyrite, and glauconite association in a Miocene deep sea biogenic ooze from the Falkland Plateau, southwest Atlantic Ocean. In: W.J. Ludwig, V.A. Krasheninnikov *et al.*, *Init. Repts. DSDP*, **71**, part 1, Washington (US Govt Printing Office), pp. 361–375.

Muza, J.P., Williams, D.F. & Wise, S.W. 1983. Paleogene oxygen isotope record for Deep Sea Drilling Project Sites 511 and 512, sub-Antarctic south Atlantic Ocean: paleotemperatures, paleoceanographic changes and the Eocene/Oligocene boundary event. In: W.J. Ludwig, V.A. Krasheninnikov *et al.*, *Init. Repts. DSDP*, **71**, part 1, Washington (US Govt Printing Office), pp. 409–422.

Ness, G., Levi, S. & Couch, R. 1980. Marine magnetic anomaly time-scales for the Cenozoic and Late Cretaceous: a precis, critique, and synthesis. *Rev. Geophys. Space Phys.*, **18**, 753–770.

Oberhänsli, H., McKenzie, J., Toumarkine, M. & Weissert, H. 1984. A paleoclimatic and paleoceanographic record of the Paleogene in the Central South Atlantic (Leg 73, Sites 522, 523 and 524). In: K.J. Hsü, J.L. La Brecque *et al.*, *Init. Repts. DSDP*, **73**, Washington (US Govt Printing Office), pp. 737–747.

Okada, H. & Bukry, D. 1980. Supplementary modification and introduction of code numbers to the 'Low-latitude coccolith biostratigraphic zonation' (Bukry, 1973; 1975). *Mar. Micropaleontol.*, **5**, 321–325.

Osborn, N.I., Ciesielski, P.F. & Ledbetter, M.T. 1983. Disconformities and paleoceanography in the Southwest Indian Ocean during the last 5.4 million years. *Geol. Soc. Am. Bull.*, **94**, 1345–1358.

Poore, R.Z. & Matthews, R.K. 1984. Late Eocene–Oligocene oxygen- and carbon-isotope record from South Atlantic Ocean, Deep Sea Drilling Project Site 522. In: K.J. Hsü, J.L. La Brecque *et al.*, *Init. Repts. DSDP*, **73**, Washington (US Govt Printing Office), pp. 725–735.

Prell, W.L. 1982. Oxygen and carbon isotope stratigraphy for the Quaternary of Hole 502B: evidence for two modes of isotopic variability. In: W.L. Prell, J.V. Gardner *et al.*, *Init Repts. DSDP*, **68**, Washington (US Govt Printing Office), pp. 455–464.

Salloway, J.C. 1983. Paleomagnetism of sediments from Deep Sea Drilling Project Leg 71. In: W.J. Ludwig, V.A. Krasheninnikov *et al.*, *Init Repts. DSDP*, **71**, part 2, Washington (US Govt Printing Office), pp. 1073–1091. ·

Savin, S.M., Douglas, R.G, Keller, G., Killingley, J.S., Shaughnessy, L., Sommer, M.A., Vincent, E. & Woodruff, F. 1981. Miocene benthic foraminiferal isotope records: a synthesis. *Mar. Micropaleontol.*, **6**. 423–450.

Savin, S.M., Douglas, R.G. & Stehli, F.G. 1975. Tertiary marine paleotemperatures. *Geol. Soc. Am. Bull.*, **86**, 1499–1510.

Schnitker, D. 1980. North Atlantic oceanography as possible cause of Antarctic glaciation and eutrophication. *Nature*, **284**, 615–616.

Shackleton, N.J. 1974. Attainment of isotopic equilibrium between ocean water and the benthonic foraminifera genus *Uvigerina*: Isotopic changes in the ocean during the last glacial. *Colloq. Int. CNRS*, **219**, 203–210.

Shackleton, N.J. & Kennett, J.P. 1975. Paleotemperature history of the Cenozoic and the initiation of Antarctic glaciation: Oxygen and carbon isotope analyses in DSDP Sites 277, 279, and 281. In: J.P. Kennett, R.E. Houtz, *et al.*, *Init. Repts. DSDP*, **29**, Washington (US Govt Printing Office), pp. 743–755.

Shackleton, N.J. & Opdyke, N.D. 1977. Oxygen isotope and paleomagnetic evidence for early Northern Hemisphere glaciation. *Nature*, **270**, 216–219.

Shackleton, N.J., Hall, M.A. & Boersma, A. 1984a. Oxygen and carbon isotope data from Leg 74 foraminifers. In: T.C. Moore, Jr, P.D. Rabinowitz *et al.*, *Init. Repts. DSDP*, **74**, Washington (US Govt Printing Office), pp. 599–612.

Shackleton, N.J. *et al.* 1984b. Oxygen isotope calibration of the onset of ice-rafting and history of glaciation in the North Atlantic region. *Nature*, **307**, 620–623.

Shaw, C.A. & Ciesielski, P.F. 1983. Silicoflagellate biostratigraphy of middle Eocene to Holocene subantarctic sediments recovered by Deep Sea Drilling Project Leg 71. In: W.J. Ludwig, V.A. Krasheninnikov *et al.*, *Init. Repts. DSDP*, **71**, part 2, Washington (US Govt Printing Office), pp. 687–737.

Sliter, W.V. 1977. Cretaceous foraminifers from the southwestern Atlantic Ocean, Leg 36, Deep Sea Drilling Project. In: P.F. Barker, I.W.D. Danziel *et al.*, *Init. Repts. DSDP*, **36**, Washington (US Govt Printing Office), pp. 519–573.

Smith, A.G. & Hallam, A. 1970. The fit of the southern continents. *Nature*, **225**, 139–144.

van Couvering, J.A., Berggren, W.A., Drake, R.E., Aguirre, E. & Curtis, G.H. 1976. The terminal Miocene event. *Mar. Micropaleontol.*, **1**, 263–286.

Vergnaud-Grazzini, C., Pierre, C. & Letolle, R. 1978. Paleoenvironment of the north-east Atlantic during the Cenozoic: oxygen and carbon isotope analyses of DSDP Sites 398, 400A, and 401. *Oceanol. Acta*, **1**, 381–390.

Weaver, F.M. 1973. Pliocene paleoclimatic and paleoglacial history of East Antarctica recorded in deep-sea piston cores. *Florida State Univ., Sedimentol. Res. Lab., Contrib.*, **36**.

Weaver, F.M. 1976. Late Miocene and Pliocene radiolarian paleobiogeography and biostratigraphy of the Southern Ocean Ph.D. dissertation, Florida State University, Tallahassee.

Weaver, F.M. 1983. Cenozoic radiolarians from the Southwest Atlantic, Falkland Plateau region, Deep Sea Drilling Project Leg 71. In: W.J. Ludwig, V.A. Krasheninnikov *et al.*, *Init. Repts. DSDP*, **71**, part 2, Washington (US Govt Printing Office), pp. 667–686.

Weaver, F.M. & Gombos, A.M. 1981. Southern high-latitude diatom biostratigraphy. *Soc. Econ. Pal. Min., Spec. Publ.*, **32**, 445–470.

Webb, P.N. 1983. Climatic, paleo-oceanographic and tectonic interpretation of Paleogene–Neogene biostratigraphy from MSSTS-1 drillhole, McMurdo Sound, Antarctica. In: R.L. Oliver, P.R. James & J.B. Jago (eds.), *Antarctic Earth Science*, Cambridge Univ. Press. p. 560.

Webb, P.N., Harwood, D.M., McKelvey, B.C., Mercer, J.H. & Stott, L.D. 1984. Cenozoic marine sedimentation and ice volume variation on the east Antarctic craton. *Geology*, **12**, 287–291.

Wise, S.W. 1981. Deep sea drilling in the Antarctic: Focus on late Miocene glaciation and applications of smear-slide biostratigraphy. In: J.E. Warme, R.G. Douglas & E.L. Winterer (eds.), The Deep Sea Drilling Project: A Decade of Progress, *Soc. Econ. Pal. Min., Spec. Publ.*, **32**, 471–487.

Wise, S.W. 1983. Mesozoic and Cenozoic calcareous nannofossils recovered by Deep Sea Drilling Project Leg 71 in the Falkland Plateau Region, Southwest Atlantic Ocean. In: W.J. Ludwig, V.A. Krasheninnikov *et al., Init. Repts. DSDP*, **71**, part 2, Washington (US Govt Printing Office), pp. 481–550.

Wise, S.W. & Wind, F.H. 1977. Mesozoic and Cenozoic calcareous nannofossils recovered by DSDP Leg 36 drilling on the Falkland Plateau, Southwest Atlantic sector of the Southern Ocean. In: P.F. Barker, I.W.D. Dalziel *et al., Init. Repts. DSDP*, **36**, Washington (US Govt Printing Office), pp. 269–492.

Wise, S.W., Ciesielski, P.F, MacKenzie, D.T., Wind, F.H., Busen, K.E. *et al.* 1982. Paleontologic and paleoenvironmental synthesis for the southwest Atlantic Ocean basin based on Jurassic to Holocene faunas and floras from the Falkland Plateau. In: C. Craddock (ed.), *Antarctic Geoscience*, Madison, Univ. of Wisconsin Press, pp. 155–163.

Woodruff, F., Savin, S.M. & Douglas, R.G. 1981. Miocene stable isotope record: a detailed deep Pacific Ocean study and its paleoclimatic implications. *Science*, **212**, 665–668.

Zinsmeister, W.J. 1978. Effect of formation of the West Antarctic Ice Sheet on shallow water marine faunas of Chile. *Antarctic J. U.S.*, **13**, 25–26.

16

Evidence from the Ostracoda of major events in the South Atlantic and world-wide over the past 80 million years

R. H. BENSON, R. E. CHAPMAN and L. T. DECK

Abstract

A global survey of 1044 samples containing about 30 000 ostracodes (156 genera) from 155 Deep Sea Drilling Project sites representing the past 80 m.y. has yielded information about 15 proposed paleoceanographic events of which the validity of six have been examined in some detail. The South Atlantic, where 329 samples from 30 sites were studied, included 208 samples from 24 sites along a west–east transect over the Rio Grande Rise and Walvis Ridge. Changes in relative ostracode abundance, generic diversity (simple and Shannon–Wiener Index), percent presence in samples, and assemblage similarity (Dice Coefficient) have shown the presence of an important 'crash' event at 65 Ma, a lesser but important event at 52 Ma, a major positive event at about 40 Ma (the origin of the psychrosphere), essentially the absence of an 'event' at 25–22 Ma (formation of the Southern Ocean Convergence System), second-order events at 14 Ma (formation of the East Antarctic Ice Cap) and at 6 Ma (Messinian Salinity Crisis), and a final downturn beginning at about 3 Ma.

Introduction

The following is a review of generic changes in the deep-sea benthic podocopid Ostracoda in response to globally significant events over the past 80 m.y., with special attention to the local importance of the South Atlantic. One hundred and fifty-six genera were identified in 1044 samples including about 30 000 specimens from 155 Deep Sea Drilling Project core sites. About 15 historical 'events', are noted with special attention to six of these events. These include the Cretaceous/ Tertiary boundary 'crash' at 65 Ma, and 52 Ma event, the origin of the psychrosphere at about 40 Ma, the Terminal Oligocene Event or the formation of the Southern Ocean Convergence System (25–22 Ma), the formation of the East Antarctic Ice Cap (about 14 Ma), and the final closure of Eurasian Tethys (12 to 6 Ma) with the development of the Messinian Salinity Crisis at 6 Ma. In addition to these events of global importance, the influence of the intrusion of North Atlantic Deep Water (NADW) into the region of the Vema Channel during the Oligocene, and the effects of the shallowing of the outer Walvis Ridge during Late Eocene, are also examined.

Although this study attempts to test the effectiveness of these events on the deep-sea ostracodes as a whole, it nevertheless remains descriptive of the 'mass properties' of the general ostracode fauna as it is distributed through time, particularly changes in diversity and taxonomic turnover. The ostracodes preserved as fossils in the DSDP cores are benthic in habit without known pelagic growth stages. They are probably the most advanced form of life with a fossil record in the deep sea. The more we study them, the more sensitive they seem to have been to changes in their environment. This study is not concerned with adaptive architectural morphology, or lesser evolutionary changes within or between species (the causes of variation within genera has been discussed in some detail elsewhere; Benson, 1975; Benson & Peypouquet, 1983; Benson, 1984a,b). It is concerned with those larger environmental changes responsible for the formation of genera, the most stable taxonomic level in the Ostracoda. It is hoped that the study described here will contribute to a broad and quantitative view of paleoceanography as well as ostracode history.

In 1975, one of us (RHB) described the basis for the discovery of the origin in the modern deep-sea psychrosphere, that cold underlayer of the present two-layer ocean water-mass system. This discovery emerged from the quantitative analysis of ostracode specimens identi-fied in the DSDP cores of the first circumnavigation of the *Glomar Challenger* and was based on data from 75 sites. This earlier work in particular examined evidence of the ostracode faunal history recovered from Leg 3, the first coring done in the South Atlantic on the Rio Grande Rise. From this study it then was claimed that there had been a major turnover in the benthic fauna at about 40 Ma. Contemporaneously, Shackleton & Kennett (1975) noted a major change in oxygen isotope data obtained from Site 277 south of the Tasman Sea at 38 Ma. Douglas (1973) had noted a change in the benthic foraminifera from the northwest Pacific occurring at about the same time (Douglas & Woodruff, 1981). Later, Corliss (1981), Keller (1982), Miller (1983, see also Miller & Curry, 1982; Miller et al, 1982), and Srinivasan & Kennett (1983), all of whom studied the Eocene/Oligocene boundary event, became concerned with the accuracy of this date or questioned its local importance. In spite of local modifications in the timing of this event, its importance as one of the two most significant changes in World Ocean history still remains. This report will emphasize the Middle/Late Eocene boundary or 40 Ma as the important datum of this event rather than the Eocene/Oligocene boundary. It is evident, however, that the major oceanographic change that caused this event did not affect all areas of the world at the same rate, nor at the same time or intensity. The present study suggests that tectonic changes in the South Atlantic seem to be near the focus of the event. The raising of the outer part of Walvis Ridge may represent part of the critical threshold system affecting it or the opening of Drake Passage may have provided the hydrodynamic drive necessary. It is in the region of this convergence that this report will ultimately concentrate.

Before the present world census of fossil deep-sea ostracodes was made, we noted some changes in local DSDP ostracode faunas at the end of the Cretaceous. These changes were not thought to be especially significant, however. Attention then was concentrated on the several genera, such as those related to *Atlanticythere* or *Abyssocythere*, that could be traced through the so-called 'K/T Boundary Event'. These showed no major morphological changes. Now, however, after pooling all of the distributional data, there does seem to be a significant and abrupt change at this time. Some of the properties of this change may prove interesting to advocates of catastrophic change (Gartner & Keany, 1978; Gartner & McGuirk, 1979; Alvarez *et al.*, 1980; Hsü, 1980; Smit & Hertogen, 1980; Officer & Drake, 1983).

Until this stage in the study of deep-sea ostracodes, we had had much more information about isolated faunas than we had about their global or general relationships. Of course we could see that the deep-sea benthos had not changed rapidly or often compared to changes in shallow marine biologic systems. Its physical regime was assumed to have been heavily buffered against the lesser climatic perturbations. Now through statistical comparison we can see how great these changes actually were. The biotic stress in the deep sea is thought to be great (Hessler & Sanders, 1967; Sanders, 1969; Hessler & Wilson, 1983), and yet the ostracodes seem to have resisted these stresses well, that is, if equitability and continued low diversity are significant measures of faunal tenacity. As stated before, we believe that when changes have occurred in the deep-sea ostracodes, these changes should reflect the more significant climatic or paleoceanographic changes that have intruded into the abyss from the upper surface world.

We would like to point out to the reader that the figures

accompanying the text are arranged with their intrinsic continuity in mind. Each is referred to numerous times in the text, and so is often cited out of numerical order.

The data base for this report

All samples (each 50 cm³) for this report were obtained from cores taken from 155 drilling sites of the Deep Sea Drilling Project by the *Glomar Challenger* from all over the World Ocean. Of these 1621 samples, 1044 had about 30 000 ostracodes representing 156 genera. Data for this report range in age from 80 Ma to the present (Fig. 1). The success in obtaining specimens was very roughly proportional to the number of samples collected at any point in time (Fig. 2), with the frequency of samples inversely proportional to their age with the notable exception of increases in the Oligocene and the Cretaceous. There is a sudden decrease in sampling opportunities immediately above the Cretaceous/Tertiary boundary (65 Ma) and a rapid increase above the Middle/Late Eocene boundary (40 Ma).

After examining the global response of the ostracode faunas through time, this report concentrates on those of the South Atlantic where 329 samples representing 30 sites were studied. Of these, 208 samples with more than ten ostracodes per sample representing 24 sites crossing the Rio Grande Rise (177 samples) and the Walvis Ridge (143 samples) were

Fig. 1. Frequency of all samples (1621) examined through time (80 million years) for ostracodes from 155 Deep Sea Drilling Sites, world-wide, compared to the frequency of those samples (1044) containing ostracodes.

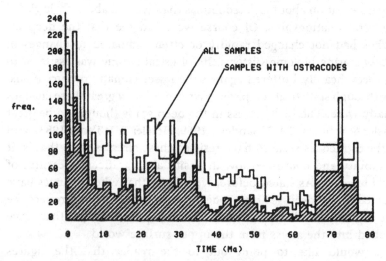

correlated by the Rho-Groups method of comparing faunal similarity (Figs. 12–17). Comparisons are also made between the faunas of the South Atlantic and those of the North Atlantic (190 samples with ostracodes from 44 sites) and the Indo-Pacific regions (509 samples with ostracodes from 77 sites).

The ostracodes of each sample averaged roughly 30 in number. One hundred and seventy-one taxa were found, of which 156 genera were considered for this analysis. The genus level was selected over the species level for this census because identification could be made much more objectively over such great areas. Morphologic stability seems greatest at this taxonomic level. All of the identifications were made or reconfirmed by one of us (RHB) over a six week period. It is thought that this study represents one of the few instances in which such a broad survey enjoyed the consistency of taxonomic judgement (with its intrinsic errors) of one person. Therefore, the variations that have been revealed in the statistical analyses are considered to be as real as the sampling and human error will allow. They are not due to differences in opinion about some of the taxa.

Of course, deep-sea ostracodes are rare with respect to other microfossils and the samples are small, but there are many samples, and lateral geographic changes in assemblage composition were not rapid. Only time, more sampling, and continued testing will show how

Fig. 2. Percent of samples in which ostracodes were found compared to the total for the intervals of time shown in Fig. 1.

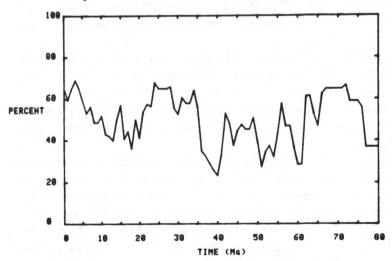

PERCENT SAMPLES WITH OSTRACODES

representative the ones of this report really are; however, the fact that some patterns through time remained consistent from area to area cannot be ignored.

Methods of study

This study is strongly oriented in two directions. The first is to examine the validity of several global 'event' hypotheses, especially the Terminal Cretaceous Event, the origin of the psychrosphere, the formation of the East Antarctic Ice Cap and the Messinian Salinity Crisis, particularly as they affect or are caused by local events in the South Atlantic. The second is the search for new evidence or new patterns of change that may emerge from a broad census and statistical analysis.

Global as well as regional and local time series were constructed (Fig. 5–11) from the total data base of samples, which included their geographic locations and positions in time over 80 my (determined by foraminifera and nannofossil zones; fide W. A. Berggren, 1977; Berggren et al., 1984). These include population counts of each species in each sample (identified to the genus level for this report), the number of taxa in each sample, the Sheldon equitability function (not shown), and the Shannon-Wiener information function used as an Index of Diversity. The data base was partitioned for these same parameters for each major subdivision (the Indo-Pacific, Fig. 6; the Atlantic, Fig. 7), and then the Atlantic was partitioned into the North and South Atlantic (Figs, 8 and 9). Finally, the South Atlantic was repartitioned into the Rio Grande Rise and Walvis Ridge (Fig. 10 and 11). Comparisons of each parameter between areas of interest were also made and used to trace differences from region to region.

The ranges of all of the ostracode genera in the DSDP samples worldwide were plotted in order of their first appearance, by their extinctions, and then tabulated in terms of frequency by age of these two characteristics (Fig. 3 center and left). The total effect of change in faunal composition with age is also tabulated as faunal turnover (Fig. 3, right).

Twenty-four DSDP sites crossing the South Atlantic (Fig. 12) were selected for comparison from among those available (those with suffiient faunas, samples with ten or more specimens). This traverse includes sites from the Sao Paulo Plateau in the west off the coast of Brazil (Site 356), east across Rio Grande Rise (Sites 518, 517, 357, 516, 22, 21, 20, 19), across the Mid-Atlantic Ridge (Sites 14, 15, 16, 519, 17) into the Angola Basin (Site 522), and then across the Walvis Ridge (Sites 527, 528, 529, 525, 526, 524) to the coast of Africa (Sites 363, 362, 532). An estimate of faunal assemblage similarity (Dice Coefficient) was made for each pair of 208 samples (Fig. 13 shows the sample locations by age). The closest tie between any single sample and the one most similar to it, irrespective of the level of that similarity, is shown in Fig. 14 (a Nearest-Neighbor analysis). For this same transect, the ties of similarity having values greater than 0.75, 0.76, and 0.77 (Rho-Groups analysis) were plotted in successive attempts (Figs. 15, 16, and 17) to test for stability in relationships having any time or regional significance. The last test was similar to the one done for Leg 3 ostracode assemblages (Benson, 1975) to demonstrate the separation of the psychrospheric fauna from the preceding thermospheric fauna.

General results of the global analysis

Examination of all of the samples (each 50 cm³) available from 155 DSDP sites from the first 50 legs, plus Legs 72–75, indicates that about two-thirds had ostracodes (1044 of 1621). Using foraminifera and nannofossil zones with ages estimated to the nearest 1 Ma interval, the frequency of samples shows (Fig. 1) that recovery in general diminishes logarithmically with age until the Cretaceous, when it then increases dramatically (due in part to a loss of resolution of small time intervals at this level).

Fig. 3. Changes in the generic composition of samples through 80 million years showing the proportions of new forms appearing for the first time, extinction of old forms, and the combined alteration of faunal composition.

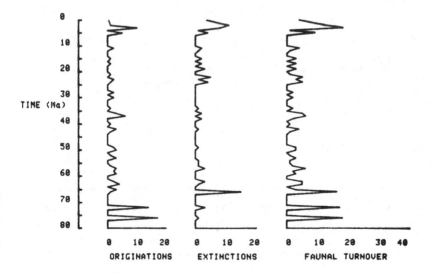

Because ostracodes have calcite exoskeletons, they rarely occur in non-carbonate sediments. Hence, to some degree their distribution here probably reflects that of the distribution of pelagic marls in the deep-sea through the past 80 Ma as was suggested by Davies & Supko (1973) (also see Davies & Worsley, 1981) for Phase I of DSDP. The Cretaceous and the Oligocene appear to be times of increased carbonate sedimentation and of ostracode abundance. The recovery record of DSDP coring seems to be biased in favor of Miocene samples and against Eocene samples. It is noteworthy how rapidly the change occurs at the end of the Cretaceous after a sustained level of high productivity and at the end of the Eocene after a sharp period of depression. Since samples were averaged for 1 Ma intervals local variation in sedimentation rates will not affect either the sample count or the specimen count.

The percentage of samples containing ostracodes fluctuates globally through time from about 25 to 70%. The oldest ostracode faunas we found in this study were in the Albian and Aptian. These faunas generally suggest restricted conditions. This study begins in the Santonian and Campanian where the ostracode frequency is higher, through the Maestrichtian, before it falls off rapidly and then fluctuates radically in the Paleocene. The percentage of samples with ostracodes rises significantly in the Eocene to drop suddenly at about 40 Ma. It then returns to its previous heights at the end of the early Oligocene to fall

Fig. 4. Standing world-wide diversity of genera. The average number of genera present globally at any 1 million year interval over the last 80 million years.

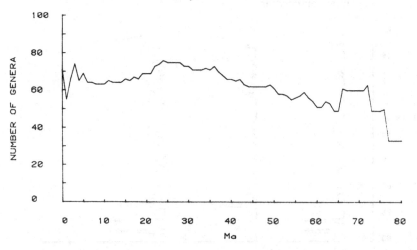

STANDING DIVERSITY OF GENERA THROUGH TIME

again in Early Miocene. It gradually builds back up to former high levels during the Pliocene (Fig. 2). Therefore, in general, the chances of finding some ostracodes in samples is directly proportional to the likelihood of the recovery of samples except at 40 Ma and during the Miocene.

The average population per million year interval of an ostracode sample varies from 20 to 40 specimens (Fig. 5). This estimate includes many individual samples with only a few specimens as well as many with over 100 and some with more than 500. In general, the populations are higher during the Cretaceous and the Oligocene with the most radical fluctuations during the Paleocene and the sustained lowest populations during the Eocene.

Fluctuations in the number of genera per sample globally and sample diversity (Shannon–Wiener Index) correlate well through 80 Ma (Fig. 5). There is a 'crash' beginning at the end of the Cretaceous sustained into the early part of the Paleocene. There are sharp variations (one of which marks the 52 Ma event) until late in the Middle Eocene when there is a spectacular increase in genera and diversity, especially in the South Atlantic and Indo-Pacific. This increase sustains itself well into the Oligocene. The dramatic global increase is thought to reflect the origin of the psychrosphere when the first young, well ventilated, nutrient-rich waters began to flow northward into the ocean basins from the newly formed Circum-Antarctic circulation. The rest of the

Fig. 5. World-wide census of the average populations of 50 cm³ DSDP samples (specimens), the average number of genera present, and the diversity (Shannon–Wiener Index) for each 1 million year interval over the last 80 million years.

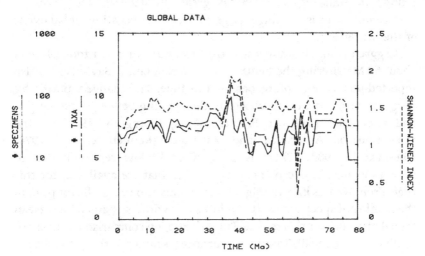

history is relatively stable with a small peak in the Middle Miocene at about 14 to 12 Ma, perhaps related to the formation of the East Antarctic Ice Cap (Kennett, 1980), and then a steady decline until about 3 Ma.

Examination of global standing diversity (the total count of genera at any one time) through time (Fig. 4) shows a surprising stability of long-lived genera over the long term. It would appear that, unlike the fluctuations of individual samples of the deep fauna, total world diversity maintains itself between 50 and 70 genera, with low points just after the Cretaceous and in the Pleistocene, and a strong sustained high through the latter part of the Oligocene.

Testing events of global importance

With the discovery of the iridium layer by Alvarez et al. (1980), the Cretaceous–Tertiary boundary event (65 Ma) has attracted considerable renewed interest (Birkelund & Bromley, 1979; Lewin, 1983). Various hypotheses have been presented for this event (McLean, 1978; Alvarez et al., 1980; Smit & Hertogen, 1980; Officer & Drake, 1983). The ostracode evidence from the deep sea strongly supports the existence of a sudden, unexpected accelerating change of global proportions. The percent of samples containing ostracodes suddenly declines as does the number of samples available. The proportion of generic extinctions without replacement (15%) is greater than at any other time included by this study. This corresponds to the estimate of extinctions of benthic foram genera made by Emiliani et al (1981). It is far below the extinction of the planktonic groups (13% survival), however (see also Raup & Sepkoski, 1982). Only during the Pliocene is the faunal turnover or the plunge in standing diversity as great. In general, the Terminal Cretaceous Event is the single largest short-term negative global event of the last 80 Ma.

Berggren (1982) has suggested that a significant event took place at about 52 Ma altering the benthic foraminifera fauna. Savin (1977) also reported an oxygen isotope peak at this time. The change reported by Berggren may have been caused by the opening between Spitzbergen and Greenland (DeGreer Fracture Zone) (Berggren, 1977) or the separation of Antarctica and Australia (Kennett et al., 1974; Oberhäsnsli, 1983). There is a significantly low representation of samples available, and ostracodes in those that are available, for this time. From what is known (Fig. 6) there seems to be a significant peak in the number of specimens in the Indo-Pacific with a sustained increase in faunal diversity for this region. The Atlantic, by comparison, is in sharp decline (Figs. 7 and 8). Taxonomic turnover is minor. If this is the time of

Fig. 6. Census for the Indo-Pacific region of the average
populations of 50 cm³ DSDP samples (specimens), the average
number of genera present, and the diversity (Shannon–Wiener
Index) for each 1 million year interval over the last 80 million
years.

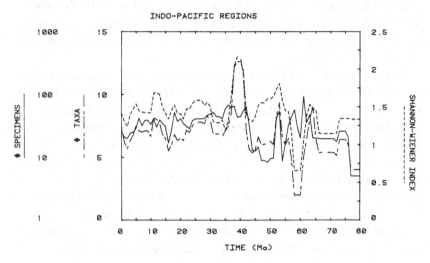

Fig. 7. Census for the Atlantic region of the average populations
of 50 cm³ DSDP samples (specimens), the average number of
genera present, and the diversity (Shannon–Wiener Index) for
each 1 million year interval over the last 80 million years.

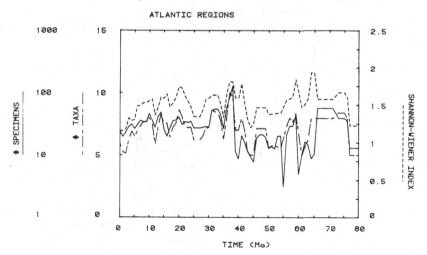

the development of the North Atlantic Deep Water (NADW), its effect is not pronounced. In summary, this event seems localized for the ostracodes. It is not as significant as some of the others.

It has been suggested that the great change at 40 Ma in ostracode faunas of the first 75 DSDP sites cored was evidence for the origin of the modern psychrosphere or two-layer ocean system (psychrosphere and thermosphere) (Benson, 1975). Other faunal, floral, and oxygen isotope evidence (Kennett & Vella, 1975; Savin, 1977; Corliss, 1981, Douglas & Woodruff, 1981; Aubry, 1983) has also pointed to an event at this time. From the present study the ostracode evidence remains strongly in support of this event.

In general, the proportion of samples with ostracodes as well as samples themselves drops noticably from the Middle to Late Eocene (40 Ma). Yet the number of genera and the diversity rises abruptly to a global all-time high. The ultimate result is a significant faunal turnover that establishes the basic stock of the modern deep-sea fauna. This is dominantly an Indo-Pacific event sharply felt in the Walvis Ridge area, but it was delayed or out of synchronization in the North Atlantic (buffered by the subsidence of the Iceland–Faroe Ridge and a surge in the NADW?) and Rio Grande Rise areas. This is as strong a positive event as the Terminal Cretaceous Event is a negative one. Its timing is mixed regionally and this diachrony is possibly due to the beginning of

Fig. 8. Census for the North Atlantic region of the average populations of 50 cm³ DSDP samples (specimens), the average number of genera present, and the diversity (Shannon–Wiener Index) for each 1 million year interval over the last 80 million years.

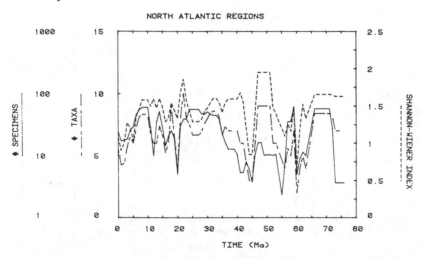

strong stratification of the water column building from deep to shallow. It is notable that shallow DSDP sites receive this signal after deeper ones (Site 357, Boersma & Shackleton, 1977; Benson & Peypouquet, 1983) and this may account for the discrepancy between its recognition at 36 to 38 Ma in some studies (Site 277, 1214 m; Shackleton & Kennett, 1975; Haq, 1982) as compared to the 40 Ma shown here. The fact that the event seems to be more sudden in some shallow regions far from a threshold can be the result of a sharply stratified water-mass structure.

The event marking the end of the Oligocene, concurrent with the suggested opening of Drake Passage (25–22 Ma; Foster, 1974) and the strong establishment of the modern deep-sea evironment (Benson, 1975), is indicated by a larger decline in the proportion of ostracodes in a rapidly decreasing number of samples, but it is not strongly indicated in global changes in the character of the ostracodes themselves. Neither the average size of populations nor the diversity of their taxa changes noticeably. There may be a short-term peaking of diversity in the Atlantic, especially in the North Atlantic, but nothing significant takes place in the Indo-Pacific.

In the Middle Miocene at about 14 Ma, there may have been a major decrease in ocean temperatures (Kennett, 1978; Shackleton & Kennett, 1975; Savin *et al.*, 1975; Savin, 1982; Berger, 1982; Vail & Hardenbol, 1979 at 15.5 Ma) associated with the beginning of ice build-up in the Antarctic. The ostracode results are mixed. Whereas there is a strong signal in the South Atlantic, it is generally weak elsewhere. Species changes can be mapped (Benson, 1984a), but in general faunal changes seem to be minor. This suggests perhaps that the circulatory patterns of the deep, cold water-mass systems were well established by this time and that the ice formation was an influence primarily on shallower systems because of eustatic changes.

The closure of the Iberian Portal (12 to 6 Ma) that sealed off the remaining vestiges of the Tethys Ocean (Benson, 1976a, 1979) did not only cause the Messinian Salinity Crisis at about 6 Ma (Hsü *et al.*, 1978; Benson, 1976b, 1978) and a sudden drop of sea level of about 70 m (Adams *et al.*, 1977); its global signal can also be seen in the deep-sea ostracode data. The number of samples and proportion of ostracodes increases sharply. A major fluctuation in standing diversity began, while the number of genera and faunal diversity generally began a downturn. As would be expected, there is a strong negative effect in the North Atlantic after a general period of high diversity lasting since 40 Ma. There seems to be an attempt at recovery after the formation of the Mediterranean, at the beginning of the Pliocene (5 Ma), only to have the diversity plunged to new lows with the closure of the Isthmus of Panama (3.5 Ma) and glaciation beginning in South America (3.5 Ma, Mercer, 1976) into the Pleistocene (beginning at 1.8 Ma).

The South Atlantic compared with the rest of the World Ocean

The formation of the water-mass controlling thresholds of the South Atlantic, by the aseismic ridges that form the Rio Grande Rise and the Walvis Ridge, seems to be the focus and probably the controlling influence of the general paleoceanographic climate of the World Ocean for the last 40 to 60 million years. Certainly the dynamics of this region, which controlled the interchange of deep psychrosphere-forming waters with the NADW, combined with that of Iberian Portal, whose closure progressively destroyed the effectiveness of the Tethys Ocean as a source of deep thermospheric water, have together been the dominant influences of World Ocean history. In this section we examine the local changes of the South Atlantic against the backdrop of the previous discussion.

First, the faunal records reflecting the water-mass histories of their particular regions in the South Atlantic (Figs. 9, 10, and 11) are different east and west of the Mid-Atlantic Ridge. Unquestionably, the benthic faunas of either region have tended to immigrate along the crests or slopes of the aseimic ridges that developed and followed the spreading of ocean floor. The dynamics of the water-mass systems of either of the regions are quite different, with the interchange through the Vema Channel of the North Atlantic Deep Water with the Antarctic Deep or Intermediate Water dominating the west, while after the Middle to Late Eocene (Sclater et al, 1977), the Walvis Ridge effectively blocked the return interchange of deep waters between the Cape and Angola Basins. Herein lie the sources of contrast for the description of local responses to, or influences on, the global events that follow.

The major differences in the response of the ostracode faunas over the last 80 million years between east (Walvis Ridge) and west (Rio Grande Rise) are the more extreme reaction of the east around the time of the 40 Ma event, and its more subdued response (recorded primarily at Site 363) to the Terminal Cretaceous Event.

The 'crash' at 65 Ma is marked at Site 357 where the iridium event is supposed to be. Ostracodes are found in samples close to this event (core 31, section 3, 47–58 cm and core 30, section 3, 97–106 cm). The number of specimens in these cores, as well as the number of taxa increases

Fig. 9. Census for the South Atlantic region of the average populations of 50 cm³ DSDP samples (specimens), the average number of genera present, and the diversity (Shannon–Wiener Index) for each 1 million year interval over the last 80 million years.

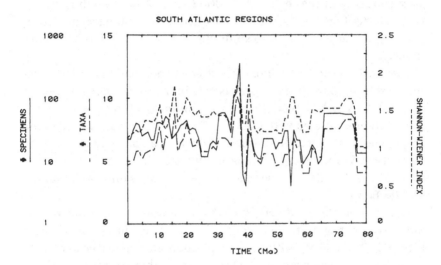

Fig. 10. Census for the Rio Grande Rise area of the average populations of 50 cm³ DSDP samples (specimens), the average number of genera present, and the diversity (Shannon–Wiener Index) for each 1 million year interval over the last 80 million years.

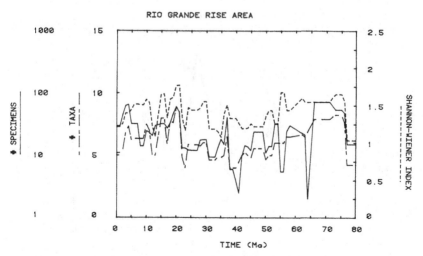

significantly (doubles). The generic composition of the assemblage changed more than 70%, while the diversity is only slightly affected. This suggests that at this paleodepth, thought to be well below 1000 m (present depth 2109 m; Benson, 1984b), the basic character of the environment remained the same, but that one water mass may have been flushed by another. All that can truly be said is that a change took place suddenly in the region of the Rio Grande Rise. Sampling on the Walvis Ridge (Site 363) is not now adequate to contribute significantly to this problem. We intend to study the changes suggested here more closely.

There are strong fluctuations in ostracode populations in both east and west from 65 to 40 Ma. A peak occurs in the Rio Grande Rise region at about 52 Ma; however, there is no meaningful change in faunal composition associated with this date. The number of taxa found in the deeper west constantly diminishes. This marks a time of unusually low diversity at Walvis Ridge. A peak in diversity occurs in both areas at about 55 Ma, but the decline that follows continues until well into the Middle Eocene.

At about 50 Ma the diversity of the ostracode fauna in the Walvis Ridge region begins to rise swiftly to culminate in the Priabonian between 40 and 38 Ma with a very high count in genera (the average is more than 15 genera per sample). This was thought to be caused by local

Fig. 11. Census for the Walvis Ridge area of the average populations of 50 cm³ DSDP samples (specimens), the average number of genera present, and the diversity (Shannon–Wiener Index) for each 1 million year interval over the last 80 million years.

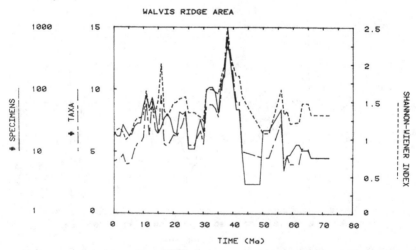

shallowing (a few specimens with eye tubercules have been found). Certain genera appear for the first time and then quickly disappear. Specimens of the shallow-water foraminifera *Lepidocyclina* were found in outcrop cores (*Vema* core V22-143; 33° 12′S 2° 17′E, 3369 m). However, the same peak is found in the Indo-Pacific and faunal turnover is seen in the cores to the west on the Rio Grande Rise even though the change in diversity is slight. Therefore, it is felt by us that the extraordinary change in the Walvis Range fauna is the result of both tectonic and water-mass character change. Furthermore, it is quite likely that this elevation of Walvis Ridge was a contributing factor to the general change in the deep-sea fauna.

From the Early Oligocene to the present there are strong and somewhat regular population oscillations in the west with a 7 My interval (at 28, 21, 14 and 7 Ma). These are not well expressed in the east except at 14 Ma. Oscillations in diversity and number of genera present are pronounced, uncoordinated between east and west and at present appear to be without an easy explanation. They are reflected to some degree in both the North Atlantic and the Indian Ocean and obviously bear further study.

The transect of 24 sites from west to east from 356 near Brazil to 362 near Walvis Bay on the coast of Africa reflects the youthfulness of the sediments penetrated by coring across the Mid-Atlantic Ridge and the increasingly older sediments penetrated approaching the continents (Figs. 12 and 13). By tying samples together according to an increase in the coefficient of the similarity of the ostracode assemblages (Dice Coefficient of Similarity), a meaningful pattern is revealed (Figs. 14–17). In the Nearest-Neighbor plot (Fig. 14), where each sample is connected to the one closest in faunal composition, two trends are evident. First, vertical ties of similarity through time or stasis are maintained *in situ*. Second, associations over long distances are most likely to be isochronous and indicate increasingly good biostratigraphic correlation from 45 to 40 Ma to present.

Tying the samples of the west–east South Atlantic transect by the same Dice Coefficient of assemblage similarity, but at increasing values as a lower threshold (Rho-Groups method; Olson & Miller, 1958; Figs. 15–17), results in an interesting pattern at the 0.76 level. Here, at the level where the dominant signal is shown, the clustering obviously divides the faunas at the 40 Ma level, with the western and deeper one maintaining its integrity best. This analysis is thought to reveal, better than all others, the origin and development of the psychrospheric fauna in the South Atlantic, succeeding a very much less coherent deep thermospheric fauna of the Early Paleogene and Cretaceous.

Not shown well in this data plot is the early establishment of the psychrospheric fauna in depths of more than 2000 m; such as at Site 357 (at about 38 Ma) near Vema Channel, and the delay of the appearance of the same fauna close by in shallower waters (about 1000 m at Site 516 at about 30 Ma). It is thought that while the deep fauna was developing in its earliest stages in a forming, cold, deep water mass (Boersma & Shackleton, 1977), an overlying intermediate water mass of new and nutrient-poor water coming from dense saline sources from the north predominated and formed an upper barrier to exclude the new

Fig. 12. Map showing the location and number of DSDP stations used for the transect across the Rio Grande Rise and Walvis Ridge areas.

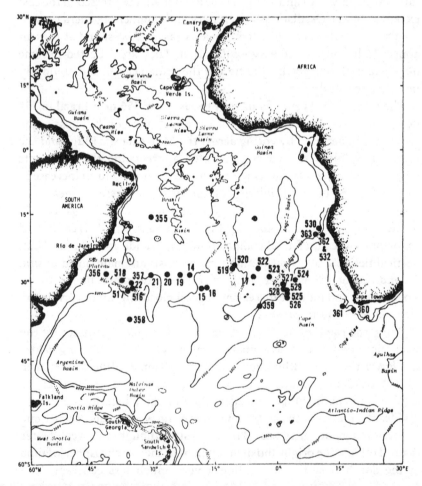

psychrospheric fauna. Only as time allowed the basins to be filled with cold water did the faunas immigrate up slope.

Conclusions

The evolution of the World Ocean over the past 80 million years has been the conversion of a latitudinal, salinity-driven, isothermal, poorly stratified, warm ocean system (thermosphere) to that of a meridional (longitudinal), thermally-driven, nearly isohaline, strongly two-layered ocean system (thermospheric gyres supported by a psychrosphere). The water masses of the thermosphere were defined, albeit poorly, by differences in salinity. Deep circulation was locally restricted with large basins poorly or well ventilated depending on their proximity to evaporating source areas. The subsequent evolution of the ocean system may have been by steps divided by stable periods as described by Berger *et al* (1981), but the size of the steps differed dramatically. The development of the psychrosphere accelerated lateral water-mass movement as regional differences in surface-water temperatures became more pronounced and basins began to be filled with denser cold water. Water-mass and marine biologic pathways of least resistance change from vertical to horizontal causing an intensification of depth stratification to occur. Reaction rates in world communication

Fig. 13. Samples used for the transect study across the Rio Grande Rise–Walvis Ridge areas plotted for transect position (by station on x-axis) versus age. Each symbol can indicate one sample, or more if very closely spaced.

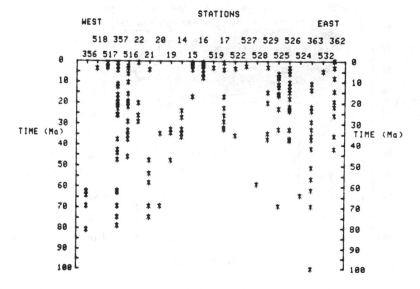

Fig. 14. Nearest-Neighbor analysis for samples in the transect along the Rio Grande Rise to the Walvis Ridge.

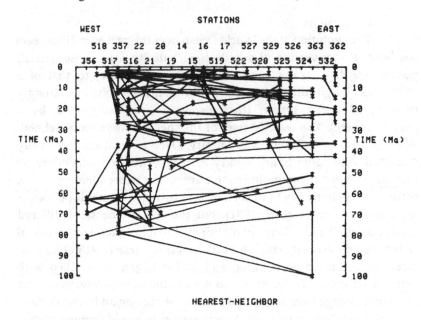

Fig. 15. Rho-Groups analysis for samples in the transect along the Rio Grande Rise to the Walvis Ridge. Critical value for the Dice Coefficient is 0.75.

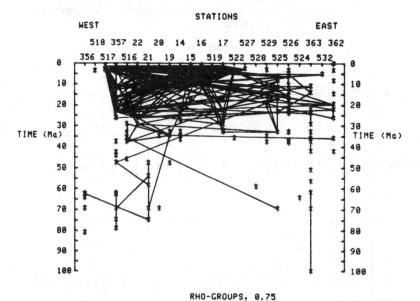

Fig. 16. Rho-Groups analysis for samples in the transect along the Rio Grande Rise to the Walvis Ridge. Critical value for the Dice Coefficient is 0.76.

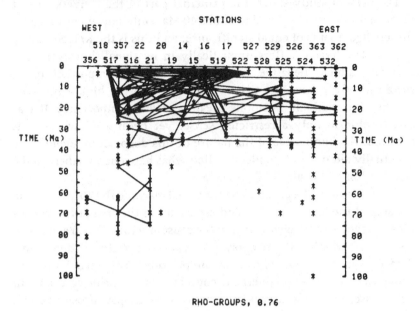

Fig. 17. Rho-Groups analysis for samples in the transect along the Rio Grande Rise to the Walvis Ridge. Critical value for the Dice Coefficient is 0.77.

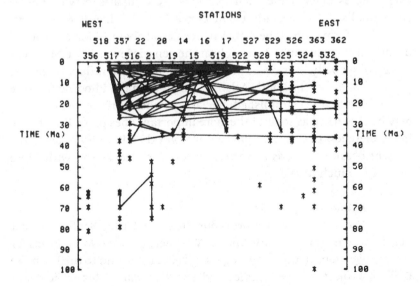

increased with more threshold effects to accelerate events and their effects over broader areas.

The present study shows the principal part of the conversion just described to have happened at about 40 Ma. Only one event precedes this change, that is of equal significance, and this is the K/T boundary 'crash'. The 'crash' differs dramatically in its signal, however. It is weaker, sudden, negative, accelerating, and less important in its sustained effect. The fact that it takes place in the highly buffered depths of a relatively immobile ocean is of great importance. It was catastrophic in only a restricted sense, yet because of where it is detected, the knowledge of the nature of this deep ocean 'event' may eventually aid others in understanding what happened farther up the water column and on the land surface.

The increase of the generic-change signal marking the 40 Ma event in the area of the South Atlantic and especially in the area of the Walvis Ridge suggests a coming together of the cause and the effect. Yet it is still not known whether the temporary elevation of Walvis Ridge at this time or the formation of the warm proto-NADW – cold Circum-Antarctic water-mass convergent couplet was the primary cause. In either case, the modern two-layer ocean with its psychrosphere was born and its effects can be traced world-wide.

Later global events are less clearly marked in the signals measured in the South Atlantic. Nothing of note happened at the end of the Oligocene as might have been expected by comparison with Vail & Hardenbol's (1979) eustatic event curve (see also Loutit & Kennett, 1981). The formation of the Antarctic Ice Cap at about 14 Ma is recorded in species changes but not strongly at the generic level. The 6 Ma event is noted globally but as expected has no special influence in the psychrosphere of the South Atlantic. Changes in the Pliocene indicate the coming of a yet unfinished glacial age whose influence on events may be as marked as some of the more important ones preceding it. The incompleteness of this last signal emphasizes the difficulty of using the present ocean system as a 'normal' model for comparison with those typical of much of the past.

Acknowledgements

We would like to express our thanks to Carlita Sanford for her help in preparing the manuscript, to W. A. Berggren and M. A. Buzas for their suggestions in the early stages of preparation, and to the Deep Sea Drilling Project for the samples and ancillary data. This project was sponsored by the Smithsonian Scholarly Studies Program.

APPENDIX 1

List of the 156 generic categories recognized including those currently in open nomenclature

Abyssocythere	Eucytheridea	Paranesidea
Abyssocythereis	Eucytherura	Pariabyssocythere
Acanthocythereis	Eximacythere	Pariceratina
Actinocythereis	Glomocythere	Parvacythere
Agrenocythere	Glossicythereis	Parviceratina
Agulhasina	Haughtonileberis	Pedicythere
Alatacythere	Hazelina	Pelecocythere
Ambocythere	Hemicytherura	Phacorhabdotus
Amphicytherura	Henryhowella	Platella
Apeteloschizocythere	Hermanites	Pokornyella
Arculicythere	Hirschmannia	Polycope
Argilloecia	Hyphalocythere	Pontocypris
Atlanticythere	Jugosocythereis	Poseidonamicus
Aurila	Kikliocythere	Pseudocythere
Australoecia	Kingerella	Pterygocythere
Bairdia	Krithe	Pterygocythereis
Bairdiacean #1	Leguminocythereis	Puriana
Bairdiid #1	Limburgina	Quadracythere
Bathycythere	Loculicytheretta	Quasibradleya
Brachycythere	Loxoconcha	Quasibuntonia
Bradleya	Loxoconchella	Radimella
Buntonia	Loxocorniculum	Rectangulocythereis
Bythoceratina	Macrocypris	Reticuloechinocythereis
Bythocyprid #1	Majungaella	Rocaleberis
Bythocypris	Mesobradleya	Rockallia
Bythocythere	Mesocythereis	Saida
Bythocytheromorpha	Monoceratina	Saipanetta
Cativella	Monoceratinid #1	Sansabella
Caudites	Murrayina	Scepticocythereis
Cestocythereis	Mutilus	Sclerochilus
Cletocythereis	Neoatlanticythere	Sphaeroleberis
Costa	Neocytherideis	Spinileberis
Cretaceratina	New Genus #1	Suhmicythere
Cyprid #1	New Genus #2	Tenedocythere
Cyprideis	New Genus #3	Thalassocythere
Cytheracean #1	New Genus #4	Trachyleberid #1
Cytheracean #2	Oblitacythereis	Trachyleberidea
Cytheralison	Oertliella	Trachyleberis
Cythereis	Ovocytheridea	Triebelina
Cythereisid #1	Oxycythereis	Unicapella
Cytherella	Paijenborchella	Unidentified #1
Cytherelloidea	Paleoabyssocythere	Unidentified #2
Cytheretta	Paleoatlanticythere	Unidentified #3
Cytheridea	Paleoposeidonamicus	Urocythereis
Cytheropteron	Paracyprid #1	Uroleberis
Cytherura	Paracypris	Veenia
Dinglella	Paracytheretta	Veeniacythereis
Dragonocythere	Paracytheridea	Verricose Form #1
Dutoitella	Paradoxostoma	Wichmannella
Echinocythereis	Paragrenocythere	Xestoleberid #1
Eocytheropteron	Parakrithe	Xestoleberis
Eucythere	Parakrithella	Zabythocypris

REFERENCES

Adams, C.G., Benson, R.H. *et al.* 1977. The Messinian Salinity Crisis and evidence of Late Miocene eustatic changes in the World Ocean. *Nature,* **269,** 383–386.
Alvarez, L.W., Alvarez, W., Asaro, F. & Michel, H.V. 1980. Extraterrestrial cause for the Cretaceous–Tertiary extinction. *Science,* **208,** 1095–1108.
Aubry, M.-P. 1983. Late Eocene to Early Oligocene calcareous nannoplankton paleobiogeography and the terminal Eocene event. *First International Conference on Paleoceanography, Zürich, 1983,* (abstracts), p. 10.
Benson, R.H. 1975. The origin of the psychrosphere as recorded in changes of deep-sea ostracode assemblages. *Lethaia,* **8,** 69–83.
Benson, R.H. 1976a. Miocene deep-sea Ostracodes of the Iberian Portal and the Balearic Basin. *Mar. Micropaleontol.,* **1,** 249–262.
Benson, R.H. (ed.) 1976b. The biodynamic effects of the Messinian Salinity Crisis. *Palaeogeogr. Palaeoclimatol. Palaeoecol.,* **20,** 1–170.
Benson, R.H. 1978. The paleoecology of the Ostracodes of DSDP Leg 42A. In: K. Hsü, L. Montadert *et al.* (eds.), *Initial Reports of the Deep Sea Drilling Project,* **42,** part 1, US Govt Printing Office, Washington, DC, pp. 777–787.
Benson, R.H. 1979. In search of lost oceans; a paradox in discovery. In: J. Gray & A.J. Boucot (eds.), *Historical Biogeography, Plate Tectonics and the Changing Environment,* Oregon State Univ. Press, pp. 379–389.
Benson, R.H. 1984a. Biomechanical stability and sudden change in the evolution of the deep-sea ostracode *Poseidonamicus. Paleobiology,* 1984.
Benson, R.H. 1984b. Estimating greater paleodepths with Ostracodes, especially in past thermospheric oceans. *Palaeogeogr. Palaeoclimatol. Palaeocol.,* 1984.
Benson, R.H. & Peypouquet, J.P. 1983. The upper and mid-bathyal Cenozoic Ostracode faunas of the Rio Grande Rise found on Leg 72 Deep Sea Drilling Project. In: P.F. Barker, R.L. Carlson, D.A. Johnson *et al. Initial Reports of the Deep Sea Drilling Project,* **72,** US Govt Printing Office, Washington DC, pp. 805–818.
Berger, W.H. 1982. Climate steps in ocean history – lessons from the Pleistocene. In: *Studies in Geophysics, Climate in Earth History,* National Academy Press, Washington, DC, pp. 43–54.
Berger, W.H., Vincent, E. & Thierstein, H.R. 1981. The deep sea record: major steps in Cenozoic ocean evolution. In: J.E. Warme, R.G. Douglas & E.L. Winterer (eds.), The Deep Sea Drilling Project: A Decade of Progress, *SEPM Spec. Publ.,* **32,** 489–504.
Berggren, W.A. 1977. Late Neogene planktonic foraminiferal biostratigraphy of the Rio Grande Rise (South Atlantic). *Mar. Micropaleontol.,* **2,** 265–313.
Berggren, W.A. 1982. Role of ocean gateways in climatic change. In: *Studies in Geophysics, Climate in Earth History,* National Academy Press, Washington, DC, pp. 118–125.
Berggren, W.A., Kent, D.V. *et al.* 1984. Cenozoic Geochronology. *Geology,* 1984.
Birkelund, T. & Bromley, R.G. (eds.). 1979. *Cretaceous–Tertiary Boundary Events,* 2 vols., Univ. Copenhagen, Denmark.
Boersma, A. & Shackleton, N.J. 1977. Tertiary oxygen and carbon isotope stratigraphy, Site 357 (mid-latitude South Atlantic). In: P. Supko & K. Perch-Nielsen (eds.), *Initial Reports of the Deep Sea Drilling Project,* **39,** US Govt Printing Office, Washington, DC, pp. 911–924.
Corliss, B.H. 1981. Deep-sea benthonic foraminiferal faunal turnover near the Eocene/Oligocene boundary. *Mar. Micropaleontol.,* **6,** 367–384.
Corliss, B.H. & Keigwin, L.D. Jr 1983. The Eocene/Oligocene event in the deep sea. *First International Conference on Paleoceanography, Zürich, 1983* (abstracts), p. 16.
Davies, T.A. & Supko, P.R. 1973. Oceanic sediments and their diagenesis: some examples from deep-sea drilling. *J. Sed. Petrol.,* **43** (2), 381–390.
Davies, T.A. & Worsley, T.R. 1981. Paleoenvironmental implications of oceanic carbonate sedimentation rates. *SEPM Spec. Publ.,* **30,** 169–180.
Douglas, R.G. 1973. Evolution and bathymetric distribution of Tertiary deep-sea benthic Foraminifera. *Geol. Soc. Amer., Abstracts,* p. 603.
Douglas, R.G. & Woodruff, F. 1981. Deep sea benthic foraminifera. In: C. Emiliani (ed.), *The Sea,* vol. 7, Wiley-Interscience, New York.

Emiliani, C., Kraus, E.B. & Shoemaker, E.M. 1981. Sudden death at the end of the Mesozoic. *Earth Planet Sci. Lett.*, **55**, 317–334.

Foster, R.J. 1974. Eocene echinoids and the Drake Passage. *Nature*, **249**, 751.

Gartner, S. & Keany, J. 1978. The terminal Cretaceous event: A geologic problem with an oceanographic solution. *Geology*, **6**, 708–712.

Gartner, S. & McGuirk, J.P. 1979. Terminal Cretaceous extinction scenario for a catastrophe. *Science*, **206**, 1272–1276.

Haq, B.U. 1982. Climatic acme events in the sea and on land. In: *Studies in Geophysics, Climate in Earth History*, National Academy Press, Washington, DC, pp. 126–132.

Hessler, R.R. & Sanders, H.L. 1967. Faunal diversity in the deep sea. *Deep Sea Res.*, **14**, 56–78.

Hessler, R.R. & Wilson, G.D.F. 1983. The origin and biogeography of Malacostracan Crustaceans in the deep sea. In: R.W. Sims, J.H. Price & P.E.S. Whalley (eds.), *Systematics Association Spec. Vol. No. 23, Evolution, Time and Space: The Emergence of the Biosphere*, Academic Press, London, pp. 227–254.

Hsü, K.J. 1980. Terrestrial catastrophe caused by cometary impact at the end of Cretaceous. *Nature*, **285**, 201–203.

Hsü, K.J. Montadert, L. *et al.* 1978. History of the Mediterranean Salinity Crisis. In: K. Hsü, L. Montadert *et al.* (eds.), *Initial Reports of the Deep Sea Drilling Project*, **42**, part 1, US Govt Printing Office, Washington, DC, pp. 1053–1078.

Keller, G. 1982. Biochronology and paleoclimatic implications of Middle Eocene to Oligocene planktic foraminiferal faunas. *Mar. Micropaleontol.*, **7**, 463–486.

Kennett, J.P. 1978. The development of planktonic biogeography in the Southern Ocean during the Cenozoic. *Mar. Micropaleontol.*, **3**, 301–345.

Kennett, J.P. 1980. Paleoceanographic and biogeographic evolution of the southern ocean during the Cenozoic, and Cenozoic microfossil datums. *Paleogeogr. Paleoclimatol. Paleoecol.*, **31**, 123–152.

Kennett, J.P. & Vella, P. 1975. Late Cenozoic planktonic foraminifera and paleoceanography at DSDP Site 284 in the cool subtropical South Pacific. In: J.P. Kennett, R.E. Houtz *et al.* (eds.), *Initial Reports of the Deep Sea Drilling Project*, **29**, US Govt Printing Office, Washington DC, pp. 769–799.

Kennett, J.P., Houtz, R.E. *et al.* 1974. Development of the Circum-Atlantic Current. *Science*, **186**, 144–147.

Lewin, R. 1983. Extinctions and the history of life. *Science*, **221**, 935–937.

Loutit, T.S. & Kennett, J.P. 1981. New Zealand and Australian Cenozoic sedimentary cycles and global sea-level changes. *Am. Ass. Petrol. Geol. Bull.*, **65** (9), 1586–1601.

McLean, D.M. 1978. A terminal Mesozoic 'greenhouse': lessons from the past. *Science*, **201**, 401–406.

Mercer, J.H. 1976. Glacial history of southernmost South America. *Quat. Res.*, **6**, 125.

Miller, K.G. 1983. Eocene–Oligocene paleoceanography of the deep Bay of Biscay: benthic foraminiferal evidence. *Mar. Micropaleontol.*, **7**, 403–440.

Miller, K.G. & Curry, W.B. 1982. Eocene to Oligocene benthic foraminiferal isotopic record in the Bay of Biscay. *Nature*, **296**, 347–352.

Miller, K.G., Gradstein, F.M. & Berggren, W.A. 1982. Late Cretaceous to Early Tertiary agglutinated benthic Foraminifera in the Labrador Sea. *Micropaleontol.*, **28** (1), 1–30.

Oberhänsli, H. 1983. Oceanographic and climatic changes recorded in Indian Ocean sediments from the latest Cretaceous through Early Neogene (DSDP Sites, 212, 217, 220, 237 and 253). *First International Conference on Paleoceanography, Zürich, 1983*, (abstracts), p. 45.

Officer, C.B. & Drake, C.L. 1983. The Cretaceous–Tertiary transition. *Science*, **219**, 1383–1390.

Olson, E.C. & Miller, R.L. 1958. *Morphological Integration*. Univ. of Chicago Press, Chicago.

Raup, D.M. & Sepkoski, J.J. Jr 1982. Mass extinctions in the marine fossil record. *Science*, **215**, 1501–1503.

Sanders, H.L. 1969. Benthic marine diversity and the stability-time hypothesis. In:

Diversity and Stability in Ecological Systems, Brookhaven Symposia in Biology, No. 22, pp. 71–81.

Savin, S. M. 1977. The history of the earth's surface temperature during the past 100 million years. *Ann. Rev. Earth Planet. Sci.*, 5, 319–355.

Savin, S. M. 1982. Stable isotopes in climatic reconstructions. In: *Studies in Geophysics, Climate in Earth History*, National Academy Press, Washington, DC, pp. 164–171.

Savin, S. M. Douglas, R. G. & Stehli, F. M. 1975. Tertiary marine paleotemperatures. *Geol. Soc. Am. Bull.*, 86, 1499–1510.

Sclater, J. G., Hellinger, S. & Tapscott, C. 1977. The paleobathymetry of the Atlantic Ocean from the Jurassic to the present. *J. Geol.*, 85 (5), 509–552.

Shackleton, N. J. & Kennett, J. P. 1975. Paleotemperature history of the Cenozoic and initiation of Antarctic glaciation: oxygen and carbon isotope analysis in DSDP Sites 277, 279, and 281. In: *Initial Reports of the Deep Sea Drilling Project*, 29, US Govt Printing Office, Washington, DC, pp. 743–755.

Smit, J. & Hertogen, J. 1980. An extraterrestrial event at the Cretaceous–Tertiary boundary. *Nature*, 285, 198–200.

Srinivasan, M. S. & Kennett, J. P. 1983. The Oligocene–Miocene boundary in the South Pacific. *Geol. Soc. Am. Bull.*, 94, 798–812.

Thierstein, H. R. & Berger, W. H. 1978. Injection events in ocean history. *Nature*, 276, 461–466.

Vail, P. R. & Hardenbol, J. 1979. Sea level changes during the Tertiary. *Oceanus*, 22, 71–79.